Marine Pollution:
Functional Responses

Academic Press Rapid Manuscript Reproduction

Proceedings of the Symposium
"Pollution and Physiology of Marine Organisms"
Held on November 14-17, 1977 at Hobcaw Barony
Georgetown, South Carolina

MARINE POLLUTION:
FUNCTIONAL RESPONSES

Edited by

Winona B. Vernberg
School of Public Health
University of South Carolina
Columbia, South Carolina

Anthony Calabrese   *Frederick P. Thurberg*
Northeast Fisheries Center
National Marine Fisheries Service, NOAA
Milford, Connecticut

F. John Vernberg
Belle W. Baruch Institute for Marine Biology and Coastal Research
University of South Carolina
Columbia, South Carolina

ACADEMIC PRESS

New York   San Francisco   London   1979
A Subsidiary of Harcourt Brace Jovanovich, Publishers

ACADEMIC PRESS, INC.
111 Fifth Avenue, New York, New York 10003

United Kingdom Edition published by
ACADEMIC PRESS, INC. (LONDON) LTD.
24/28 Oval Road, London NW1 7DX

Library of Congress Cataloging in Publication Data
Main entry under title:

Marine pollution.

    Includes index.
    1. Marine ecology.   2. Marine pollution.
3. Marine fauna—Physiology.   I. Vernberg,
Winona B.,   Date.
QH541.5.S3M285   574.5'2636   79-10451
ISBN 0-12-718260-8

30,807

PRINTED IN THE UNITED STATES OF AMERICA

79 80 81 82   9 8 7 6 5 4 3 2 1

CONTENTS

PART II METALS

PART III PESTICIDES AND PCBs

PART IV   MULTIPLE FACTOR
INTERACTIONS

CONTRIBUTORS

Numbers in parentheses indicate the pages on which authors' contributions begin.

J. W. *Anderson* (3, 69), Battelle Pacific Northwest Laboratories, Marine Research Laboratory, Sequim, Washington 98382

Joel E. *Bodammer* (223), NOAA, National Marine Fisheries Service, Northeast Fisheries Center, Oxford Laboratory, Oxford, Maryland 21654

Anita C. *Brannon* (307), Faculty of Biology, University of West Florida, Pensacola, Florida 32504

Elaine M. *Caldarone* (271), Department of Fisheries and Wildlife, Oregon State University, Marine Science Center, Newport, Oregon 97365

Richard S. *Caldwell* (271), Department of Fisheries and Wildlife, Oregon State University, Marine Science Center, Newport, Oregon 97365

Angela C. *Cantelmo* (307), Department of Biology, Ramapo College of New Jersey, Mahwah, New Jersey 07748

Philip J. *Conklin* (307), Faculty of Biology, University of West Florida, Pensacola, Florida 32504

John D. *Costlow, Jr.* (355), Duke University Marine Laboratory, Beaufort, North Carolina 28516

D. G. *Crosby* (291), Environmental Toxicology Department, University of California at Davis, Davis, California 95616

Patricia A. *Cunningham* (183), Department of Biology, North Carolina Central University, Durham, North Carolina 27707

Margaret A. Dawson (171), National Marine Fisheries Service, Northeast Fisheries Center, Milford Laboratory, Milford, Connecticut 06460

A. L. DeVries (53), Department of Physiology and Biophysics, University of Illinois, Urbana, Illinois 61801

Ronald Eisler (111), U.S. Environmental Protection Agency, Environmental Research Laboratory, Narragansett, Rhode Island 02882

David W. Engel (239), National Marine Fisheries Service, NOAA, Southeast Fisheries Center, Beaufort Laboratory, Beaufort, North Carolina 28516

Bruce A. Fowler (239), Laboratory of Environmental Toxicology, National Institute of Environmental Health Sciences, Research Triangle Park, North Carolina 27709

Ferris R. Fox (307), Faculty of Biology, University of West Florida, Pensacola, Florida 32504

K. W. Fucik (85), Scientific Applications, Inc., 2760 29th Street, Boulder, Colorado 80302

Wayne S. Gardner (23), Skidaway Institute of Oceanography, P.O. Box 13687, Savannah, Georgia 31406

R. L. Garnas (291), Environmental Toxicology Department, University of California at Davis, Davis, California 95616

Richard F. Lee (23), Skidaway Institute of Oceanography, P.O. Box 13687, Savannah, Georgia

Robert J. Livingston (389), Department of Biological Science, Florida State University, Tallahassee, Florida 32306

J. M. Neff (85), Department of Biology, Texas A & M University, College Station, Texas 77843

DelWayne R. Nimmo (259), U.S. Environmental Protection Agency, Environmental Research Laboratory, Sabine Island, Gulf Breeze, Florida 32561

Richard M. Philpot (23), National Institutes of Health, National Institute of Environmental Health Sciences, Phamacology Branch, Research Triangle Park, North Carolina 27709

K. Ranga Rao (307), Faculty of Biology, University of West Florida, Pensacola, Florida 32504

Stanley D. Rice (39), Northwest and Alaska Fisheries Center Auke Bay Laboratory, National Marine Fisheries Service, NOAA, P.O. Box 155, Auke Bay, Alaska 99821

John L. Roberts (365), Department of Zoology, University of Massachusetts, Amherst, Massachusetts 02882

G. Roesijadi (69), Battelle Pacific Northwest Laboratories, Marine Research Laboratory, Sequim, Washington 98382

Barbara A. Rosene (271), Department of Fisheries and Wildlife, Oregon State University, Marine Science Center, Newport, Oregon 97365

Norman I. Rubinstein (341), Faculty of Biology, University of West Florida, Pensacola, Florida 32504

Geoffrey I. Scott (415), Environmental Protection Agency, National Marine Water Quality Laboratory, Bear's Bluff Field Station, P.O. Box 368, John's Island, South Carolina 29445

J. R. Sharp (85), Department of Biology, Texas A & M University, College Station, Texas 77843

Carl J. Sindermann (437), U.S. Department of Commerce, NOAA, National Marine Fisheries Service, Northeast Fisheries Center, Sandy Hook Laboratory, Highlands, New Jersey 07732

Sara C. Singer (23), Skidaway Institute of Oceanography, P.O. Box 13687, Savannah, Georgia 31406

Kenneth R. Tenore (23), Skidaway Institute of Oceanography, P.O. Box 13687, Savannah, Georgia 31406

Robert E. Thomas (39), Department of Biological Science, Chico State University, Chico, California 95926

Winona B. Vernberg (415), School of Public Health and Belle W. Baruch Institute, University of South Carolina, Columbia, South Carolina 29208

Judith S. Weis (151), Department of Zoology and Physiology, Rutgers-The State University of New Jersey, Newark, New Jersey 07102

Peddrick Weis (151), Department of Anatomy, New Jersey Medical School, Newark, New Jersey 07102

PREFACE

The papers in this volume were presented at a symposium entitled "Pollution and Physiology of Marine Organisms" held at Hobcaw Barony, Georgetown, South Carolina in November 1977. This symposium, which was sponsored by the School of Public Health and the Belle W. Baruch Institute for Marine Biology and Coastal Research of the University of South Carolina and the Northeast Fisheries Center, National Marine Fisheries Service, is the third one organized around the effects of pollutants on marine organisms.

There has been a veritable explosion in interest in this subject area in the past few years, and we are grateful to Carl Sindermann for his thought-provoking and insightful overview of our present level of knowledge as well as charting where our future needs lie. It is clear that we have long passed the point where death is the only criterion for determining whether a substance is harmful or not and the large number of papers on sublethal effects in this volume attest to this present trend in physiological pollution studies.

The editors are indebted to the authors for their cooperation, and to the staff of the University of South Carolina for their assistance in many ways, especially Ms. Kathi Hicks, Ms. Dorothy Knight, and Ms. Lois Lawrence.

PART I
Petroleum Hydrocarbons

AN ASSESSMENT OF KNOWLEDGE CONCERNING THE FATE AND EFFECTS OF PETROLEUM HYDROCARBONS IN THE MARINE ENVIRONMENT

J. W. Anderson

Battelle
Pacific Northwest Laboratories
Marine Research Laboratory
Sequim, Washington

INTRODUCTION

From August 1976 to October 1977, there have been four major North American symposia regarding the fate and effects of petroleum hydrocarbons in the marine environment. The proceedings from these are currently available (AIBS, 1976; API, 1977; Wolfe, 1977; and Journal of the Fisheries Research Board of Canada, 1978). In addition, much of the recent literature has been reviewed in a two volume publication edited by D.C. Malins (1977). This extensive amount of information provides ample evidence of our progress in understanding the biological effects of these hydrocarbons and their behavior in marine ecosystems. Indeed, there has probably been an order of magnitude increase in our knowledge during the past three years. Sufficient data are available on the lethal (short-term) concentrations of various oils and specific petroleum hydrocarbons for a variety of marine organisms (Rice *et al.*, 1977). The matter of tissue accumulation and magnification of specific hydrocarbons has been reviewed (Varanasi and Malins, 1977), and Anderson (1977) has attempted to relate tissue uptake and release with abnormal physiological response. Much present research is directed toward finding sensitive biological response parameters, which will allow us to evaluate the state of health of organisms exposed in the field or under experimental conditions. In the former case, the intention is to assess the impact of a spill or effluent under "natural" conditions; the latter evaluation is an attempt to define the lower "safe" threshold concentration of a species, such that the level of damage to a species or population can be predicted from the composition and concentration of hydrocarbon contamination. The latter approach may be the only feasible method of assessing damage, due to the extreme variability of organisms in the natural environment.

Recent emphasis has been placed on field investigations to validate the significance of laboratory data or to determine the recovery of a population or community after a spill (Journal of the Fisheries Research Board of Canada, 1978). It is an extremely difficult and time consuming process to follow the recovery of a local region but, in a few cases such as Chedabuto Bay and West Falmouth, etc., interdisciplinary studies have compiled sufficient data for assessment of impact and the time scale for recovery. Analyses of petroleum hydrocarbons in contaminated sediment sometimes provides interesting correlations with benthic populations (Armstrong et al., 1977). Field experiments in the natural environment concerning rate of change of petroleum contamination in sediments aid in the interpretation of both field and laboratory data (Anderson et al., 1978).

With the mass of data now available, it has been possible to review specific aspects of the fate and effects of hydrocarbons (Malins, 1977a). An overall assessment of our state of knowledge should be attempted to help formulate hypotheses and identify gaps in our understanding. Of course, an integral part of any assessment is a recommended course of future research designed to produce answers to the most significant problems remaining to be solved. This paper will attempt to examine several hydrocarbon research topics, pointing out apparent strengths and weaknesses in our current understanding. The following sections will be devoted to an overall examination of findings in an attempt to fit the pieces of the puzzle together to form a unified picture.

DISCUSSION

Toxicity

Lethal Concentrations. As explained by Rice et al., (1977), there are numerous problems associated with the exposure of marine organisms to petroleum hydrocarbons in bioassay experiments. The variables encountered often lead to differences in results which may or may not be interpreted correctly. They concluded, after reviewing recent literature, that oil toxicity is due to the chemical toxicity of soluble

aromatics, rather than the physical toxicity of dispersed oil
droplets. Hence one can predict in general terms the toxicity
of a given oil, based on the relative concentration of toxic
aromatics within the oil and the physical/chemical characteristics of the oil/water mixture.

Both Neff *et al.* (1976b) and Rice *et al.*, (1977) agree
regarding the comparative toxicities of oil components. Monoaromatics are least toxic, and acute toxicity increases with
increasing molecular size up to the 4- and 5-ring aromatic
compounds, which have very low water solubility. With increasing degrees of alkylation of the aromatic nucleus of 1-, 2-,
3-ring compounds, there seems to be an increase in the toxicity of the compounds. The position of alkylation on such
monoaromatics as xylene may influence toxicity. The most
toxic compounds in refined and crude oils are the di- and tri-
aromatic compounents. Neff *et al.* (1976a) reported 96 hr
LC50 values of 0.3 to 0.6 ppm for 1-methylphenanthrene, fluo-
ranthrene, and phenanthrene, indicating the high toxicity of
these polynuclear aromatics for the polychaete, *Neanthes
arenaceodentata*. The latter compounds are less soluble and
present in the parent oils at lower concentrations than mono-
and di-aromatics. Therefore, the compounds contributing most
to the toxicity of refined and crude oils are probably the
diaromatic hydrocarbons, characterized by naphthalenes, with
alkylbenzenes falling second in importance.

Examination of data on sensitivities of adult organisms
to crude oil and No. 2 fuel oil tested at various temperatures
provide estimates of ranges of toxicity to marine organisms.
The range of about 1 to 20 ppm crude oil in water include the
96 hr LC50 tests with a wide variety of marine fish and
invertebrates in static exposures (Rice *et al.*, 1977). Comparable values for No. 2 fuel oil range between 0.4 and 6 ppm.
Tests conducted at the lower temperatures of 4 to 10°C
generally result in LC50 values near the lower portion of the
range, probably due to the longer retention time of monoaromatics in the static exposure water.

While there are limited data available on quantitative
comparisons between adult sensitivities and those of early
life stages, the evidence indicates that larval stages may be
the most sensitive. Eggs of fish and invertebrates are often
quite tolerant, and juveniles may or may not be more sensitive
than adults. In many cases, the differences between adult and
early life stages may only be a factor of 2 or 3 and not
orders of magnitude. One very significant difference, which
will be addressed later in this paper, is the duration of
exposure for adults and early life stages. It is common to
rear young under conditions of constant exposure for as much
as 60 days, while adults are often exposed for only 4 days.
Investigators do not generally take into account the

differences in their time-dose exposure when comparing tolerances to hydrocarbons. The most sensitive organisms appear to
be small crustaceans or crustacean larvae. The larval,
juvenile, and adult crustaceans seem to be most sensitive during molting, but the differences between species and the
mechanisms of pollutant action are not well understood.

The need for toxicity tests using flowing systems has been
generally recognized, but little information is available.
Vanderhorst et al. (1976) reported a 96 hr LC50 concentration
of 0.8 ppm for coon stripe shrimp tested with a flowing dispersion of No. 2 fuel oil. They also noted the difficulty of
actually deriving an LC50 value from static exposures when
concentrations are rapidly declining. Anderson and Kiesser
(unpublished) have recently used the same exposure system as
Vanderhorst et al. to test the sensitivity of the mysid,
Neomysis awatschensis, to extracts of Prudhoe Bay crude oil.
Juveniles of this species, hatched in clean water in the laboratory, exhibited a static 24 hr LC50 value of 2.9 ppm total
hydrocarbons, while a value of 0.18 ppm was derived from the
constant flowing system. The ratio of these two values is
0.06 which is near the values produced when sublethal response
concentrations (static) are divided by static acute toxicity
(96 hr LC50) values. This provides evidence that values from
flowing exposure systems more accurately define the sensitivity of an organism to extracts of a crude oil.

The subject of toxicity from oil-contamined sediment has
not been thoroughly investigated, but all indications are that
very high concentrations associated with sediment are required
to produce mortality. Several studies have noted the lack of
significant mortality during exposures to concentrations in
excess of 1000 ppm (Wells and Sprague, 1976; Anderson et al.,
1977; Anderson and Kiesser, unpublished data). However, Krebs
and Burns (1977) noted the mortality and adverse responses of
Uca living in sediment containing 1000-7000 ppm. Shaw et al.
(1976) reported mortality of the clam, Macoma balthica, from
sediment containing 640-3890 ppm (dry weight). The peak in
mortality occurred at 29 days, and at 44 and 60 days mortality
was reduced, perhaps indicating a loss of toxic components
from the substrate. By 60 days, sediment concentrations were
approximately equal to background levels. It is not known
whether toxicity of sediment is the result of contact, ingestion of particles, or from interstitial water containing compounds leached from particles or freed by metabolic conver-
sions of microorganisms.

Sublethal Concentrations

This topic has also been reviewed recently (Anderson, 1977), and only salient aspects of our knowledge will be considered in this paper. There are two general questions which should be associated with studies on effects from sublethal concentrations of hydrocarbons. The first concern is whether or not the parameter under study represents a reliable and sensitive method of determining the condition or state of health of the organism. Secondly, could this approach be used to monitor the health of organisms collected in polluted locations? In relation to this last question, Anderson (1977) has suggested that tissue content of some petroleum components (primarily aromatics) might be indicative of state of health, and thus provide another means of monitoring the marine environment.

Much of the research effort in the U.S. over the past three years has been directed toward finding a sensitive measure of organismic condition which reflects the extent of stress in its environment. In many of these studies, hydrocarbon uptake and release rates also have been measured to evaluate correlations between presence and extent of abnormal response and the level of tissue contamination. There is no absolute reason to believe that a pollutant must be accumulated in the tissues to produce an adverse effect, but, to my knowledge, every environmental pollutant studied has been found in tissues of exposed animals at levels above those of the surrounding water. Cell and membrane tissue damage could result from pollutant contact alone, but accumulation seems to occur due to tissue lipid vs. seawater solubilities. We will again turn to the topic of accumulation in a later section, but first, sublethal response parameters and associated tissue contamination must be examined.

Approaches to evaluate the effects of sublethal concentrations on marine organisms cover subject areas as broad as the range of research fields within the biological sciences. Of course, this is no coincidence, since biologists with expertise ranging from ultrastructure to population dynamics have applied their talents to the study of pollutant effects. The same problems exist in sublethal studies as acute toxicity research regarding the lack of uniformity in oils tested, exposure conditions and analytical methods. I must draw heavily upon results of studies using similar exposures, oils and methods of quantitation in attempting to compare findings on various organisms. Much of this has already been presented in diagrammatic or tabular form (Anderson, 1977).

Respiration is the metabolic parameter which has been used most frequently in the past to determine the extent of stress from pollutants. Responses of organisms to hydrocarbon exposures may be abnormally high, low or unaffected. While

the magnitude of response is seldom dependent on hydrocarbon
concentrations, the mysid, *Mysidopsis almyra*, did exhibit
increased respiratory rate with increased concentrations of
No. 2 fuel oil (Anderson, J.W., 1975). The highest level of
oxygen consumption was correlated with a concentration of 0.4
ppm total naphthalenes, from both a water-soluble fraction and
an oil-water dispersion of the oil. Additional tests with
another crustacean, *Penaeus aztecus*, showed that exposure for
4 hr to 0.83 ppm total naphthalenes produced an elevated
respiration rate, which was still exhibited 26 hr after transfer to clean seawater. Examination of tissue naphthalenes
content of *P. aztecus* showed that about 7 ppm was still present in the body, and a period of 96 hr was required for
complete depuration. Following this line of investigation,
the grass shrimp, *Palaemonetes pugio*, which showed a depression in respiratory rate after 5 hr of exposure, were allowed
to depurate for 7 days before respiratory rate was again
measured. Tissue naphthalenes returned to background levels
in about 3 days, and the 7 day respiratory rate was equal to
control values. These data indicate that there may be a
cause and effect relationship between tissue content of naphthalenes and abnormal respiratory response.

Another physiological parameter which has been used to
examine the effects of a variety of pollutants, is the ability
of an organism to regulate internal concentrations of solutes.
Both total osmotic and ionic regulation has been studied in
fish and invertebrates. The research on effects of petroleum
hydrocarbons has not clearly demonstrated major alterations in
an organism's capacity to adapt to new salinities. In studies
on both oysters (Anderson, R.D., 1975; Anderson and Anderson,
1976) and shrimp (Anderson *et al.*, 1974; Cox, 1974), there was
a lag in the rate of chloride ion adjustment to new salinities
but the effect was only transitory. By about 96 hr after
exposrue to hydrocarbons and transfer to new salinities,
Penaeus aztecus had depurated tissue naphthalenes and returned
to control levels of chloride ion concentration. *Crassostrea
virginica* exposed to much higher levels of oil required
between 3 and 10 days to adjust to a new steady state level of
blood chloride, and the major decrease in content of tissue
naphthalenes occurred by 7 days. Again, these data suggest a
relationship between tissue content of aromatics, represented
by naphthalenes, and the ability of organisms to regulate
metabolic functions.

An approach which is assumed to be a sensitive measure of
organism "health" is behavior, but few investigations have
coupled quantitative measures of behavioral responses with
good analytical chemistry. If these two areas of expertise
could be combined in the study of behavior, the results might
provide a good estimate of the lowest level of hydrocarbon

contamination producing adverse effects on marine organisms. Percy and Mullin (1977) demonstrated suppression in the loco-motor activity of an amphipod and a coelenterate medusa after exposure to extracts of four oils. Rather high concentrations (15 ppm and greater) were used to produce these effects and, in some cases, the organisms exhibited recovery during 24 hr in clean water.

Even correlation of gross behavioral abnormalities with tissue and organ content of contaminants may provide useful information. To provide some preliminary data on the feasi-bility of this approach, Dixit and Anderson (1977) studied the relationship between naphthalenes content in organs of acutely exposed *Fundulus similus* and the stage of abnormality from "normal" to death. As expected from earlier studies (Lee *et al.*, 1972; Roubal *et al.*, 1977) high concentrations of naph-thalenes were found in the gall bladder and liver. A point of interest was the relatively consistent and longer retention time of naphthalenes in the brain tissue of the fish. Levels of the compounds at or above 200 ppm in the brain may inter-fere with regulatory mechanisms and lead to death. The first sign of abnormal behavior occurred at the point that brain tissue contained about 200 ppm, and as more adverse effects were exhibited the concentration increased. After 23 hr in clean water, 140 ppm was still present in the brain, but by 15 days, depuration was complete.

Oil-contaminated sediment or food have been used in recent studies (Percy, 1976; Shaw *et al.*, 1976; Taylor and Karinen, 1977) to evaluate the survival and/or behavior of benthic invertebrates. The Arctic amphipod, *Onisimus affinis*, avoided oil-tainted food as well as oil masses, but this response diminished with preexposure to oil and when weathered oil was used. The isopod, *Mesiodotea entomon*, was more resistant to oil and exhibited no preference for clean food (Percy, 1976). Taylor and Karinen (1977) studied the behavior of the detriti-vore, *Macoma balthica*, in response to oil-contaminated sub-strate and in response to oil extracts flowing over clean substrate. They reported that at least 95% of these clams burrowed to the surface before dying, and this response was elicited within 3 days by 50% of the animals exposed to 0.37 ppm naphthalene equivalents. At a calculated dose of 0.2 ppm naphthalene equivalents, 50% of the clams would not burrow into the sediment within 60 minutes. A concentration of 0.67g oil/cm^2 was found to stimulate 50% of the clams to move to the surface within 24 hr. No death was observed in these 4 to 6 day experiments, while in the field studies of Shaw *et al.* (1976) doses of 5 μl oil/cm^2 applied to surface sediment daily for 5 days produced significant mortality. The behavior of these clams is a response that can be readily quantitated, and it appears that in some cases it may be sensitive to

relatively low levels of petroleum hydrocarbons. Oil sorbed
to sediment can be tolerated to a greater extent than hydrocarbons dissolved or suspended in the water column. Krebs and
Burns (1977) noted the impairment of *Uca pugnax* behavior
during field observations after the West Falmouth spill and in
laboratory exposures to 100 ppm fuel oil in sediment. Abnormal
responses were most apparent at colder temperatures (13-15°C),
and mortalities were greatest during overwintering. While
behavior is still possibly a sensitive measure of the response
or organisms to sublethal levels of toxicants, there has been
no clear demonstration of effects from low levels of petroleum
hydrocarbons on or in sediment. It is quite possible that the
effects observed are due to the hydrocarbons released (by
various means) from the sediment to the interstitial water,
where they are readily available for tissue accumulation.

Growth of marine organisms under clean and oil-contaminated
conditions has been studied by a number of investigators using
a variety of techniques (reviewed by Anderson, 1977). The
findings indicate that growth is a parameter which demonstrates
the sensitivity of larval or juvenile stages to low hydrocarbon concentrations. At levels of 0.1 and 0.6 ppm total naphthalenes, organisms ranging from fish to polychaetes exhibit
reduced growth and/or survival. These studies involved exposure times from 4 to 60 days, and this time factor is worth
further discussion. To evaluate the sensitivity of a biological parameter as it relates to pollutant response, the exposure conditions, particularly time, should be considered on a
cumulative basis. For example, a very comprehensive study of
lobster larvae conducted by Wells and Sprague (1976) showed
that the 4 day LC50 value for 3rd- and 4th-stage larvae was
4.9 ppm, while the 30 day LC50 concentration for larvae and
the concentrations suppressing growth was 0.14 ppm, which
provided a ratio of "safe" to acute concentration of about
0.03. This is the conventional manner of calculating differences between 4 day lethal doses and the results of long-
term tests. I believe a more realistic approach to examing
differences would include the product of dose and exposure
time, since hydrocarbons are constantly taken up by marine
organisms, and a critical level of contamination is probably
associated with death. When exposure time (in days) is
multiplied by the LC50 concentration (in ppm), values for the
ratio of the product (ppm-days) suppressing growth to the
product producing mortality in 96 hr, range from 0.2 to 0.8
for *Homarus americanus, Palaemonetes pugio,* and *Neanthes
arenaceodentata* (Wells and Sprague, 1976; Anderson, 1977).
These ratios for the product of LC50 concentration and time
of exposure do not indicate such a wide difference between the
results of 96 hr lethal tests and those of long-term growth
studies. This is not to say that growth and reproduction

studies are not more sensitive approaches to the problem, but for northern species they may be very time consuming, and a consistent oil exposure system is difficult to construct and expensive to maintain and monitor. It would be surprising if the ratios were not even larger when the 96 hr tests were con- ducted in flowing exposure systems. If data from both short- term (acute) exposures and long-term exposures to concentra- tions producing abnormal physiological or developmental responses are compared on a time specific basis, using the same oils and delivery systems, we will have an excellent opportunity to evaluate the significance and sensitivity of a response parameter. Based on the available information, growth would seem to be a meaningful and sensitive measure of the effects of hydrocarbons on populations. Most of the results relate to chronic environmental exposure as might be produced by effluents. Data on oil spills indicate that the levels of hydrocarbons in the water drop to quite low levels within a matter of days. Long-term exposures, therefore, are simulating a constant input observed near effluent outfalls or in areas where the oil-contaminated substrate is leaching hydrocarbons to the water column.

Other approaches to the determination of responses to sub- lethal concentrations of hydrocarbons might be lumped into two large categories. The first is morphological and histo- logical effects, and the second would be the very broad topic of biochemical alterations in response to hydrocarbons. Hawkes (1977) summarized much of the available information on the effects of hydrocarbons on fish tissue and presented some recent findings from her own research. Tissue and ultra- structural damage has been demonstrated, but it is too early to formulate general patterns associated with specific levels of oil or hydrocarbon component exposures. It appears to the author that this is also the case for studies related to bio- chemical alterations of organisms exposed to hydrocarbons. Both of these fields show potential for use as sensitive bio- logical parameters, but the amount of information generated thus far, does not provide us with a suitable base for making evaluations.

Included in this symposium (Roesijadi and Anderson, 1978) is an example of biochemical alterations that respond to hydrocarbon exposure. The detritivore, *Macoma inquinata*, exposed in the natural environment to oil-contaminated sedi- ment exhibited decreases in condition index and free amino acid content, as compared to control animals. A variety of parameters can be measured after realistic exposures of this type to determine threshold levels of sediment contamination for benthic organisms.

The above information, especially the growth and repro- duction findings, show that levels of 0.1 to 0.3 ppm total naphthalenes, regardless of the oil, appear to be the lowest concentrations in water producing abnormal and deleterious responses during long-term, sublethal exposures to petroleum hydrocarbons.

Uptake and Release

Water Exposure. Much of the recent data on uptake and release kinetics exhibited by various marine organisms has already been reviewed (Neff *et al.*, 1976a, b; Anderson, 1977). In addition, Lee (1977) has described data on bivalves and their relation to field monitoring of petroleum contamination. Therefore, this section summarizes these data by taking a broad overview of our knowledge and pointing out trends. Bivalve molluscs accumulate hydrocarbons from both the particulate and dissolved phase. Differences in uptake of n-alkanes reported in various studies may be due to the presence or absence of dispersed oil droplets or oil sorbed to particles. When only water-soluble hydrocarbons are available, uptake of alkanes is relatively small. In all cases, these paraffins seem to be the compounds which are most rapidly released, and in some cases the rapid decrease may be due to incorporation of the compounds in feces or pseudofeces and subsequent elimination.

A comparison of bioaccumulation factors and short-term release rates for the clam, *Rangia cuneata*, tested with 2- and 5-ring aromatics was presented by Neff *et al.* (1976b). The lowest tissue magnification was exhibited by naphthalene, the greatest by phenanthrene, and the 4- and 5-ring compounds were accumulated to intermediate levels. Only 13-16% of phenanthrene and benzo(a)-pyrene were released after 24 hr in clean water, while 66% of naphthalene was depurated, and 26% of chrysene (4 rings) remained. Neff *et al.* (1976a) summarized extensive data on fish and shrimp, as well as bivalves. *Rangia* exposed to a water-soluble fraction of No. 2 fuel oil for 24 hr were analyzed for content of specific naphthalenes at the end of exposure and after 24 hr in clean water. Concentrations in the exposure water decreased from naphthalene through methyl- and dimethylnaphthalenes to the trimethylnaphthalenes, but the bioaccumulation factors increased in this order. Values ranged from a low of 2.3 for naphthalene to a high of 26.7 for trimethylnaphthalenes. Release during 24 hr was greatest for naphthalene (79%), but there was no apparent pattern for the release of other compounds which ranged from 32 to 51%.

One cannot extrapolate from one species to another regarding uptake and release rates of aromatic hydrocarbons. While bivalves tend to slowly but constantly accumulate hydrocarbons, fish and shrimp take these compounds up very rapidly, reaching maximum levels within a few hours.   Longer exposures often produce lower tissue concentrations of the parent compounds. For a review of metabolite production, transformation and retention times in marine organisms, the reader is referred to Malins (1977a, b) and Varanasi and Malins (1977).   The accumulation of parent petroleum hydrocarbons as well as metabolites in the liver, gall bladder and hepatopancreas (crustaceans) has been demonstrated, and there is no doubt that this activity contributes significantly to the release rates exhibited by fish and crustaceans.   The retention times of a few days for fish and crustaceans, no doubt give evidence of more active systems for ventilation, blood circulation, excretion and hydrocarbon metabolism, as compared to bivalves.   Malins (1977b) has pointed out the need to better understand the transport and effects of metabolites of petroleum hydrocarbons in organisms and the marine environment.

Induction of these enzymatic degradation systems also occurs during chronic exposure to petroleum or other organic pollutants.   There are instances where this process may help explain findings from sublethal effects studies.   There have been several instances where sublethal responses, including respiration and growth, have returned to normal levels during constant exposure of organisms.   One specific demonstration of induced hydrocarbon resistance was reported by Rossi and Anderson (1978) studying the increased tolerance of the polychaete, *Neanthes arenaceodentata*, chronically exposed to No. 2 fuel oil water-soluble fractions.   While larvae and juveniles did not exhibit increased tolerance to acute levels of these extracts after chronic exposure, adult animals increased in tolerance such that 96 hr LC50 values were consistently greater by a factor of between 2 and 3.   The resistance was not dependent upon the number of generations reared under chronic exposure (within 3 generations) nor the concentration of the exposure water.   Resistance was reduced by transferring the first generation organisms to clean water 14 days before LC50 tests, but a 7 day period had no effect, and these same transfers in the next two generations did not reduce the tolerance of the worms.   Lee *et al.* (1977) have demonstrated the presence of an aryl hydrocarbon hydroxylase system in the polychaete, *Nereis* sp., and Lee's paper (1978) in this volume will also address induction.

While the capabilities of individual species alter the rates of hydrocarbon uptake and release, there are some basic patterns associated with the chemical structure of specific compounds.   At least a portion of the depuration process appears to be dependent upon the partitioning of hydrocarbons between the water phase of tissues and tissue lipids.   When animals are returned to clean seawater, the lipid/water

partition coefficients for the hydrocarbons favor their gradual
release from the tissues, and such exchange is dependent upon
the lipophilic nature of the compound. There are numerous
examples where tissues first release naphthalene, followed by
methylnaphthalenes and then the di- and trimethyl isomers.
This pattern has been observed in bivalves lacking enzymatic
degradation pathways, as well as fish and crustaceans. There
is still a question regarding the possible longer retention
of some contaminants which might have been incorporated in a
more stable tissue compartment of bivalves as a result of
longer exposure periods.

   Sediment Exposure. Until very recently, little information
was published regarding the bioavailability of sediment-bound
petroleum hydrocarbons. Rossi (1977) reported on the comparative aspects of naphthalene uptake from water and sediment by
N. arenaceodentata. Biomagnification of C^{14}-naphthalene from
water was approximately 40 times in 24 hr, but after 28 days
in sediment containing 9 ppm naphthalenes, there was no
significant uptake. When C^{14}-methylnaphthalene was used to
contaminate artificial detritus on which the worms fed for 16
days, no accumulation was observed. A period of 24 hr in clean
water was sufficient for the organisms to egest the apparent
tissue contamination via fecal pellets. While the sediment
studies of Rossi (1977) involved only a fine layer of substrate
with no interstitial water, Anderson *et al.* (1977) utilized an
exposure system with a 10 cm depth of sediment containing
sipunculid worms, *Phascolosoma agassizii*. When exposed to
rather high concentrations of hydrocarbons on/in sediment,
these animals exhibited quite low levels of tissue contamination, and even 2 days of depuration resulted in a substantial
release of naphthalenes. One and two week depuration periods
were sufficient for the organisms to completely release naphthalene. It was not possible to separate the uptake via
ingested sediment from the possible uptake from interstitial
water using the above exposure system. Anderson and Kiesser
(unpublished data) designed a system to allow approximation of
the contribution of interstitial water, released to the water
column, to the contamination of amphipods burrowing in oiled
substrate. Amphipods (*Anonyx laticoxae*) living in the oiled
sediment accumulated 8 ppm total naphthalenes after 4 days,
while those in clean substrate suspended above the bottom contained 4 ppm. After 18 days of exposure, neither group
contained a significant amount of contamination, while the
sediment still possessed 80% of the initial total hydrocarbons
and 30% of the naphthalenes. Very similar findings were
reported by Roesijadi *et al.* (1978b) studying the bioavailability of sediment-bound naphthalenes for the detritivore,
Macoma inquinata.

Utilizing C^{14}-labelled aromatics, Roesijadi *et al.* (1978a)
studied the comparative uptake of 3- through 5-ring aromatic
hydrocarbons by *M. inquinata.* They found that after an
initially rapid uptake, phenanthrene (3 rings) concentrations
in tissue constantly decreased during the 56 day exposure to
contaminated sediment. Exposure to chrysene (4 rings) showed
initial uptake to about 20 days, followed by a decrease during
the next 30 days. Benzo(a)pyrene tested in the same system
was constantly accumulated during the entire 42 day exposure
period.

These data lend further evidence to the hypothesis that
tissue accumulation and retention are the result of relative
partitioning between the water and lipid phases of tissue.
Based on all the above experimentation, it appears that polynuclear aromatics, such as benzo(a)pyrene, may be taken up by
deposit-feeding benthic organisms. Naphthalenes and phenanthrenes which are rapidly accumulated from solution are
apparently not taken up (or at least retained) to a significant extent from oil-contaminated sediment. An important
factor to consider in this evaluation is the hydrocarbons in
interstitial water and the amount of organism contact with
this fluid before it is diluted by the water column. The
adverse effects on *Uca*, reported by Krebs and Burns (1977),
may well be the result of long contact with interstitial water
derived from partially anaerobic sediment which contains
rather high amounts of aromatics.

CONCLUSIONS AND RECOMMENDATIONS

Toxicity of crude and refined oils appears to be primarily
the result of concentrations in the water extracts or disper-
sions of mono- and diaromatic petroleum hydrocarbons. Crude
oils contain a higher proportion of monoaromatics than refined
products, and these are the most volatile compounds found in
water extracts. The length of time organisms are exposed to
these low molecular weight compounds will be temperature
dependent, and bioassays conducted in cold water ($10^{o}C$ or less)
appear to verify these assumptions. There is a general pattern
of toxicity related to the degree of alkylation and the water
solubility of the compounds. Four- and five-ring aromatic
compounds, such as chrysene and benzo(a)pyrene, have very low
solubility in seawater and, thus, are relatively nontoxic
during short exposures. With the compound classes known as
benzenes (1 ring), naphthalenes (2 rings) and phenanthrenes
(3 rings), toxicity appears to increase with the degree of
alkylation. Most of the pertinent information in this area of
study has been obtained, but very few long-term flowing

exposure systems have been used to accurately evaluate the
threshold levels of chronic exposure. It is necessary to
obtain time-concentration data to determine "safe" levels of
contamination under conditions of exposure, which may have
durations of one week to constant contamination. These long-
term exposures should incorporate multidisciplinary approaches
such that, in addition to deriving information on mortality
of larvae, juveniles and adults, sublethal responses could be
evaluated, and tissue contamination levels can be correlated
with both lethal and sublethal findings. The differences
observed in short-term tests on adults and long-term exposures
with larvae or juveniles will likely be more clearly defined
using this approach.

Biological parameters which are likely to be useful in
assessing the effects of sublethal hydrocarbon concentrations
on marine organisms include morphological effects, biochemical
alterations, and studies on the growth, reproduction and
behavior of animals. In no case should an investigation be
conducted or results accepted if sound analytical chemistry
is not used for description of the exposure conditions and
resulting tissue contamination. Within these investigations
there should be consideration of the hydrocarbon metabolizing
capabilities of the test species and the attempt to determine
the fate and possible effects of these compounds. During
realistic long-term exposures to fish, crustaceans, and polychaetes, there is evidence of resistance induction. It is
possible that the duration of abnormal responses may be closely
linked to the capabilities of a given species to maintain a
constant level of internal contamination when exposure conditions are also constant. This hypothesis should be further
tested as it has important implications regarding the tolerance limits of chronically exposed organisms, which may aid
in explaining the results of both field and laboratory
experiments.

Recent findings on the uptake and release kinetics of
hydrocarbons exhibited by a variety of marine organisms must
be carefully considered in interpreting the results of sublethal studies. While many questions regarding the uptake,
production and release of hydrocarbon metabolites are still
unanswered, data on parent compounds seem to fit a rather
consistent pattern. Retention times of hydrocarbons in tissues, regardless of the species involved, tend to be dependent
upon the relative ratios of water and lipid solubilities.
Factors such as the level of organ-system development in a
species and its hydrocarbon metabolizing capabilities can
certainly affect the rates of depuration, but the pattern of
specific compound release is basically the same. Naphthalene
is usually the first compound to be depurated from contaminated tissues, followed by methylnaphthalenes. The di- and

Taylor, T. L. and J. F. Karinen. 1977. Response of the clam, *Macoma balthica* (Linnaeus), exposed to Prudhoe Bay crude oil as unmixed oil, water-soluble fraction, and oil-contaminated sediment in the laboratory. Pp. 229-237. In: Fate and Effects of Petroleum Hydrocarbons in Marine Ecosystems and Organisms. D. A. Wolfe (ed.), Pergamon Press, Oxford.

Vanderhorst, J. R., C. I. Gibson and L. J. Moore. 1976. Toxicity of No. 2 fuel oil to coon stripe shrimp. Mar. Poll. Bull. 7:106-107.

Varanasi, U. and D. C. Malins. 1977. Metabolism of petroleum hydrocarbons: Accumulation and biotransformation in marine organisms. Pp. 175-262. In: Effects of Petroleum on Arctic and Subarctic Marine Environments and Organisms. D. C. Malins (ed.), Academic Press, New York.

Wells, P. G. and J. B. Sprague. 1976. Effects of crude oil on American lobster (*Homarus americanus*) larvae in the laboratory. J. Fish. Res. Bd. Canada 33:1604-1614.

Wolfe, D. A. 1977. Fate and Effects of Petroleum Hydrocarbons in Marine Ecosystems and Organisms. Pergamon Press, Oxford.

DETOXIFICATION SYSTEM IN POLYCHAETE WORMS: IMPORTANCE IN THE DEGRADATION OF SEDIMENT HYDROCARBONS

Richard F. Lee
Sara C. Singer
Kenneth R. Tenore
Wayne S. Gardner

Skidaway Institute of Oceanography
P. O. Box 13687
Savannah, Georgia

Richard M. Philpot

National Institutes of Health
National Institute of Environmental Health Sciences
Pharmacology Branch
Research Triangle Park, North Carolina

INTRODUCTION

Coastal sediments near urban areas and around oil spills contain high concentration of hydrocarbons. The sources of these hydrocarbons, particularly polycyclic aromatic hydrocarbons, include wastewater effluents, spills of petroleum and its derivatives, and hydrocarbons associated with atmospheric particulates resulting from fossil fuel combustion (Hites, 1976; Ocean Affairs Board, 1975; Van Vleet and Quinn, 1977). Hydrocarbons can be incorporated into the bottom after sedimentation from surface waters. Once incorporated into the sediment these hydrocarbons can persist for many years (Blumer and Sass, 1972).

Polychaete annelids are important members of the benthic community in most marine areas. Studies around oil spills, refinery effluents and oil field brine effluents have shown that polychaetes are often the dominant animals in these areas (Armstrong *et al.*, 1976; Baker, 1976; Reish, 1971; Sanders

23

et al., 1972). The cosmopolitan species, *Capitella capitata*,
dominated the benthos after the West Falmouth oil spill
(Sanders *et al.*, 1972) and in parts of Long Beach harbor
receiving refinery wastes (Reish, 1971). Carr and Reish (1977)
determined 28 day-LD_{50} values for five polychaete species
exposed to water extracts of fuel oil and crude oil. There
was a wide range of LD_{50} values for the different species with
Capitella capitata the species most resistant to the toxic
effects of oil. Species, such as *Arenicola marina* and *Nereis
diversicola*, although reduced in numbers after oil spills re-
establish in the sediments in spite of the continued presence
of oil (Prouse and Gordon, 1976; Levell, 1976; Baker, 1976).

Many polychaetes are deposit feeders and are directly
exposed to sediment hydrocarbon which they can take up and
degrade (Prouse and Gordon, 1976; Lee, 1976; Lee *et al.*, 1977;
Rossi and Anderson, 1977, 1978). The mixed function oxygenase
(MFO) system, which is responsible for the metabolic modification of many foreign compounds in animals, is present in
polychaetes (Lee, 1976; Lee *et al.*, 1977).

This paper discusses the hydrocarbon metabolic system in
polychaetes and its induction after exposure to crude oil and
benz(a)anthracene. Evidence is presented which suggests an
important role for polychaetes in both directly degrading
hydrocarbons and in exposing hydrocarbons in sediments to
microbial action as a result of feeding activity.

MATERIALS

Enzyme and P-450 Studies

Tissues of *Nereis virens*, collected in mudflats in Maine,
were dissected, weighed and homogenized in 0.15 M KCl buffered
with 0.05 M Tris buffer (pH 7.5) using a Potter-Elvehjem
homogenizer. Cell debris and nuclei were removed by centrifugation at 700 x g for 10 minutes at 4^0C. Mitochondria were
collected by centrifugation of the supernatant at 8,000 x g
for 10 minutes. The supernatant from the mitochondrial fraction was centrifuged at 140,000 x g for 60 minutes in an
ultracentrifuge (Beckman Model L-40) to sediment microsomes.
The protein content of each fraction, after resuspension in
the buffer KCl solution, was determined by the method of Lowry
et al. (1951) using bovine serum albumin as the reference
standard.

The assay for mixed function oxygenase activity was similar to that of Wattenberg *et al.* (1962) with modifications
described by Nebert and Gelboin (1968). Each assay mixture,
in a total volume of 1 ml, contained 50 μmoles of Tris buffer,

pH 7.5, 0.6 μmoles of NADPH, 3 μmoles of $MgCl_2$, 0.2 ml of cell homogenate (1-2 mg protein) and 0.01 μmoles of benzo(a)pyrene (added in 50 μl of methanol). The mixture was shaken at 30^0C for 30 minutes. The reaction was stopped by addition of 1 ml of cold acetone followed by 3 ml hexane. The organic phase was extracted with 3.0 ml of 1 N NaOH. The concentration of hydroxylated benzo(a)pyrene in the alkali phase was determined with a fluorometer (Turner Model 430) with activation of 396 nm and emission at 522 nm. For some experiments the fluorescent spectra of the alkali phase at these activities and excitation wavelengths was recorded. Enzyme activites were determined in triplicate and compared with a boiled enzyme control. One unit of hydroxylase activity was defined as that amount catalyzing the formation in a 60 minute incubation at 30^oC of hydroxylated product causing fluorescence equivalent to that of 1×10^{-12} moles of 3-hydroxybenzo(a)pyrene.

For the lipase experiment a microsomal homogenate from the intestine containing 2 mg of protein, was incubated for 30 minutes in the buffered KCl solution (pH 7.4) with 4 units of phospholipase C (phosphatidyl choline phosphohydrolase EC 3.1.4.3, *Clostridium welchii,* type I, Sigma Chemical Co., specific activity 5 units per mg, where 1 unit liberates 1 μmole of water-soluble organic phosphorus from egg yolk lecithin per min. at pH 7.3 at 37^0C). After the preincubation with lipase the normal assay for mixed function oxygenase was carried out.

For the pH studies tissues were extracted with the normal pH 7.5 buffer. Microsomes were prepared from aliquots of the extract and the pellets were resuspended in buffers of different pH. These were 0.05 M potassium phosphate (pH 6.5 and 7.0) and 0.05 M Tris (pH 7.5, 8.0 and 8.5).

For determination of cytochrome P-450 cuvettes contained between 1 and 2 mg microsomal protein in 0.05 M Tris buffer at pH 7.5 containing 0.15 M KCl. The sample cuvette was bubbled for 20 seconds with carbon monoxide and then reduced by addition of a few grains of sodium dithionite. The reference cuvette was reduced by sodium dithionite addition. Spectra were run at room temperature (26^0C) on an Aminco DW-2A spectrophotometer at a full scale readout of 0.01 absorbance unit. The amount of P-450 was determined by the method of Omura and Sato (1964).

Induction of MFO Activity in Capitella

Culture trays containing 2000 g of sand sediment (< 0.3 mm grain size) were seeded with 25 *Capitella capitata* which were grown on a diet of mixed cereal (350 mg per day). After ten weeks a large culture of *Capitella* occupied the trays. Five grams of Kuwait crude oil was mixed into the sediment in

one tank (tank 1) while 300 mg of benz(a)anthracene were mixed
into the sediment of a second tank (tank 2). After 4 weeks,
5 mg of benz(a)anthracene was mixed into 1 ml of corn oil and
added to the food three times a week to tank 2. A total volume
of 5 liters of seawater was maintained in each tank with a flow
rate of about 20 ml per minute. Extracts of whole worms assayed
for mixed function oxygenase activity by the methods described
above.

Degradation of Oil by Microbes and Capitella

Five ml of Kuwait crude oil, enriched with 0.1 g each of
anthracene, fluoranthene and benz(a)anthracene and 0.05 g of
benzo(a)pyrene were mixed with sediment collected from a near-
by estuary. Six trays containing 2000 g of sediment received
100 ml/min. of filtered (1 µ), temperature regulated (20^0C)
seawater (salinity ca. 25o/oo). Four days after the water
flow was started, half of the trays were seeded with *Capitella
capitata* (20 worms per tray). At various time intervals three
randomly selected cores were taken from each tray and mixed
to form a composite for the extraction of hydrocarbons and
determination of dry weight.

Samples of sediment (0.2 g) were extracted with hexane.
Hexane extracts were concentrated to dryness and dissolved in
10 µl of methanol. Anthracene, fluoranthene, benz(a)anthra-
cene and benzo(a)pyrene were measured by passing 5 µl of the
methanol solution through an assembled liquid chromatograph
equipped with a fluorometric detector (Fluoromonitor-American
Instrument Co.). A standard curve was prepared for each compound. The chromatograph was equipped with a LiChrosorb RP-2
(Merck) column (25 cm long and 2.0 mm inner diameter). The
sample was eluted with solvent mixtures added sequentially
(solvent 1 -methanol:water, 1:1; solvent 2 -methanol:water, 2:3;
solvent 3 -methanol:water, 3:2).

Microbial degradation of the polycyclic aromatic hydrocarbons was determined by adding ^{14}C-labelled hydrocarbons to
a sediment sea water slurry (1 g sediment per 50 ml seawater)
in a 125 ml glass bottle capped with a silicon stopper. During
incubation the sediment was kept in suspension by shaking
(Junior Orbit Shaker, Lab-Line Co.). Controls were bottles
containing mercuric chloride. At the end of the incubation
2N sulfuric acid was added and the released [14]CO_2 was collected
in phenethylamine and counted in a liquid scintillation counter
(Beckman LS 100C). The radioactive hydrocarbons used were
anthracene -9 10-[14]C (33 mci/mM-California Bionuclear Corp.),
[14]C-12-benz(a)anthracene (48 mci/mM-Amersham), benzo(a)pyrene-3,
6-[14]C (21 mci/mM-Amersham).

RESULTS

Mixed Function Oxygenase in Nereis virens

Mixed function oxygenase activity was associated with the lower portion of the intestine with little or no enzyme activity in the pharynx, esophagus or upper portion of the intestine. Subcellular fractionation of the intestine indicated that the enzyme system was associated with the microsomal fraction. Specific activity of the washed microsomes were 21,000 enzyme units/mg microsomal protein. Specific activity of crude homogenates were 150 ± 50 enzyme units/mg protein. Activity was shown to be linear with time and protein in the limits of the assay conditions. Full enzyme activity required addition of NADPH, Mg^{+2} and oxygen (Table 1). Enzyme activity was not effected by the detergent Tween-80 at 0.1% but was completely inhibited at this concentration by sodium dodecyl sulfate and Triton X-100. A 30 minute incubation of the microsomal preparation with phospholipase C completely inhibited mixed function oxygenase activity. The temperature optima curve showed a peak at 30^0C while the pH curve showed highest activity between pH 7 and 8 (Figures 1 and 2).

Carbon monoxide (bubbled for 15 seconds), SKF 525-A (1 x 10^{-5} M) and 7, 8-benzoflavone (1 x 10^{-4} M), all inhibitors of cytochrome P-450, reduced MFO activity by 55, 30 and 21%, respectively. SKF 525-A is 2(diethylamino) ethyl 2,2-diphenyl-pentanoate. An assay for cytochrome P-450 in microsomes of the intestine showed the presence of this cytochrome (Figure 3). Even after extensive washings the preparation shows the presence of blood respiratory pigments (peak at 424 nm and trough at 437 nm). The concentration calculated from the spectrum was 6.6 picomoles of P-450/mg of microsomal protein.

TABLE 1. Requirements for mixed function oxygenase activity in microsomal preparations from the intestine of Nereis virens

Incubation system	Relative activity (%)
Complete (MgCl$_2$, NADPH, O$_2$)	*100*
Minus MgCl$_2$	*73*
Minus NADPH	*0*
Minus NADPH plus NADH	*35*
Nitrogen atmosphere	*10*

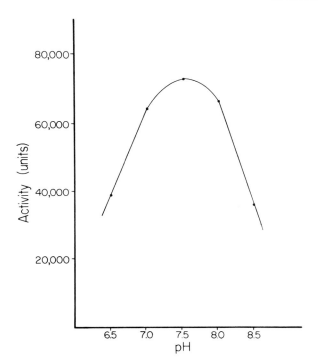

FIGURE 1. *Optima pH for mixed function oxygenase in micro-somes of* Nereis virens *intestine.*

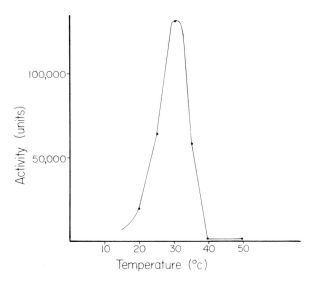

FIGURE 2. *Temperature optima for mixed function oxygenase in microsomes of* Nereis virens *intestine.*

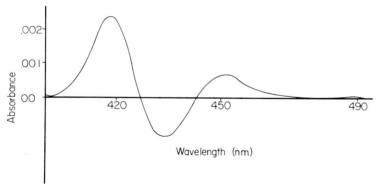

FIGURE 3. Carbon monoxide difference spectrum of a micro-somal preparation of Nereis virens *intestine.*

Mixed Function Oxygenase Activity in Capitella *after Exposure to Crude Oil and Benz(a)anthracene*

Kuwait crude oil or benz(a)anthracene were mixed into sediment containing a culture of *Capitella capitata*. At time intervals of 3, 6 and 10 weeks approximately 100 worms were removed from each tray for an assay of MFO activity (Table 2). Whole worms were extracted and assayed since their small size prevented dissection of specific tissues. The generation time was approximately 3 weeks so that after 10 weeks the trays contained primarily the third generation of worms exposed to oil.

Worms exposed to benz(a)anthracene increased in activity from zero to 200 units/mg protein after 3 weeks and finally to 500 units/mg protein after 10 weeks. In worms exposed to crude oil very low MFO activity was observed after 6 weeks of exposure and somewhat higher activity after 10 weeks. The association of the enzyme with the microsomes (collected at 140,000 x g) indicated that bacteria were not responsible for the observed hydrocarbon degradation since bacteria were removed by the preliminary low speed centrifucation (800 x g).

Degradation of Hydrocarbons in Sediments by Capitella *and Microbes*

Crude oil enriched with anthracene, fluoranthene, benz(a)-anthracene and benzo(a)pyrene was mixed with sediment in trays. Half of the trays contained in addition to the normal micro-fauna, a culture of *Capitella capitata,* while the remaining trays contained no worms. The concentrations of anthracene, fluoranthene, and benz(a)anthracene showed a linear decrease during the 31 week experiment with a greater decrease in trays with worms. In Figure 4 the decreasing concentration of

TABLE 2.   Induction of mixed function oxygenase activity
in Capitella capitata

Compounds	Time of exposure (weeks)	MFO activity (enzyme units/ mg protein)
Benz(a)anthracene	0	0
	3	200
	6	200
	10	500
Kuwait crude oil	0	0
	3	0
	6	10
	10	100

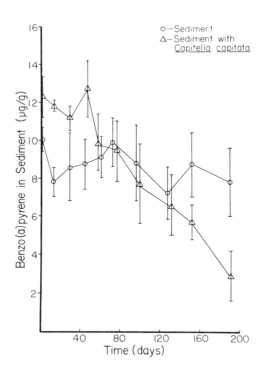

FIGURE 4.   Changes in sediment benzo(a)pyrene concentrations
with and without Capitella capitata.

benzo(a)pyrene in trays containing *Capitella capitata* can been
seen and contrasts with little or no decrease in this compound
in trays without worms. The microbial degradation rate of
added ^{14}C-benzo(a)pyrene at a concentration of 1.0 µg/g sediment in oil treated trays was 0.3 ng/g sediment/day. At the
start of the experiment the concentration of benzo(a)pyrene was
25 µg/g sediment. After 160 days microbes degraded only 0.05 µg
of benzo(a)pyrene to carbon dioxide. Thus it appears that
Capitella was responsible for most of the observed degradation
of benzo(a)pyrene in the sediments.
    Figure 5 compares the microbial degradation of ^{14}C-
benz(a)anthracene in the top and bottom of oil-treated sediments with and without worms. The surface sediments without
worms show very high activity whereas little or no activity
was observed in subsurface sediments of these trays. In trays
with worms microbial degradation was seen in both surface and
subsurface sediments. These results suggest that in trays
without worms most of the hydrocarbon degradation takes place
on the surface, while the presence of worms allows microbes to
attack subsurface hydrocarbons. The worms may provide aeration
and nutrient to hydrocarbon-degrading microbes in subsurface
sediments.

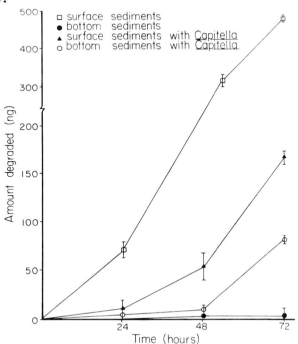

*FIGURE 5. Degradation of ^{14}C-benz(a)anthracene (2.5 µg/g
sediment) in sediment with and without* Capitella capitata.

DISCUSSION

Mixed function oxygenase (MFO) activity in *Nereis virens* associated with the intestine had specific activity of 150 units/mg protein which can be compared with MFO activity in blue crab (*Callinectes sapidus*) stomach where specific activity ranged from 1400 to 2500 units/mg protein (Singer and Lee, 1977). Purification of the hepatic mixed function oxygenase system in fish and mammals has shown it to be composed of cytochrome P-450, NADPH-cytochrome P-450 reductase and phospholipid (Lu *et al.*, 1976; Philpot *et al.*, 1976). The evidence we have presented suggests that the MFO system in polychaetes is composed of similar components. A phospholipid requirement was suggested by the inhibition of enzyme activity by detergents and phospholipase C. Spectral evidence of P-450 was demonstrated and an indication of its importance was shown by inhibition of MFO activity by P-450 inhibitors, including carbon monoxide, SKF 525-A and benzoflavone. The cofactor requirements, pH optima and temperature optima of the MFO in polychaetes was similar to those observed in crustaceans, insects, fish and mammals (Khan *et al.*, 1972; Nebert and Gelboin, 1968; Pedersen *et al.*, 1974; Pohl *et al.*, 974; Singer and Lee, 1977; Terriere and Yu, 1976; Wilkinson and Brattsten, 1972).

MFO activity in *Capitella capitata* could be detected only after exposure of the organisms to either petroleum or its components. To demonstrate the basal level of MFO activity in *Capitella* would probably require dissection and assay of intestinal tissue, rather than using extracts of whole worms. It would be difficult to carry this out because of the small size of *Capitella* (ca. 1 mg dry wt./worm).

After *Capitella* cultures were exposed for long periods to oil or its components the third generation worms had much higher MFO activity than first or second generation worms. Rossi and Anderson (1978) exposed the polychaete, *Neanthes arenaceodentata,* to fuel oil and found that third generation worms were significantly more resistant, *i.e.*, higher TL_m values, than the first generation. Possibly, the increase resistance of polychaetes to oil after three generations was due to an increase to the MFO system which acts to detoxify the aromatic hydrocarbons. Grassle and Grassle (1976,1977) have shown by a study of the electrophoretic patterns that *Capitella capitata* is actually a complex of at least six sibling species. Thus, it may be that exposure to oil results in selection for species or strains resistant to oil.

In insects pesticide-resistant strains have higher MFO activity than less tolerant strains (Brooks, 1977; Walker and Terriere, 1970; Wilkinson and Brattsten, 1972). This increase in enzyme activity appears to allow oxidative detoxification of pesticides.

The importance of microbes in degrading sediment hydrocarbon has been well documented but only recently has the role of polychaetes in this process been recognized (Lee, 1976; Prouse and Gordon, 1976; Rossi and Anderson, 1977). The results we have presented show that *Capitella capitata* can metabolize hydrocarbons with a MFO system and also can enhance microbial degradation of subsurface hydrocarbons. Some of the polychaetes belong to the meiofauna (larger than microscopic but less than 1 mm), which is composed of harpactoicoid copepods, nematodes, turbellarians and polychaetes (Marc, 1942). Benthic animals larger than meiofauna are placed in the macrofauna category. Many of these benthic species are deposit feeders that are important in the oxidation and recycling of sediment organic matter (Tenore *et al.*, 1977). In undisturbed sediments most microbial activity is restricted to the surface, but feeding processes of benthic animals results in sediment turnover to depths as great as 15 cm (Rhoads, 1967) thus allowing microfauna to degrade organic matter from deeper sediments. We conclude that petroleum degradation in marine sediments is a result of interactions between microfauna, meiofauna, and macrofauna.

SUMMARY

The polychaetes, *Nereis virens* and *Capitella capitata,* have a mixed function oxygenase (MFO) system which metabolizes polycyclic aromatic hydrocarbons. In *Nereis virens* this enzyme system was associated with microsomes of the lower intestine. Similarities between the MFO of polychaetes and other animals, both vertebrate and invertebrate, were noted with respect to cytochrome P-450 and phospholipid components, cofactor requirements, effect of inhibitors, pH optima and temperature optima. MFO activity was found in *Capitella capitata* only after exposure to either petroleum or its components with worms, primarily of the third generation, having higher MFO activity than the first or second generation of worms exposed to petroleum. Selection of a genotype with increased MFO activity was suggested. Polychaetes were important in degradation of crude oil both by directly degrading hydrocarbons and indirectly by turning over sediments during feeding, thus allowing petroleum-degrading microbes to metabolize subsurface hydrocarbons.

ACKNOWLEDGMENTS

These studies were supported by the National Science Foundation, Office for the International Decade of Ocean Exploration, Grant No. OCE74-05283 A01 and Georgia Sea Grant Program, supported by NOAA Office of Sea Grant, U. S. Department of Commerce, under Grant No. 10-32-RR273-077. Hydroxy benzo(a)pyrenes were provided by A. R. Patel through the National Cancer Institute Carcinogenesis Research Program. SKF 525-A was kindly donated by Smith Kline and French Labs.

LITERATURE CITED

Armstrong, H. W., K. Fucik, J. W. Anderson, and J. W. Neff. 1976. Effects of oil field brine effluent on benthic organisms in Trinity Bay, Texas. Unpublished manuscript.

Baker, J. M. 1976. Investigation of refinery effluent effects through field surveys. In: Marine Ecology and Oil Pollution, pp. 201-205, ed. by J. M. Baker. New York: John Wiley and Sons.

Blumer, M. and J. Sass. 1972. Indigenous and petroleum-derived hydrocarbons in a polluted sediment. Mar. Pollut. Bull. 3: 92-94.

Brooks, G. T. 1977. Chlorinated insecticides: retrospect and prospect. In: Pesticide Chemistry in the 20th Century, pp. 1-20, ed. by J. R. Plimmer. Washington, D. C.: American Chemical Society.

Carr, R. S. and D. J. Reish. 1977. The effect of petroleum hydrocarbons on the survival and life history of polychaetous annelids. In: Fate and Effects of Petroleum Hydrocarbons in Marine Organisms and Ecosystems, pp. 168-173, ed. by D. A. Wolfe. New York: Pergamon Press.

Hites, R. A. 1976. Sources of polycyclic aromatic hydrocarbons in the aquatic environment. In: Sources, Effects, and Sinks of Hydrocarbons in the Aquatic Environment, pp. 325-332. Washington, D. C.: American Institute of Biological Sciences.

Khan, M. A. Q., W. Coello, A. A. Khan and H. Pinto. 1972. Some characteristics of the microsomal mixed-function oxidase in the freshwater crayfish, *Cambarus*. Life Sci. 11: 405-415.

Lee, R. F. 1976. Metabolism of petroleum hydrocarbons in
    marine sediments. In: Sources, Effects, and Sinks of
    Hydrocarbons in the Aquatic Environment, pp. 333-344.
    Washington, D. C.: American Institute of Biological Sciences.
Lee, R. F., E. Furlong and S. Singer. 1977. Metabolism of
    hydrocarbons in marine invertebrates. Aryl hydrocarbon
    hydroxylase from the tissues of the blue crab, *Callinectes
    sapidus,* and the polychaete worm *Nereis* sp. In: Pollutant
    Effects on Marine Organisms, pp. 111-124, ed. by C. S. Giam.
    Lexington, Mass.: D. C. Heath.
Levell, D. 1976. The effect of Kuwait crude oil and the dis-
    persant BP 1100X on lugworm, *Arenicola marina* L. In:
    Marine Ecology and Oil Pollution, pp. 131-158, ed. by J. M.
    Baker. New York: John Wiley and Sons.
Lowry, O. H., N. J. Rosenbrough, L. A. Farr and R. J. Randall.
    1951. Protein measurement with the Folin phenol reagent.
    J. Biol. Chem. 193: 265-275.
Lu, A. Y. H., W. Levin, M. Vore, A. H. Conney, D. R. Thakker,
    G. Holder and D. M. Jerina. 1976. Metabolism of benzo(a)-
    pyrene by purified microsomal cytochrome P-448 and epoxide
    hydrase. In: Polynuclear Aromatic Hydrocarbons: Chemistry,
    Metabolism and Carcinogenesis pp. 115-124, ed. by R. I.
    Freudenthal and P. W. Jones. New York: Raven Press.
Grassle, J. P. and J. F. Grassle. 1976. Sibling species in
    the marine pollution indicator *Capitella* (polychaete).
    Science 192: 567-569.
Grassle, J. F. and J. P. Grassle. 1977. Temporal adaptations
    in sibling species of *Capitella*. In: Ecology of Marine
    Benthos, pp. 177-189, ed. by B. C. Coull. Columbia, South
    Carolina: University of South Carolina Press.
Marc, M. F. 1942. A study of a marine benthic community with
    special reference to the microorganisms. J. Mar. Biol.
    Assoc. U. K. 25: 517-554.
Nebert, D. W. and H. V. Gelboin. 1968. Substrate-inductible
    microsomal aryl hydroxylase in mammalian cell culture.
    I. Assay and properties of inducible enzyme. J. Biol.
    Chem. 243: 6242-6249.
Ocean Affairs Board, Commission on Natural Resources, National
    Research Council. 1975. Petroleum in the Marine Environ-
    ment. 107 pp. Washington, D. C.: National Academy of Science.
Omura, T. and R. Sato. 1964. The carbon monoxide-binding
    binding pigment of liver microsomes. J. Biol. Chem. 239:
    2370-2378.
Pedersen, M. G., W. K. Hershberger and M. R. Juchau. 1974.
    Metabolism of 3,4-benzpyrene in rainbow trout (*Salmo
    gairdneri*). Bull. Environ. Contam. Toxicol. 12: 481-486.

Pohl, R. J., J. R. Bend, A. M. Guarino and J. R. Fouts. 1974.
    Hepatic microsomal mixed-function oxidase of several marine
    species from coastal Maine. Drug Metabolism and Disposi-
    tion 2: 545-555.
Philpot, R. M., M. O. James and J. R. Bend. 1976. Metabolism
    of benzo(a)pyrene and other xenobiotics by microsomal
    mixed-function oxidases in marine species. In: Sources,
    Effects and Sinks of Hydrocarbons in the Aquatic Environ-
    ment, pp. 184-189. Washington, D. C.: American Institute
    of Biological Sciences.
Prouse, N. J. and D. C. Gordon. 1976. Interactions between
    the deposit feeding polychaete *Arenicola marina* and oiled
    sediment. In: Sources, Effects and Sinks of Hydrocarbons
    in the Aquatic Environment, pp. 407-422. Washington, D. C.:
    American Institute of Biological Sciences.
Reish, D. J. 1971. Effect of pollution abatement in Los
    Angeles harbours. Mar. Pollut. Bull. 2: 71-74.
Rhoads, D. C. 1967. Biogenic reworking of intertidal and
    subtidal sediments in Barnstable Harbor and Buzzards Bay,
    Mass. J. Geol. 75: 61-76.
Rossi, S. S. and J. W. Anderson. 1977. Accumulation and
    release of fuel-oil-derived diaromatic hydrocarbons by
    the polychaete *Neanthes areanceodentata*. Mar. Biol. 39:
    51-55.
Rossi, S. S. and J. W. Anderson. 1978. Petroleum hydrocarbon
    resistance in the marine worm *Neanthes arenaceodentata*
    (polychaetea:annelida), induced by chronic exposure to No.
    2 fuel oil. Bull. Environ. Contam. Toxicol. 20 (in press).
Sanders, H. L., J. F. Grassle and G. R. Hampson. 1972. The
    West Falmouth Oil Spill. I. Biology. Woods Hold OceanO-
    graphic Institution, Technical Report No. 72-20.
Singer, S. C. and R. F. Lee. 1977. Mixed function oxygenase
    activity in the blue crab, *Callinectes sapidus:* tissue
    distribution and correlation with changes during molting
    and development. Biol. Bull. 153: 377-386.
Tenore, K. R., J. Tietjen and J. Lee. 1977. Effect of meio-
    fauna on incorporation of aged eelgrass, *Zostera marina,*
    detritus by the polychaete *Nephthys incisa*. J. Fish. Res.
    Bd. Canada. 34: 563-567.
Terriere, L. C. and S. J. Yu. 1976. Microsomal oxidases in
    the flesh fly (*Sarcophaga bullata* Parker) and the black
    blow fly (*Phorima regina meigen*). Pest. Biochem. Physiol.
    6: 223-228.
Van Vleet, E. S. and J. G. Quinn. 1977. Input and fate of
    petroleum hydrocarbons entering the Providence River and
    upper Narragansett Bay from wastewater effluents. Environ.
    Sci. Technol. 11: 1086-1092.

Walker, C. R. and L. C. Terriere. 1970. Induction of microsomal oxidases by dieldrin in *Musca domestica*. Entomologia Experimentalis et Applicata 13: 260-274.

Wilkinson, C. F. and L. B. Brattsten. 1972. Microsomal drug metabolizing enzymes in insects. Drug Metabolism Reviews 1: 153-228.

Wattenberg, L. W., J. L. Leong and P. J. Strand. 1962. Benzpyrene hydroxylase activity in the gastrointestinal tract. Cancer Res. 22: 1120-1125.

THE EFFECT OF EXPOSURE TEMPERATURES ON OXYGEN CONSUMPTION
AND OPERCULAR BREATHING RATES OF PINK SALMON FRY EXPOSED
TO TOLUENE, NAPHTHALENE, AND WATER-SOLUBLE FRACTIONS
OF COOK INLET CRUDE OIL AND NO. 2 FUEL OIL

Robert E. Thomas

Department of Biological Science
Chico State University
Chico, California

Stanley D. Rice

Northwest and Alaska Fisheries Center Auke Bay Laboratory
National Marine Fisheries Service, NOAA
P. O. Box 155, Auke Bay, Alaska

INTRODUCTION

Changes in the respiratory activity of fishes have been
observed during exposures to pollutants such as DDT (Schaumburg
et al. 1967), kraft pulp mill effluent (Walden *et al.* 1970;
Davis 1973), zinc (Sparks *et al.* 1972), copper (Drummond *et al.*
1973), combinations of copper and zinc (Sellers *et al.* 1975),
benzene (Brocksen and Bailey 1973), and refined and crude oils
(Anderson *et al.* 1974; Thomas and Rice 1975; Rice *et al.* 1977a).
An immediate increase in opercular breathing rates and coughing
rates (back flushing of gills) was observed in pink salmon
(*Oncorhynchus gorbuscha)* fry exposed to sublethal concentrations
of the water-soluble fractions (WSF) of Prudhoe Bay and Cook
Inlet crude oils and No. 2 fuel oil (Thomas and Rice 1975;
Rice *et al.* 1977a). Although the breathing and coughing rates
decreased after the initial 24-h exposure, elevated breathing
and coughing rates of fry exposed to a stable concentration
of the Cook Inlet crude oil WSF remained higher than control
levels for the duration of a 72-h exposure (Rice *et al.* 1977a).

In the previous study, Rice *et al.* (1977a) found that increased breathing rate correlated well with the presence of aromatic hydrocarbons in the tissues and they postulated that the increased breathing rate was coupled with an increased energy demand for detoxifying and eliminating the hydrocarbons from the tissues. However, they did not measure oxygen consumption, and it is possible that oxygen transfer across the gills was impaired enough to stimulate increased opercular breathing for compensation without an actual increase in energy demand.

The primary objective of this study was to measure breathing rates in pink salmon fry exposed to equivalent concentrations of aromatic hydrocarbon toxicants at 4^0 and 12^0C to determine if the response at the lower temperature differs from that at the higher temperature. This experiment was of interest because spills in Alaskan waters will probably occur at lower temperatures than have been studied elsewhere, and little is known about the effects of aromatic hydrocarbons at these lower temperatures. Before we could study the effects of temperature on the breathing rate response of fry exposed to aromatic hydrocarbons, we conducted several experiments at one temperature (12ºC). First, we compared opercular breathing and oxygen consumption in fry exposed to aromatic hydrocarbons in order to determine the validity of using changes in opercular breathing rates as indicators of altered metabolism. Second, we measured the effects of several different concentrations of two oil-derived hydrocarbons, toluene and naphthalene, on breathing rates, to determine if the response is proportional to concentration. Toluene and naphthalene were chosen as representative of mononuclear and dinuclear aromatic hydrocarbons, the classes of hydrocarbons in water-soluble fractions of oil believed to be most responsible for the actue toxicity of oil. Third, we compared breathing rates in fry exposed to equivalent concentrations of toluene and naphthalene to determine the relative effectiveness of a mononuclear (toluene) and a dinuclear (naphthalene) aromatic hydrocarbon in stimulating a change in breathing rate response.

MATERIALS AND METHODS

These studies were conducted at the Northwest and Alaska Fisheries Center Auke Bay Laboratory using pink slamon fry raised in gravel incubators (Bailey and Taylor 1974). The fryemerged in April 1976 and were kept in running seawater aquaria until used in the study. Temperatures in the rearing aquaria during the experimental periods, June-August 1976,

ranged from 8^0 to 12^0C. The fry were fed daily with Oregon
Moist Pellets[1] and grew to 1-2 g and a length of 4.5-5.5 cm
before use in the experiments.

Water-soluble fractions were prepared with Cook Inlet crude
oil and No. 2 fuel oil as described by Rice *et al.* (1977a).
The WSF's were analyzed for parts per million of total aromatic
hydrocarbons by gas chromatography to the methods described by
Cheatham *et al.*[2] and had aromatic hydrocarbon profiles similar
to that reported by Korn *et al.* (in press). Naphthalene and
toluene stock solutions that approached saturation were prepared as described by Korn *et al.* (in press). The concentrations of toluene or naphthalene in the exposure solutions
were determined by the ultraviolet spectrophotometric methods
of Neff and Anderson (1975) according to the procedures outlined by Rice *et al.* (1977a).

Acute toxicities were determined for each toxicant so that
the concentrations during the sublethal exposures (24-h or less)
could be compared on an equated basis, as a percent of the 24-h
median tolerance limit (TLm) for that toxicant. Bioassays
for WSF's of Cook Inlet crude oil and No. 2 fuel oil were conducted according to methods described by Rice *et al.* (1977a)
and for toluene and naphthalene according to methods described
by Korn *et al.* (in press). Median tolerance limits with 95%
fiducial limits were calculated by probit statistics (Finney
1971). The 24-h TLm's and associated difucial limits of pink
salmon fry exposed at 12^0C to each toxicant were as follows
(TLm's to toluene and naphthalene are expressed in parts per
million and TLM's to the WSF's of Cook Inlet crude oil and
No. 2 fuel oil are expressed in parts per million of total
aromatic hydrocarbons):

Toxicant	*24-h TLm*	*95% fiducial limits*
Toluene	*5.38*	*4.42-6.54*
Naphthalene	*0.92*	*0.78-1.08*
Cook Inlet crude oil WSF	*1.73*	*1.54-1.92*
No. 2 fuel oil WSF	*0.651*	*0.540-0.729*

The test chambers containing the pink salmon fry during
breathing rate and oxygen consumption measurements were located
in a heat- and sound-insulated room to protect the fry from
extraneous stimuli. The recording equipment and reservoirs

[1]*Reference to trade name does not imply endorsement by the
National Marine Fisheries Service, NOAA.*
[2]*Cheatham, D. L., R. S. McMahon, S. J. Way, J. W. Short,
and S. D. Rice. The relative importance of evaporation and
biodegradation and the effect of lower temperature on the loss
of some mononuclear and dinuclear aromatic hydrocarbons from
seawater. Manuscript in preparation.*

containing the test solutions were located outside this room.
Water was pumped from the reservoirs to headtanks inside the
room. The headtanks maintained a constant hydrostatic pressure
and provided waterflow to each animal compartment at about 1
liter per hour. A flowthrough system was used to provide continuous and stable concentrations of toxicant and water passing
from the animal chambers was discarded. Concentrations of the
test solutions were determined by ultraviolet spectrophotometry
at 1-h intervals and sufficient quantities of concentrated
stock solutions were added to the reservoir's tanks to maintain the desired toxicant concentration. In the exposures
involving simultaneous measurements of oxygen consumption and
breathing rate, the concentrations were not maintained in the
stock reservoir and consequently the exposure concentrations
declined during the test. The temperature of the test solutions was controlled by thermostatically-regulated flow of
ethylene glycol solutions through stainless steel cooling
coils submerged in the reservoirs; the temperature was maintained within 0.4^0C of the desired temperature.

To determine oxygen consumption and breathing rate simultaneously, each pink salmon fry was confined in an airtight
PVC cylinder (3 X 9 cm), threaded at each end and capped with
PVC screw caps with a fitting mounted in the middle that permitted the passage of water. Oxygen consumption was determined by measuring the difference in oxygen concentrations
between inflowing and outflowing water with a Radiometer
electrode (E5046). Opercular breathing movements were detected
by stainless steel mesh electrodes filling the cross-sectional
area of each end cap. For each experimental situation (two
concentrations of naphthalene and two concentrations of toluene
at 12^0C, and one concentration of toluene at 4^0C), six new fish
(previously acclimated to 4^0 or 12^0C) were placed in the chambers (one fish per chamber) and permitted to acclimate to the
new environmental conditions for 18 h prior to recording the
basal or control breathing rate and oxygen consumption. After
beginning the exposure to the toxicant, breathing rate and
oxygen uptake were recorded at 2-h intervals for a period of
10-12 h. Six fish each were exposed to two concentrations of
naphthalene and two concentrations of toluene at 12^0C and
one concentration of toluene at 4^0C. The concentration of the
toxicant was not maintained in the stock tanks and as a result
the concentration declined during the exposure to 20% or less
of the initial concentration by the end of the exposure.

To determine breathing rates of large numbers of pink salmon fry, we recorded the opercular movements detected by stainless steel mesh electrodes using an apparatus previously described (Thomas and Rice 1975). Thirty new fish (50 ± 5 mm
long) were used for each toxicant concentration and at each
temperature tested. Seven of these fish were not exposed to

the pollutant and were used to determine if external stimuli
were affecting the test results or if there were differences
in opercular breathing rates at the selected times. The fish
were placed in the recording chambers (one fish per chamber)
at 1430, following 30 h of acclimation at the test temperature,
and breathing rate was first recorded at 0630 the following
day. This first recorded rate was used as the normal or
control breathing rate and all subsequent breathing rates
during exposure were compared to this value. The test solutions
were introduced to the experimental animals at 0830 and
breathing rates were recorded at 3, 6, 9, 12, and 15 h post-
exposure. Measurements were taken precisely at the same times
each day, even though previous experiments (Thomas and Rice
1975) did not detect a diurnal rhythm. The differences in the
initial breathing rates and the rates during the exposure were
tested by analysis of variance and Student's t test.

RESULTS

   Oxygen consumption and breathing rates of fry exposed to
toluene and naphthalene began to increase immediately upon
exposure and declined in later hours during exposure (Figure 1).
Breathing rate reached maximum response values at 2 or 4 h,
while the oxygen consumption rates were greatest at 6 or 8 h
of exposure. Sample sizes were limited because of the difficulty in making simultaneous measurements of oxygen consumption and opercular breathing rates. For the experiment in
which oxygen consumption and breathing rates were measured
simultaneously, the concentrations of the toxicants were not
maintained constant and the fish were exposed to decreasing
concentrations. Both oxygen consumption and breathing rate
returned to nearly pre-exposure levels during the 10-h exposures in all except the low-temperature toluene test. The
concentrations of toxicants at the end of the 10-h exposure
period were less than 20% of the 24-h TLm for all toxicants
tested. Two of six fry exposed to toluene at 4^0C were dead
prior to the 6-h recording, while no deaths occurred at
approximately the same concentration of toluene at 12^0C.
   All three concentrations of naphthalene (107%, 70%, and
45% of the 24-h TLm) resulted in significant increases in the
opercular breathing rate of pink salmon fry (P < 0.01), while
of the four toluene concentrations (94%, 69%, 45%, and 30% of
the 24-h TLm), only the two highest resulted in a significant
increase in the breathing rate (Figure 2). The two higher
naphthalene concentrations resulted in significant increases
in breathing rate for the entire 25-h exposure but only the
highest toluene concentration caused a significantly elevated

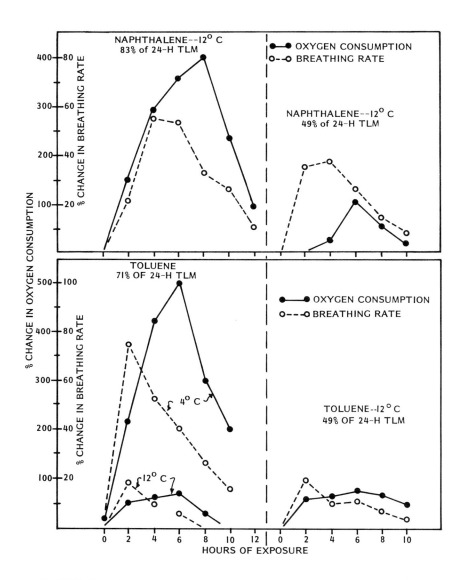

FIGURE 1.   Percent changes in oxygen consumption and breathing rate of pink salmon fry exposed to naphthalene and toluene at 12⁰C and toluene at 4⁰C.   Sample size was 4-6. Control (0 hour) oxygen consumption and breathing rates at 12⁰C were 202 µl/g/h and 67 beats/min.   Exposure concentrations declined to about 20% of the initial concentration at 10 h.

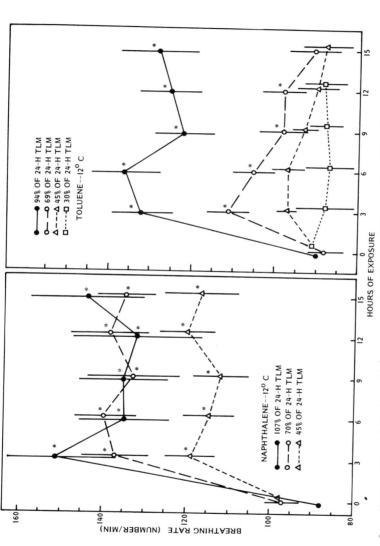

FIGURE 2. Mean opercular breathing rate (± 5% confidence limit) of pink salmon fry exposed to several different concentrations of toluene and naphthalene at 12°C. Exposure concentrations are expressed as a percent of the 24-h TLm and were stable during exposure. Asterisk indicates significant difference at the 0.01 level between the mean given and the mean at 0 hour. Sample size occurring at the later recording periods at the highest concentrations due to fry mortality.

respiratory rate for the entire 15-h period, even though the
exposure concentrations in all of these tests were stable.
Some fry died during the exposure to the highest concentrations
of naphthalene and toluene. Eight of 23 fry were dead at the
end of the 15-h exposure to 107% of the 24-h TLm for naphthalene, while 6 of 23 fry were dead at the end of 15-h exposure
to 94% of the 24-h TLm for toluene.

Both toluene and naphthalene caused linear increases in
breathing rate with increasing exposure concentration, but
the threshold concentration of toluene (expressed as a percent of the 24-h TLm) needed to cause response was higher than
the threshold for naphthalene (Figure 3). For example, exposure to naphthalene at an estimated concentration of 20% of
the 24-h TLm caused a 10% increase in breathing rate for pink
salmon fry while a similar increase in rate after exposure to
toluene occurred at an estimated exposure concentration of
46% of the 24-h TLm.

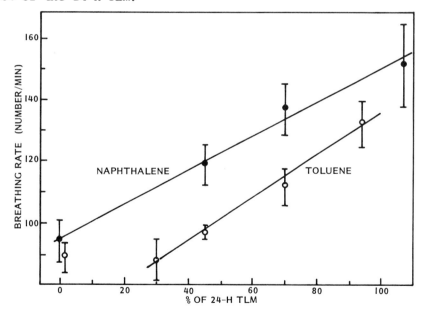

*FIGURE 3.   Linear relation between mean opercular breathing
rate (±95% confidence limit) and exposure concentration after
3 h of exposure at 12⁰C where concentrations are expressed as
a percent of the 24-h TLm of toluene and naphthalene.   The
regressions for breathing rates of pink salmon fry exposed to
naphthalene (y = 0.54x + 95) and toluene (y = 0.7X + 63) had
correlation coefficients of 0.995 and 0.994, respectively.
The regression line and coefficient for toluene do not include
the initial response at 0 hour.*

The increases in breathing rate after exposure to the four toxicants at 4^0 and 12^0C were not uniform between temperatures or toxicants (Figure 4), even though the concentrations used in the exposures were all equivalent (80-84% of the 24-h TLm determined at 12^0C). The effect of temperature on breathing rate response was most dramatic for exposures to naphthalene and No. 2 fuel oil (Figure 4a and d)--increases at 4^0C were twice those observed at 12^0C (100% increase versus 47% to 48%). In contrast, temperature had little effect on breathing rates during exposures to toluene (the percent increases were between 51% and 55% for exposures both at 4^0 and 12^0C; Figure 4b). Temperature had some effect on the breathing rate during exposure to the Cook Inlet crude oil WSF (56% at 4^0C and 38% at 12^0C; Figure 4c). During exposure at both temperatures breathing rates were significantly different from the control rates for the entire 15-h period with all pollutants except toluene at 4^0C. For toluene at 4^0C, the measurement was not significantly different at 15-h because the sample size had been reduced drastically by death of fry (17 of 24 were dead at 15-h). No fry died during exposure to the same concentration of toluene at 12^0C or during exposure to similar concentrations (82% of the 24-h TLm) of the other three toxicants at either 4^0 or 12^0C.

DISCUSSION

Although relatively few fish were used in the studies of oxygen consumption, the pattern of increased oxygen consumption along with increased breathing rates in each of the exposures indicates that increased breathing rate of pink salmon fry reflects increased energy demands. Oxygen consumption was greatest at about 6-h of exposure, about 4 h after the occurrence of maximum breathing rates. Apparently, as the fry became physiologically acclimated to the stress of the toxicant they increased the efficiency of oxygen extraction, thus decreasing the need to move water across the gills and subsequent expenditure of energy.

If the fish are metabolizing the toxicants, energy demands of pink salmon fry should increase as the concentration of toxicant increases. This increased energy demand would be reflected by increased breathing rates and oxygen consumption. Earlier studies have shown increases in breathing rate. As the concentrations of WSF's of Prudhoe Bay and Cook Inlet crude oils and No. 2 fuel oil increased in test exposures the rate of breathing of pink salmon fry increased (Thomas and Rice 1975; Rice *et al.* 1977a). Naphthalene and toluene are seen to have a similar concentration-dependent effect on pink salmon fry breathing

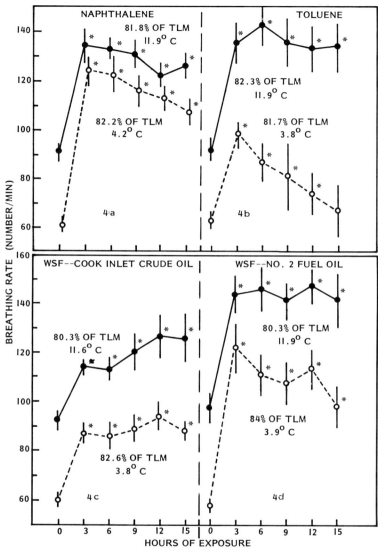

FIGURE 4. Effect of temperature on mean opercular breathing rate (±95% confidence limit) of pink salmon fry exposed to equivalent concentrations of toluene, naphthalene, and water-soluble fractions of Cook Inlet crude oil and No. 2 fuel oil. Concentrations were stable during the exposure and consisted of 80-84% (average = 82%) of the 24-h TLm determined at 12°C. Asterisk indicates significant difference at 0.01 level between the mean given and the mean at 0 hour. Sample sizes ranged from 11 to 23, except for the exposure to toluene at 4°C where deaths toward the end of the exposure decreased the sample size to 6.

THE EFFECT OF NAPHTHALENE ON SYNTHESIS OF PEPTIDE ANTIFREEZE
IN THE BERING SEA SCULPIN, *MYOXOCEPHALUS VERRUCOSUS*

A. L. DeVries

Department of Physiology and Biophysics
University of Illinois
Urbana, Illinois

INTRODUCTION

Over the past several years there has been much interest
in the effects shown by marine organisms upon exposure to petroleum hydrocarbons. This interest has been expressed in the
form of studies where organisms have been exposed to the water
soluble fraction of petroleum or the soluble fractions of some
of its refined products such as fuel oil (NAS, 1975). Generally
the approach has involved determining how much of the water
soluble fraction is necessary to cause death in an organism,
in a specified period of time. In general the hydrocarbon
toxicants responsible for death were not identified and if so
the concentrations were not accurately determined because of
the lack of good analytical techniques (NAS, 1975).

Recently there has been more emphasis on identifying the
toxic components of the water soluble fraction of petroleum
and attempting to elucidate the mechanisms by which they dis-
rupt behavior and cause death (Winters *et al.*, 1976; Boylan
and Taylor, 1971). Of the many petroleum hydrocarbon pollutants studied, recent results indicate that the aromatic compounds, and in particular naphthalene and naphthalene type
compounds are probably the most toxic (Boyland and Tripp, 1971).
In addition, naphthalene is of particular interest because it
has been demonstrated that it is rapidly taken up by organisms
and in the case of fishes it has been shown to concentrate
in the liver where it is metabolized (Lee *et al.*, 1972).
Recently, studies have also shown that this hydrocarbon com-
prises a significant portion of some crude oils and fuel oil
fractions (Boyland and Tripp, 1971). Because of its toxicity

and relatively high concentration in certain petroleums, we
decided to investigate the toxic effects of naphthalene on the
physiology and biochemistry of selected Bering Sea fishes at
lethal and sublethal concentrations. The fact that naphthalene
is rapidly taken up and appears to accumulate in the liver, at
low levels of exposure (Lee *et al.*, 1972), suggested to us
that this toxicant might affect liver protein synthesis. The
liver protein synthetic systems selected for study were the
secreted plasma proteins (albumins and globulins) and the biological antifreeze peptides which have been demonstrated to
be present in certain members of the fish families Gadidae
and Cottidae which inhabit the Bering Sea. The plasma protein
system was selected because it is a relatively easy system
to work with (See review by Haschemeyer, 1976). The species
selected for this study was the cottid, *Myoxocephalus verrucosus*
(Bean) because it was found to be a very hardy fish which sur-
vives with few problems in captivity. It also lacks scales
and thus can be repeatedly sampled in an experimental study
without damaging the integument. The biological antifreeze
which protects this fish at subfreezing temperatures consists
of small peptides which are composed of approximately 40%
alanine (Raymond *et al.*, 1975). A recent protein synthesis
study in antarctic nototheniid fishes indicates that their
antifreeze compounds (glycopeptides) are synthesized in the
liver (Hudson, in press). No studies have been done to demonstrate whether the peptide antifreeze is synthesized in the
liver of the sculpin, however a recent study of peptide antifreeze synthesis in the winter flounder, *Pseudopleuronectes
americanus*, presents evidence that the site of synthesis of
its peptide is the liver (Hew and Yip, 1976). Thus there is
little reason to believe that the site of peptide synthesis
would be different in *Myoxocephalus*.

MATERIALS AND METHODS

    About 200 specimens of the sculpin, *Myoxocephalus verrucosus*
were collected by the fishing crew aboard the R/V *Miller
Freeman* while on Leg II of cruise OCSEAP RP-4-MF-76-A in the
eastern Bering Sea. The fish were caught with a 100 foot otter
trawl which was towed for 30 minutes at a depth of 50 m near
St. George Island in the Bering Sea during the month of May,
1976. The water temperature near the bottom was 0^0C as indicated by an XBT probe trace. The fish were held on deck of
the ship in 1220 liter fiberglassed circular tanks while at
sea. The tanks were closed with resined plywood covers which
sealed the tanks and water was introduced through a stand-
pipe. This design ensured that the tanks were full and that

there was little water movement even in rough seas. Seawater
from the ship's fire main was continually circulated through
the tanks and few of the sculpins died while at sea. The
water temperature of the fire main varied between -0.5 and
+4^{0}C depending on how near the ship was to the edge of the ice.
    Short term acute exposures to naphthalene were done aboard
the ship in tanks holding 160 liters of seawater containing
naphthalene at concentrations of 3, 4 and 5 ppm. Naphthalene
was introduced into the water by first solubilizing it in
ethanol and then fusing the ethanol into a stream of high
velocity seawater which ensured adequate mixing. When sea
conditions permitted, 10 sculpins were transferred to each
tank which contained a different amount of naphthalene. The
times at which they first showed stress and when they died
were recorded. Cessation of opercular movement coincided
with death as indicated by the fact that when the exposed
fish were removed to fresh seawater they failed to recover.
Only preliminary short term studies were done aboard the ship
because of the lack of aquarium space, the large size of the
specimens (average weight = 500 g) and the short duration of
the cruise. These preliminary naphthalene toxicity determinations were intended only to provide a rough approximation of
the TLm (concentration at which 50% of specimens survive) so
that sublethal concentrations could be selected for long term
exposure studies to be conducted at the aquarium facility at
Scripps Institution of Oceanography (SIO) using a flow-through
system.

   Long Term Sublethal Exposure. Forty specimens of the sculpin collected on this cruise were shipped by air freight to
the aquarium facility at SIO. They were packaged in heavy
plastic bags in Igloo coolers and the water aerated using a
portable battery powered air pump. Some ice was put in the
coolers and after 30 hours of air travel the temperature of
the water was +4^{0}C. Of the 40 specimens shipped all survived
the air shipment. After a week in the aquarium at +7^0 the
fish began feeding on pieces of yellow tail tuna. The fish
were treated once a week with the antibiotic, furacin (30 g/100
l of seawater) to prevent bacterial infections. After two
weeks acclimation at +7^{0}C they were transferred to +14^{0}C
water and held at that temperature for 2 months. At this temperature the fish were fed twice a week and each week thereafter a few of the specimens were selected for blood samples.
Blood plasma samples were assayed for ion content as well as
for the disappearance of the thermal hysteresis which is a
measure of the peptide antifreeze content. Sculpin sampled
in May near the ice edge had low blood freezing points (-1.9^{0}C)
indicating their blood contained a "full" complement of antifreeze. After warm acclimation most of the peptide antifreeze

disappeared at which time 12 specimens were transferred to two
60 liter tanks where they were cold acclimated to +0.5^{0}C
seawater. One 60 liter tank contained naphthalene at a concentration of 1 ppm while the other tank served as a control.
Naphthalene concentration in the tank was determined by measuring the absorbance at 276 mμ and the absorbance agreed with
that obtained for a standard solution of 1 ppm. The naphthalene was introduced by solubilizing it in 95% ethanol and
infusing it through an 18 gauge needle at the rate of 0.4 ml
per minute into a stream of seawater flowing at the rate of
800 ml per minute. For metering the alcoholic naphthalene
and seawater, two variable speed peristaltic pumps were used.
Silicone tubing was used and it was changed once a week. The
calibration of the pumps was checked daily as well as the concentration of the naphthalene in the seawater by measuring
the absorbance at 276 mμ.

Blood samples were taken periodically from the caudal vein
of both the control and exposed fish using a 30 gauge hypoder-
mic needle while the fish were under light anesthesia. The
plasma levels of sodium and potassium were determined using a
Corning 450 flame photometer and the chlorides determined using
a Buchler chloridometer. Freezing and melting points were
determined according to the method of DeVries (1971a). The
difference between the freezing and melting point is referred
to as a thermal hysteresis and is a reasonably accurate estimate of concentration of the peptide antifreeze in the blood
(DeVries, 1971b).

Naphthalene Uptake Measurements. Two specimens of *Myoxocephalus* weighing approximately 200 g each were used in two
separate naphthalene uptake experiments. The experiments
involved solubilizing 5 μc of Naphthalene-1-C-14 (specific
activity 39.8 μCi/mg) in one ml of ethanol and slowly infusing
it through a 30 gauge needle into 5 liters of filtered seawater.
The seawater was stirred with a magnetic stirrer to ensure
adequate mixing. The water was gently aerated and radioactivity
measurements of a water sample indicated that loss of naphthalene through evaporation from the surface of the water was
insignificant. After the sculpin was put into the container,
water samples were withdrawn and analyzed for radioactivity
as follows. One ml of seawater was diluted to 3.5 ml with
distilled water and then shaken with 11.5 ml of Aquasol (New
England Nuclear). The resulting stiff clear gels were counted
in a Beckman liquid scintillation counter. After 22 hours,
the sculpins were washed with methanol and the radioactivity

determined in 10 mg samples of its tissues. The tissue samples
were digested with Protosol (New England Nuclear), and after
digestion was complete they were neutralized with Tris-HCl.
They were counted in Aquasol before and after addition of an
internal standard.

   Measurement of Resting Metabolism. Oxygen consumption
measurements were done in a 6 liter glass jar at $+1^0C$. Water
was circulated through a Rank electrode chamber and back to
the respirometer. The electrode potential was displayed on a
Houston strip chart recorder and a span of one millivolt indicated an oxygen concentration change from 0 to 7.9 ml per
liter of seawater. At $+1^0C$ the response time of the electrode
was about 10 minutes. Runs usually lasted 2 hours. Date were
recorded only after the first 15 minutes to ensure that the
rate of response of the electrode was constant.
   When fish are put into a small chamber they often exhibit
elevated rates of oxygen consumption for several hours due to
hyperactivity resulting from strange surroundings and handling.
Therefore, the sculpins were held in the respirometer several
hours before their oxygen consumption was determined. During
this acclimation period water was slowly circulated through
the respirometer. The oxygen consumption is expressed as
ml O_2 consumed/g/hour.

   Measurement of Plasma Protein Synthesis. After 6 weeks
of low temperature acclimation, 4 control fish and 4 exposed
fish were injected each with 40 µCi of L-leucine-C-14(U)
(specific activity: 320 mCi/mmol) and the incorporation into
secreted plasma protein determined. Forty µCi of the isotope
was made up to a volume of 0.4 ml with a buffered, balanced
salt solution and injected into the caudal vein. The solution
was injected through a 10 cm length of polyethylene tubing
(PE-10) attached to a 30 gauge needle. Once the needle pierced
the vein the injection could be made without disturbing the
needle. The fish were lightly anesthetized with MS 222
(0.1 g/l) both before the injection was given and before blood
samples were drawn. The arousal time was usually about 3 minutes. At various times after the injection, 300 µl samples
of blood were withdrawn from the caudal vein and immediately
centrifuged before clotting occurred. A 100 µl aliquot of
the plasma was transferred to a 2.3 cm Whatman No. 3 mm filter
paper. The paper disc was allowed to dry for 5 minutes then
washed twice for 10 minutes in each of the following solutions:
cold 10% trichloracetic acid (TCA), cold 3% perchloric acid,
cold 95% ethanol and ether. After drying at room temperature
for 15 minutes the disc was put into a scintillation vial
containing 10 ml of toluene containing 4 g per liter of 2,5-
diphenyl-oxazole (PPO), 0.05 g per liter of 1,4-bis (2-phenyl-

oxazoyl)benzene (POPOP). Another 100 µl aliquot of the blood plasma sample was added to 100 µl of 10% TCA, shaken and centrifuged after having been immersed for 1 hour in ice. A 100 µl aliquot of the TCA supernatant was counted in 10 ml of Aquasol.

When it was apparent that the secretion of leucine labeled plasma protein was occurring at a constant rate, the fish were anesthetized, weighed and sacrificed. The liver was removed, weighed and homogenized in one volume of a buffer containing 0.35 $\underline{M}$ sucrose, 0.05 $\underline{M}$ Tris, pH 7.4, 0.025 $\underline{M}$ KCl and 0.01 $\underline{M}$ MgCl$_2$ for 3 minutes at low speed in a Waring Blender. The homogenate was centrifuged at 1000 x g for 3 minutes to sediment the cellular debris and rid the homogenate of bubbles. One hundred µl aliquots were assayed for radioactivity using the filter disc technique described above and the free radioactivity determined in TCA soluble supernatants.

The radioactive samples were counted on a Beckman liquid scintillation counter. The recoveries of radioactivity in the forms of labeled protein and free radioactivity in the plasma and liver were calculated according to the method outlined by Haschemeyer (1973).

Effect of Naphthalene on Liver Morphology. Upon completion of the leucine incorporation experiment but before the livers were homogenized, small sections of liver were preserved in 10% formalin for histological examination. The liver samples were embedded, sectioned and stained with hematoxylin and eosin. They were examined by light microscopy.

RESULTS AND DISCUSSION

Acute Exposure Studies. Of the several species of Bering Sea fishes that have been exposed to naphthalene at a temperature of +1^{0}C, the cottids and pleuronectids appear to be more resistant to exposure than the gadids. Concentrations of 4 ppm caused death in the cod, *Gadus macrocephalus* and pollack, *Theragra chalcogramma* within 2 hours, but not in the sculpin *Myoxocephalus* and the rock sole *Lepidopsetta bilineata* until 20 hours of exposure had occurred (Devries, unpublished data). At a concentration of 3 ppm cods fail to maintain their equilibrium after 3 hours of exposure and die after 13 hours. At this concentration sculpins show signs of loss of equilibrium only after 12 hours of exposure and after 48 hours only 10% die. No toxic lethal doses are given for this study because only 10 specimens were exposed at each concentration. The results of the acute exposures are considered to be preliminary because the exact naphthalene concentrations could not be determined at sea. Collection of water samples and analysis of water samples at a later time was considered unreliable.

The acute exposure experiments done at sea did however permit a reasonable estimate of the naphthalene dose which could be tolerated for long periods of time. On the basis of these short term exposure studies, a concentration of 1 ppm was selected for the sublethal long term exposures done in the laboratory at SIO and at this level of exposure no fish died during a period of exposure which lasted 6 weeks.

The cause of death from short term exposures to high concentrations of naphthalene is not known. Exposed fish usually lost their ability to retain their equilibrium in the water column very quickly, and shortly thereafter stopped ventilating. Such behavior suggests that the function of the nervous system may be impaired. The high concentration of radioactivity in the brain (Table 1) of sculpin exposed to radioactive naphthalene lends some support to this hypothesis.

Short Term Naphthalene Uptake in Sculpin. Exposure of the sculpin *Myoxocephalus* to low levels of naphthalene (0.025 ppm) indicates that even at levels well below 1 ppm fish rapidly take up naphthalene from seawater. (Figure 1) illustrates that within two hours 75% of the naphthalene in 5 liters of seawater had been taken up by a 200 g *Myoxocephalus*. After several hours of exposure significant amounts of radioactivity (presumably naphthalene or its metabolites) were found in the various body fluids and tissues. Most of this radioactivity was associated with the liver (Table 1). Similar rates of uptake and concentrations of radioactivity in the liver have also been found when specimens of the temperate Pacific sculpin *Oligocottus maculosus* were exposed to low levels of radioactive naphthalene (Lee et al., 1972). Studies indicate that

TABLE 1. Distribution of radioactivity in various tissues and fluids of Myoxocephalus verrucosus *after 22 hours exposure in five liters of seawater containing 1 μCi (0.025 ppm) naphthalene C-14 per liter.*

Tissue	dpm/100 mg tissue
Liver	140,000
Bile	119,000
Brain	40,200
Gut	14,900
Kidney	7,900
Gill	4,300
Muscle	2,750
Blood Serum	1,670

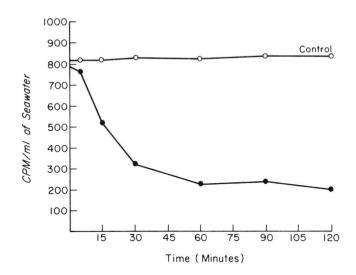

FIGURE 1. *Uptake of naphthalene by a 200 g* Myoxocephalus *verrucosus. The fish was introduced into 5 liters of seawater which contained 5 μCi of naphthalene-1-C-14. The concentration was 0.025 mg per liter. The decline in radioactivity of one ml seawater samples was followed as a function of time.*

the liver metabolizes naphthalene to more water soluble products such as 1,2-dihydro-1,2-dihydroxynaphthalene which are excreted via the bile (Lee *et al.*, 1972). In rats, some of the naphthalene is metabolized to 1,2-dihydro-1-naphthyl glucosiduronic acid and excreted via the urine (Boyland and Solomon, 1955). There appears to be no data available concerning the toxicity of these metabolites.

Long Term Naphthalene Exposure Studies. Although none of the six sculpins, *Myoxocephalus,* exposed to seawater containing 1 ppm naphthalene died, their condition appeared to deteriorate over the course of exposure. Previously all of these fish fed on pieces of yellow tail fillet and although the controls continued to feed during the course of cold acclimation, the naphthalene exposed fish refused to feed. During the 6 week exposure period they were lightly anesthetized twice and small pieces of fish forced into their stomach. This "force feeding" did not appear to have any adverse effects on the specimens. The reduced food intake undoubtedly had some influence on their condition, however their condition appeared worse than that of the starved controls. The exposed fish also appeared to be less active when transferred to a containers of anesthetic.

Effect on Blood Chemistry. Periodic sampling of both the
control and naphthalene exposed fish during cold acclimation
showed that the concentrations of ions in their blood did not
change significantly after the initial increase which resulted
from transferring them from warm to cold water (Table 2). The
change in concentrations of Na+ and Cl⁻ associated with cold
acclimation are in accord with what has been observed with
other marine fishes upon cold acclimation (DeVries, 1971b).
On the basis of the data presented in this study, it appears
that naphthalene does not affect the capability for ionic and
osmotic regulation in *Myoxocephalus*.

One important difference between the control and exposed
fish that was noted was the large decrease in the hematocrit
value following exposure. It should be pointed out that each
time the fish were sampled about 2.5% of their blood volume
was removed. The drop in hematocrit suggests that replacement
of red blood cells did not occur during exposure or if it did,
the rate was much slower than that of control fish. The decrease in hematocrit did not occur in a few of the control fish
which were starved and sampled as often as the exposed fish.
The decrease in hematocrit value is similar to that observed
for other higher vertebrates which have been exposed to naphthalene (Marks, 1965).

Effect of Naphthalene on Concentration of Peptide Antifreeze.
When the warm acclimated control sculpin were cold acclimated
for 6 weeks at +0.5⁰C, they produced only a small amount of
peptide antifreeze. The plasma freezing point dropped from
-1.01⁰C to -1.36⁰C during this time. The change in the difference between the freezing point and melting point which is
an estimate of the amount of peptide antifreeze synthesized,
amounted to only 0.28⁰C (Table 2) during the course of cold
acclimation. A similar change (0.22⁰C) was also observed with
the plasmas of the naphthalene exposed fish during the 6
week period of cold acclimation (Table 2). The slightly
smaller change associated with the exposed fishes was found
to be insignificant when a statistical comparison was made
with the controls. These results nevertheless indicate that
peptide antifreeze was being synthesized during exposure to
naphthalene. Since naphthalene exposure appeared to have no
effect on the amount of or the rate at which the peptide antifreeze was synthesized, its effect on total protein synthesis
in the liver was investigated using incorporation of C[14]
leucine as a measure of protein synthesis (see section on
Effect on Synthesis of Secreted Liver Proteins).

The difference between the freezing and melting points
observed with winter fish caught in the Bering Sea was 1.3⁰C.
The 0.6⁰C freezing-melting point difference observed after
cold acclimation indicates that only about half of the usual

TABLE 2. Physiochemical properties of blood plasma of two groups of Myoxocephalus verrucosus cold acclimated to +0.5°C. One of the groups was exposed to 1ppm naphthalene during the course of cold acclimation. The number of specimens analyzed are given in parentheses and the values represent the mean ± the standard error.

Number of specimens and acclimation conditions	Freezing Point °C	Melting Point °C	Melting Point minus Freezing Point	Na mM/l	Cl mM/l	K mM/l	Hematocrit %
CONTROLS							
60 days at +12°C (8)	-1.01 ± 0.03	-0.68 ± 0.02	0.33 ± 0.03	184 ± 2.12	171 ± 1.93	3.95 ± 0.37	18.0 ± .37
15 days at +0.5°C (6)	-1.25 ± 0.03	-0.85 ± 0.03	0.40 ± 0.03	207 ± 4.84	190 ± 3.83	2.98 ± 0.29	
27 days at +0.5°C (6)	-1.28 ± 0.02	-0.79 ± 0.02	0.49 ± 0.02	203 ± 0.72	185 ± 0.87	2.88 ± 0.29	
50 days at +0.5°C (4)	-1.36 ± 0.04	-0.75 ± 0.04	0.61 ± 0.04	207 ± 0.36	193 ± 2.27	3.00 ± 0.33	17.6 ± 0.43
NAPHTHALENE EXPOSURES							
15 days at +0.5°C (6)	-1.03 ± 0.02	-0.85 ± 0.03	0.17 ± 0.02	205 ± 2.05	183 ± 1.54	3.92 ± 0.19	9.0 ± 0.38
27 days at +0.5°C (6)	-1.11 ± 0.04	-0.84 ± 0.02	0.27 ± 0.03	205 ± 4.10	178 ± 4.25	2.39 ± 0.27	4.6 ± 0.30
50 days at +0.5°C (4)	-1.34 ± 0.04	-0.78 ± 0.04	0.56 ± 0.04	214 ± 5.97	187 ± 4.37	3.43 ± 0.17	2.5 ± 0.45

amount of antifreeze was synthesized. This small change in
thermal hysteresis observed with this extended period of cold
acclimation was somewhat surprising, and it was less than the
change observed when the closely related sculpin, *Myoxocephalus
scorpius* was cold acclimated in the laboratory during the warm
early autumn months of August and September (Duman and DeVries,
1974). With *M. scorpius* the amount of antifreeze synthesized
during cold acclimation was significantly more than with *M.
verrucosus*. In the two studies the acclimation regimes were
the same except that the study described in this paper was
done between mid- and late summer, while the other was done
between late summer and early autumn. The slow rate of peptide antifreeze production observed during acclimation in
this study suggests to us that the experiment should have been
conducted late in the autumn season rather than during the
summer. The difference in antifreeze production during cold
acclimation in these two species suggests to us that the control
of its production in the sculpin is a seasonal phenomenon and
involves more than just exposure to low temperature and short
days. It is apparent that in order to examine the effects of
naphthalene exposure on peptide antifreeze synthesis, the
acclimation experiments must be done during the late autumn,
a time during which the sculpins normally produce their peptide antifreezes.

   Effect of Naphthalene on Metabolism. The oxygen consumption rates for the exposed and control fish are given in Table
3. Data are for fish of similar weights and are therefore
comparable. They clearly show that exposed fish have lower
rates of oxygen consumption. This low rate is not entirely
unexpected in view of their reduced food intake and apparent
poor condition.

   Effect on Liver Cellular Structure. Prior to preparation
of the liver homogenates livers were examined for gross changes
in appearance and there appeared to be no difference between
the control and exposed fish with the exception that the livers
of the exposed fish appeared to be slightly hardened and contained more pigment. Histological examination of H and E
stained sections of the livers of exposed fish revealed that
some of the hepatocytes were shrunken and a reduction in the
volume of cytoplasm present. In these cells, thickening of the

TABLE 3. Rates of oxygen consumption for the sculpin
Myoxocephalus verrucosus *after 6 weeks acclimation to +0.05°C
and exposure to 1 ppm naphthalene*

	Milliliters of $O_2/g/hour$
Controls	0.044, 0.059, 0.045
Exposed to Naphthalene	0.029, 0.025, 0.032, 0.025

wall had occurred and in some there were deposits of fibrous material. These morphological changes were similar to some of those noted in trout exposed to the water soluble fraction of crude oil (Hawkes, 1977).

Effect on Synthesis of Liver Proteins. The time course of the appearance of labeled plasma protein at -0.5^0C after injection of radioactive leucine into the caudal vein of control and exposed sculpin is shown in Figure 2. Examination of the respective curves reveals that there is no difference in the rates of incorporation between the control and exposed specimens. The shapes of the curves are similar to those obtained for the incorporation of labeled amino acids into plasma protein at 20^0C in the toad fish (Haschemeyer, 1973).

The recoveries of labeled plasma ranged between 3% and 10% for both groups. These values are slightly lower than those reported for the toad fish at 10^0C (Haschemeyer, 1973), and for the antarctic cod, *Dissostichus mawsoni* at -1.5^0C (Hudson, in press). The recoveries of radioactivity in liver protein (63-76%) however were similar to those reported for the toad fish and antarctic cod. Recoveries of free radioactivity in the liver were significantly higher than those in the toad fish (Table 4), however this is most likely due to the fact that twice as much isotope per unit of body weight

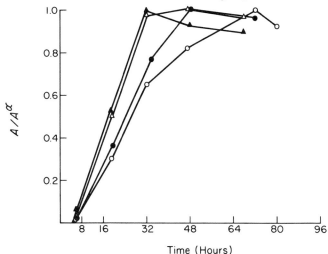

FIGURE 2. *Time course of appearance of radioactive label in the TCA insoluble fraction of the plasma of the sculpin,* Myoxocephalus verrucosus, *at +0.5^{O}C after 40 µCi of leucine C-14 was injected into the caudal vein. The control values are given by (Δ) and (o). The data are given as dpm divided by the maximum dpm value of the plateau (A/A$^\alpha$).*

Table 4. Recoveries of radioactivity in the form of plasma and liver protein and TCA soluble radioactivity after injection of 40 µCi leucine C-14 into specimens of Myoxocephalus verrucosus.

Specimen	Plasma protein µCi	Plasma protein %[a]	Liver protein µCi	Liver protein %[a]	Plasma free radioactivity µCi	Plasma free radioactivity %[a]	Liver free radioactivity µCi	Liver free radioactivity %[a]
CONTROL								
1	0.22	17	2.1	63	0.06	2	0.94	29
2	0.21	11	1.5	76	0.04	2	0.25	12
EXPOSED TO NAPHTHALENE								
1	0.19	3	3.5	64	0.14	3	0.164	30
2	0.19	8	1.8	73	0.06	2	0.40	17

[a] Percentages of recovery are percentages of total radioactivity recovered in the TCA soluble and insoluble fractions.

was administered to the sculpin as was to the other species. Free radioactivity recovered from the plasma was less than 3% of the total, a value which is similar to that recovered from the plasma of the toad fish. There appear to be no differences in the recoveries of both the labeled protein and free radioactivity between the control and exposed fish. The lack of a difference suggests that 6 weeks exposure to 1 ppm naphthalene has little if any detectable effect on the rate at which liver proteins are synthesized and secreted into the blood.

In view of the fact that some of the liver hepatocytes appeared non-functional, the normal pattern of protein synthesis was unexpected. Such a result however might be explained by the fact that the normal cells have a large capability for increasing their synthesis to the point where they can handle the synthetic requirement of both the mixed function oxidase system needed for detoxification and the secreted plasma proteins. Recent studies done on trout exposed to the water soluble fraction of crude oil have shown loss of hepatocytes occurs and that the remaining hepatocytes exhibit a proliferation of the rough endoplasmic reticulum which indicates a high level of protein synthesis (Hawkes, 1977). Such studies suggest that this cellular state is a reflection of the increased synthesis of the enzymes associated with the mixed function oxidase system (Hawkes, 1977) which is induced in the presence of many toxicants (Burns, 1976).

The normal rate of synthesis of secreted plasma proteins in fishes exposed to naphthalene suggests that the fraction of protein synthesis in the liver destined for export is small and the liver has a large synthetic capacity that can handle the synthetic requirements of both the mixed function oxidase system and the plasma proteins at the same time. This appears to be the case even when many of the hepatocytes show signs of cellular deterioration.

LITERATURE CITED

Boyland, D. B. and B. W. Tripp. 1971. Determination of hydrocarbons in seawater extracts of crude oil and crude oil fractions. Nature 230:44-47.

Boyland, E. and J. B. Solomon. 1955. Metabolism of polycyclic compounds: acid-labile precursors of naphthalene produced as metabolites of naphthalene. Biochem. J. 59:518-522.

Burns, K. A. 1976. Microsomal mixed function oxidases in an estuarine fish, *Fundulus heteroclitus*, and their induction as a result of environmental contamination. Comp. Biochem. Physiol. 53B:443-446.

DeVries, A. L. 1971a. Glycoproteins as biological antifreeze agents in antarctic fishes. Science 172: 1152-1155.

DeVries, A. L. 1971b. Freezing resistance in fishes, In: Fish Physiology, W. S. Hoar and D. J. Randall, eds. Academic Press, London Vol. 6, pp. 157-190.

DeVries, A. L., (unpublished observations).

Duman, J. G. and A. L. DeVries. 1974. The effects of temperature and photoperiod on antifreeze production in cold-water fishes. J. Exp. Zool. 190: 89-98.

Haschemeyer, A. E. V. 1973. Kinetic analysis of synthesis and secretion of plasma proteins in a marine teleost. J. Biol. Chem. 248: 1643-1649.

Haschemeyer, A. E. V. 1976. Kinetics of protein synthesis in higher organisms *in vivo*. Trends in Biochemical Sciences 1: 133-136.

Hawkes, J. W. 1977. The effects of petroleum hydrocarbon exposure on the structure of fish tissues. In: Proceedings of Fate and Effects of Petroleum Hydrocarbons in Marine Ecosystems and Organisms, D. A. Wolfe, ed., Pergamon Press, New York, pp. 115-128.

Hew, C. L. and C. Yip. 1976. The synthesis of freezing-point-depressing protein of the winter flounder *Pseudopleuronectus americanus* in *Xenopus laevis* oocytes. Biochem. and Biophy. Research Comm. 71: 845-850.

Hudson, A. P. 1976. Comparative studies of plasma protein synthesis in temperate and antarctic fishes. Biol. Bull. 151: 414.

Lee, R. F., R. Sauerheber and G. H. Dobbs. 1972. Uptake, metabolism and discharge of polycyclic aromatic hydrocarbons by marine fish. Marine Biol. 17: 201-208.

Marks, P. A. 1965. Drug induced hemolytic anemias associated with glucose-6-phosphate dehydrogenase deficiency; a genetically heterogeneous trait. Ann. N. Y. Acad. Sci. 123: 198-206.

National Academy of Sciences, Ocean Affairs Board, Commission on Natural Resources, National Research Council, 1975. Petroleum in the marine environment, Washington, D. C., pp. 1-107.

Raymond, J. A., Y. Lin, and A. L. DeVries. 1975. Glycoprotein and protein antifreezes in two Alaskan fishes. J. Exp. Zool. 193: 125-130.

Winters, K., R. O'Donnel, J. C. Batteron and C. Van Baalen. 1976. Water-soluble components of four fuel oils. Marine Biol. 36: 269-276.

CONDITION INDEX AND FREE AMINO ACID CONTENT
OF *MACOMA INQUINATA* EXPOSED TO
OIL-CONTAMINATED MARINE SEDIMENTS

G. Roesijadi
J. W. Anderson

Battelle Marine Research Laboratory
Sequim, Washington

INTRODUCTION

It is evident that experimental approaches in the study of
biological effects of petroleum contamination of the marine
environment will require use of chronic, long-term exposure
and examination of the responses of organisms to sublethal
concentrations of petroleum (Wilson, 1975). With most of the
physiological parameters which have been considered to date,
(e.g., oxygen consumption, osmoregulation, breathing rate,
coughing rate) (reviewed by Neff *et al.*, 1976; Anderson, 1977),
responses following petroleum exposure have varied with test
species, petroleum concentrations, or duration of exposure.
While deviations from control responses were observed in many
studies, interpretations regarding the stressful nature of the
exposures have been difficult to make. An exception to this
trend of variable responses was that of growth rate inhibition
reported for several crustacean species (Neff *et al.*, 1976;
Anderson, 1977) and a polychaetous annelid (Rossi, in press).
Growth, when examined in terms of energy balance, has been
shown to be a sensitive indicator of stress in bivalve molluscs
(Bayne, 1975; Bayne *et al.*, 1976a). Bayne, for example, has
shown that scope for growth (=assimilated calories-respired
calories) in mussels *Mytilus edulis* is affected by temperature,
food availability, salinity, and oxygen concentration. Using
a related approach, Gilfillan (1975) and Gilfillan *et al.* (1976)
estimated carbon flux (=assimilated carbon-respired carbon)
in the clam *Mya aremaria* and reported a reduction in carbon
flux following petroleum exposure. A sustained alteration in
energy balance can result in changes in biomass and, thus, a
change in a parameter such as condition index (defined as the
ratio of dry tissue weight to shell cavity volume or length)

in bivalves. Fluctuations in the condition of bivalves are
considered normal and have been correlated to seasonal changes
in nutrient storage and utilization (Walne, 1970; Trevallion,
1971). Furthermore, periods of stress can lead to reductions
in condition (Trevallion, 1971). Condition index in bivalves
has been used as a measure of relative health, nutritional state
and growth.

Changes in concentrations of free amino acids (FAA) in
tissues of marine animals have also been shown to occur in
response to both natural and pollutant stresses (Jeffries,
1972; Roesijadi et al., 1976; Bayne et al., 1976b). Jeffries
(1972) and Bayne et al. (1976b) expressed FAA content as
taurine: glycine ratios and reported that elevated taurine:
glycine ratios of the bivalves Mercenaria mercenaria and
Mytilus edulis were indicative of stressed individuals, primarily due to decreases in glycine content. Although the
underlying mechanisms for the changes in free amino acid
content are not presently understood, it appears that determination of taurine:glycine may be useful in identifying
organisms which have experienced physiological stress.

The positive correlations observed to date between presumably stressful environmental conditions and both bivalve
condition index and free amino acid content indicate that both
parameters warrant additional study as physiological indicators
of pollutant stress. At the present time, neither parameter
has been extensively studied in this regard. To assess their
applicability as physiological indicators, condition index and
tissue free amino acid content were measured in the detriti-
vorous clam Macoma inquinata exposed to sediment contaminated
with Prudhoe Bay crude oil. Exposures were conducted in both
the laboratory and field. The results are presented in this
paper.

MATERIALS AND METHODS

Animals

Clams were collected from the intertidal region of Sequim
Bay, Washington (U.S.A.) and held in a water table containing
flowing seawater of approximately 10°C and 30°/oo. Clams were
used in the experiment within 2 days of collection.

Experimental

Sand from the upper intertidal region of Sequim Bay was
passed through a 6 mm sieve, and portions were weighed for use
as control and exposure sediment. The latter was prepared as

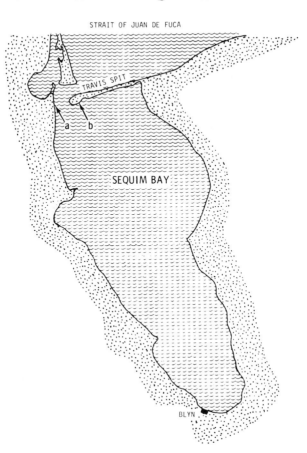

FIGURE 1. Sequim Bay, Washington, U.S.A. Points a and b are study areas (48° 5'; 123° 3' W). Scale = 1:40,000

described by Anderson *et al.* (1977) using Prudhoe Bay crude oil at 1,000 µg oil/g sediment nominal concentration. Exposures were conducted in fiberglass boxes divided into 3 equal compartments, each with a fiberglass mesh bottom (described in detail by Anderson *et al.*, 1977).

Two experiments were conducted. The first was a coordinated laboratory and field experiment in which roughly 5 kg sediment and 10 clams were placed in each compartment of an exposure tray, i.e., 30 clams per tray. One exposure box was inserted into the sediment in the intertidal zone of Sequim Bay opposite the tip of Travis Spit (point a of Figure 1.), while another was submerged in a holding tank in the laboratory. In the latter case, cement blocks were used to prop the boxes off the bottom of the holding tank. Each holding tank contained flowing seawater (750 ml/min) with a simulated tidal flux which

was achieved by draining seawater from the holding tank. Tidal
fluxes in the laboratory were timed to coincide with each
negative tide in the field. The mesh on the bottoms of the
exposure boxes allowed seawater to drain through the sediments
at low tide. Control boxes containing clams and non-oiled
sediment were prepared for both laboratory and field exposures.
Exposure began on November 22, 1976, and lasted 55 days. Sediments were analyzed for total petroleum hydrocarbons at the
beginning and the end of exposure. Clams were sampled at the
end of exposure. Some were analyzed for aliphatic and diaromatic hydrocarbons and the remainder for condition index.

The second experiment consisted of a single field exposure.
Exposure sediment was prepared as described above and placed
in 2 exposure boxes. Twenty clams were placed in each compartment of the 2 exposure boxes, i.e., a total of 120 clams. The
boxes were inserted into the sediment in the intertidal region
of Sequim Bay on the south side of Travis Spit (point b of
Figure 1). Controls in non-oiled sediment were also prepared
and placed at the same location. The exposure was initiated
June 18, 1977, and lasted 38 days. Sediments were analyzed
for total petroleum hydrocarbons at the beginning and end of
exposure. Clams were sampled at the end of exposure and
analyzed for aliphatic and diaromatic hydrocarbons, condition
index, or free amino acids (FAA) of selected tissues.

Analytical

Sediments were extracted in carbon tetrachloride and
analyzed for total petroleum hydrocarbons by infrared spectrophotometry (American Petroleum Institute, 1958). Clam
tissues were analyzed for aliphatic and diaromatic hydrocarbons by gas chromatography (Warner, 1976).

Condition index was determined as:

$$\frac{\text{ash-free dry weight}}{\text{length}^3} \times 1000 \quad \text{(de Wilde. 1975)}$$

Clams were shucked, dried at 65°C for at least 1 week, weighed,
ashed at 450°C for 24 hr, then reweighed. Ash-free dry weight
was calculated as the difference between dry and ash weights.

For free amino acid analysis, adductor muscle, foot, and
mantle were dissected from an individual clam, pooled, then
homogenized in distilled water using a 2:1 ratio of distilled
water:wet weight. Homogenates were deproteinized with 7%
trichloroacetic acid (TCA) as described by Roesijadi *et al.*
(1976). TCA supernatants were extracted 3 times with diethyl

ether to remove TCA, then lyophilized. Lyophilized extracts were taken up in 0.2 sodium citrate and adjusted to pH 2.2 by the addition of 0.1 N NaOH. Amino acids were analyzed by automated ion-exchange chromatography.

RESULTS

In the coordinated laboratory and field exposure, experiment 1, the behavior of clams in the laboratory indicated stress in oil-exposed individuals. During the exposure, 5 of 30 (i.e., 17%) exposed clams came to the sediment surface, a response previously reported for stressed *Macoma balthica* (de Wilde, 1975; Taylor and Karinen, 1977). Five clams (17%) in the contaminated sediment were also found dead and buried. In control sediment, no clams surfaced, and only one (3%) died during the experiment. Field-exposed clams were not examined until the end of the experiment. At that time, it was evident that there was considerable scouring due to wave action at the experimental site. Some sediment in the boxes had been washed away and partially replaced by new material. Dead or surfaced clams were not present in control or oil-exposed boxes in the field. However, pieces of broken shell from *M. inquinata* were present in those boxes. Nine of 30 (30%) control and 13 of 30 (43%) exposed clams were recovered for subsequent analysis.

Measurements of condition index in control and exposed clams in the first experiment were not significantly different (Student's t test $p<0.05$) from each other, although mean values for exposed clams were lower than for controls in both the laboratory and field (Table 1). On the other hand, condition index of animals following transfer to the field were significantly lower than those held in the laboratory (Student's t test $p<0.05$). Additionally, total numbers of dead and unrecovered clams (controls plus exposed) in the field (38 or 63%) greatly exceeded that in the laboratory exposure (6 or 10%). Shifting sand as a result of wave action can prevent settling and feeding and, thus, lead to reduced condition index (Trevallion, 1971). Such a process may explain our observations with *Macoma inquinata* in the field. However, stress due to oil was evident in the laboratory exposure since 10 or 33% of exposed clams, as opposed to one or 3% of controls, either died or surfaced during the exposure.

Although differences in condition index between oil-exposed and control clams, described above, were not statistically significant, mean values for exposed animals were ~10% lower than controls in both laboratory and field exposures. The possibility existed that the decreases in condition index of

TABLE 1. Condition index of Macoma inquinata exposed to oil contaminated sediment in the coordinated laboratory and field exposures (experiment 1).

Treatment	Sample size	Condition index $\bar{x} \pm S.E.$
Lab		
Control	20	11.5 ± 0.5 n.s.
Exposed	14	10.3 ± 0.6
Field		
Control	7	9.1 ± 0.8 n.s.
Exposed	8	8.1 ± 0.6

n.s. = Not significant at $p < .05$; Student's t test.

oil-exposed clams may have been real but obscured by individual variation (coefficient of variation ~20%). In order to demonstrate statistically significant differences under such conditions, we calculated (see Sokal and Rohlf, 1969, p. 246) that a sample size n of at least 65 per treatment would be required. Therefore, in the second experiment (conducted in the field only), we exposed 120 clams to oil-contaminated sediment with an additional 120 control clams. The increase in n above 65 was allowed to account for mortalities, as well as clams required for hydrocarbon and free amino acid analyses.

In the second experiment, clams were not examined until the end of exposure. At that time, we recovered 71 live clams, 24 empty shells, and 25 missing (i.e., not accounted for) from oil-contaminated sediment. In control sediment, 110 control clams were recovered alive, 1 as empty shell, and 9 were missing. Missing clams may have resulted from individuals which came to the sediment surface and were either consumed by predators or washed away by wave action. Those clams were pooled with empty shells in determining the total percent mortality of 41% for exposed and 8% for control clams. The percent of clams missing in oil-contaminated sediment of 21% corresponded closely with the 17% which came to the surface in oil-contaminated sediment of the previously described laboratory exposure.

Oil-contaminated sediment appeared to be anaerobic below the top 0.5 cm; the color was black and smelled of H_2S. Occasional small anaerobic pockets were observed in control sediment.

Condition index of clams in the second experiment was significantly decreased by oil exposure (Table 2) (Student's t test, p<0.001). It is interesting to note that the decrease in mean values by 12% due to oil exposure and coefficient of variation of 23% were similar to values for the first experiment described above. Thus, increasing the sample size was sufficient to demonstrate a highly statistically significant effect of oil-exposure on condition index.

When free amino acids (FAA) in muscle and mantle were examined (Table 3), it was determined that levels of arginine, glycine, lysine, threonine and total FAA were significantly reduced (Student's t test, p<0.001; except arginine, p<0.02) in oil-exposed clams. Reductions in glycine had the largest effect on the total FAA pool since glycine was present at the highest concentrations and the decreases were relatively large. The decreases in arginine, lysine, and threonine may represent a nutritional effect since these have been shown to be essential amino acids in animal groups which have been studied to date (Mahler and Cordes, 1971). The increase in taurine:glycine ratios from 0.55 ± 0.04 (S.E.) in controls to 0.89 ± 0.12 (S.E.) in exposed clams were also significant (Student's t test, p<0.02).

TABLE 2.   *Condition index of* Macoma inquinata *exposed to oil-contaminated sediment in the second experiment. Exposure was conducted in the field only*

Treatment	Sample size	Condition index $\bar{x} \pm S.E.$
Control	91	8.9 + 0.2
		***
Exposed	50	7.5 + 0.3

***Significant at p<0.001; Student's t test.

TABLE 3. Free amino acid content of Macoma inquinata
*exposed to oil-contaminated sediment in the second experiment.
Conducted in the field only   (n=8)*

	Concentration (μ moles/g)	
	Control	Exposed
Amino acid	*x ± S.E.*	*x ± S.E.*
Alanine	*22.53 ± 1.82*	*16.80 ± 2.02*
Arginine	*6.89 ± 0.40*	*4.56 ± 0.73 a*
Aspartate	*1.44 ± 0.17*	*0.90 ± 0.27*
Glutamate	*2.72 ± 0.22*	*2.21 ± 0.19*
Glycine	*70.25 ± 4.51*	*43.56 ± 4.99 b*
Histidine	*0.28 ± 0.03*	*0.22 ± 0.04*
Isoleucine	*0.37 ± 0.02*	*0.32 ± 0.04*
Leucine	*0.63 ± 0.05*	*0.48 ± 0.05*
Lysine	*0.59 ± 0.04*	*0.41 ± 0.04 b*
Methionine	*0.20 ± 0.03*	*0.13 ± 0.02*
Phenylalanine	*0.22 ± 0.02*	*0.19 ± 0.02*
Proline	*0.57 ± 0.04*	*0.74 ± 0.25*
Serine	*4.09 ± 0.41*	*3.24 ± 0.39*
Taurine	*37.06 ± 1.94*	*35.08 ± 2.39*
Threonine	*1.16 ± 0.05*	*0.87 ± 0.07 b*
Tyrosine	*0.37 ± 0.03*	*0.33 ± 0.03*
Valine	*0.54 ± 0.04*	*0.44 ± 0.16*
Total	*150.57 ± 8.15*	*110.48 ± 8.24 b*
Taurine:Glycine	*0.55 ± 0.04*	*0.89 ± 0.12 a*

[a]*Significant at p < 0.02, Student's t test.*
[b]*Significant at p < 0.01, Student's t test.*
 All others not significant at p < 0.05, Student's t test.

Values for sediment hydrocarbon concentrations are summarized in Table 4. At the beginning of exposure in the coordinated laboratory and field exposure (the first experiment),
contaminated sediment contained 1232.9 ± 82.0 (S.E.) μg/g
total petroleum hydrocarbons. After 55 days, hydrocarbon concentrations for sediment in the laboratory decreased to 616.1
± 84.9 (S.E.) μg/g, 50% of the initial value, while those in
the field decreased to 87.8 ± 27.3 (S.E.) μg/g, 7% of the
initial level. Control sediment contained undetectable levels,

TABLE 4. Total petroleum hydrocarbons in exposure sediment in both experiments. x̄ ± S.E. of 3 replicate samples.

Treatment	Hydrocarbon concentration (μg/g)	
	Initial	*Final*
Experiment 1[a]		
Lab control	*n.d.*[b]	*n.d.*
exposed	*1232.9 ± 82.0*	*616.1 ± 84.9*
Field control	*n.d.*	*n.d.*
exposed	*1232.9 ± 82.0*	*87.8 ± 27.3*
Experiment 2[c]		
control	*n.d.*	*n.d.*
exposed	*1144.0 ± 47.3*	*364.0 ± 66.4*

[a]*Duration of experiment 1 - 55 days.*
[b]*n.d. = not detectable (<7.5 μg/g).*
[c]*Duration of experiment 2 = 38 days.*

both initially and at 55 days. In the second experiment, initial total petroleum hydrocarbons in exposure sediment were 1144 ± 47.3 (S.E.) μg/g, and levels after 38 days were 364 ± 66.4 (S.E.) μg/g. Again, levels in control sediment were undetectable.

Hydrocarbon analyses of clam tissue indicated relatively low levels in exposed clams relative to petroleum hydrocarbons in the exposure sediment (Table 5). In both experiments, control clams possessed undetectable levels of diaromatic hydrocarbons and small amounts of aliphatics ($\leq$ 0.030 μg/g). In the first experiment, levels of aliphatics in clams exposed to oil in both the laboratory and field exposures were similar, approximately 0.2 μg/g. However, diaromatic hydrocarbons, primarily dimethylnaphthalenes, in laboratory-exposed clams ($\bar{x}$ = 3.80 μg/g) were approximately 100 times higher than those exposed in the field ($\bar{x}$ = 0.03 μg/g). Hydrocarbon concentrations of clams exposed to oil-contaminated sediment in the field in the second experiment were relatively similar to levels detected in field-exposed clams in the first experiment.

TABLE 5. *Aliphatic and diaromatic hydrocarbons in Macoma inquinata exposed to oil-contaminated sediments in both experiments*

Treatment	Hydrocarbon concentrations ($\mu g/g$) [a,b]		
	C_{12-28} Aliphatics	Methylnaphthalenes	Dimethylnaphthalenes
Experiment 1			
Lab control	0.030	<0.001	<0.005
exposed	0.140	1.153	5.209
exposed	0.422	<0.010	2.387
Field control	0.020	<0.001	<0.005
exposed	0.140	<0.050	0.022
exposed	0.058	<0.010	0.031
Experiment 2			
control	<0.004	<0.004	<0.004
control	<0.004	<0.004	<0.004
exposed	0.46	0.02	0.06
exposed	0.42	0.01	0.07

[a] Each sample consisted of pooled tissue of at least 2 clams.
[b] Levels of naphthalene were not dectable in all cases (<0.004 $\mu g/g$).

DISCUSSION

Our study has shown that both condition index and certain free amino acids in *Macoma inquinata* were significantly affected by exposure to oil-contaminated sediments. Adequate sample size was an important factor in demonstrating statistically significant reductions in clam condition. The reduction in condition index in exposed clams, although small (~10%), provided evidence of a deterioration in physiological state. A decrease in bivalve condition index is an indication that affected clams may have been in a state of negative energy balance; in other words, metabolized energy exceeded energy consumed as food. Utilization of storage products such as tissue protein, lipid, and carbohydrates may have been necessary to provide the balance of the energy for metabolism under such conditions (Gabbott, 1976). Condition index of oysters and mussels have been closely correlated with tissue glycogen content (Walne, 1970; Gabbott and Stephenson, 1974; Gabbott and Bayne, 1973).

Using the criteria proposed by de Wilde (1975) for *Macoma balthica,* a value of > 10 represents good condition, ~ 8 moderate, and < 6 poor. If these criteria are applicable to *M. inquinata,* the clams in this study possessed mean values for condition index which ranged from good to moderate. Condition in bivalves is known to undergo seasonal variations which are reflective of reproductive and physiological state (Walne, 1970; Trevallion, 1971; de Wilde, 1975). Periodic sampling of clams from our collection site during the course of this study indicated that condition index of our experimental clams, especially those used in field exposures, were lower than those of the natural population. In general, clams in this study possessed lower condition index (Table 1, 2) than those collected freshly at the times of experiment termination (13.8 ± 1.6 and 17.0 ± 0.8 for experiments 1 and 2, respectively). Thus, experimental manipulation (e.g., collection, handling, placement in boxes) also was a factor which influenced condition index of clams in this study.

Since arginine, lysine and threonine are considered to be essential amino acids (Mahler and Cordes, 1971), decreases in their concentrations in oil-exposed clams were also suggestive of alterations in physiological state. Increased utilization of these amino acids, possibly in protein systhesis, by oil-exposed clams or decreased ingestion during feeding may have accounted for our observations. The large decrease in glycine content in our oil-exposed clams was consistent with previous studies which examined free amino acid levels in marine animals subjected to pollutant or natural stresses (Jeffries, 1972; Roesijadi *et al.,* 1976; Bayne *et al.,* 1976b). As a consequence of the decrease in glycine, the taurine:glycine ratio was elevated in oil-exposed clams. The actual values of 0.55 ± 0.04 (S.E.) for control clams and 0.89 ± 0.12 (S.E.) for exposed clams in this study were not directly comparable to those reported by Jeffries (1972) or Bayne *et al.* (1976b) since taurine levels in *Macoma inquinata* were much lower than those in the bivalves *Mercenaria mercenaria* and *Mytilus edulis* used in the other studies. Although taurine:glycine ratios may prove useful in identifying bivalves which have experienced stressful environmental conditions, it is evident that the cause of the change in the ratios is due primarily to alterations in glycine content. This pattern has been consistent in the studies conducted to date. Examination of glycine metabolism would certainly be useful in understanding this apparent stress response.

Hydrocarbon concentrations in exposure sediments in this study were high, but consistent with values reported in marine sediments containing high levels of petroleum (Gilfillan *et al.*, 1976; Krebs and Burns, 1977). The relatively low levels of hydrocarbons detected in tissue of exposed clams agreed well with previous studies in our laboratory regarding bioavailability of sediment-sorbed hydrocarbons (Anderson *et al.*, 1977; Roesijadi *et al.*, 1978). Observations of stress in *Macoma inquinata* exposed to oil-contaminated sediment were also consistent with previous reports that either survival or carbon flux was reduced in bivalves exposed to oil-contaminated sediment (Dow, 1975; Gilfillan *et al.*, 1976; Shaw *et al.*, 1976). However, Shaw *et al.* (1977) did not observe changes in dry weight:shell length ratios in *M. balthica* exposed to Prudhoe Bay crude oil. Additional studies are required to reconcile discrepancies between our findings and those of Shaw *et al.* (1977).

SUMMARY

Stress in *Macoma inquinata* exposed to oil-contaminated sediment was evident at 3 levels of examination. These were reduced survival, reduced condition index, and reduced levels of free amino acids, primarily glycine, with concomitant elevations in the taurine:glycine ratio. This study has provided evidence that free amino acid content, as well as condition index, may be useful indicators of physiological stress in marine bivalves exposed to petroleum. These parameters may also be useful as general indicators of pollutant stress in vivalves and, possibly, other marine organisms.

PHYSIOLOGICAL BASIS OF DIFFERENTIAL SENSITIVITY OF FISH
EMBRYONIC STAGES TO OIL POLLUTION

J. R. Sharp, K. W. Fucik[1] and J. M. Neff

Department of Biology
Texas A & M University
College Station, Texas

INTRODUCTION

Because of their mobility, adult fish can avoid or emigrate
from polluted areas. Such is not the case, however for the
embryonic and larval stages which are either planktonic or
demersal. Demersal eggs, such as those of the killifish,
Fundulus heteroclitus, may be particularly susceptible to those
pollutants that are incorporated into bottom sediments and
gradually leached out into the water column. Blumer *et al.*
(1970) reported that, following a spill of #2 fuel oil near
West Falmouth, Mass., large amounts of petroleum were incorporated into the sediments. Two months after the spill,
essentially unchanged oil was still being released from the
sediments and surviving bivalve molluscs from the area were
still heavily contaminated, especially with the aromatic fractions of the oil. Such areas of chronic hydrocarbon pollution
could pose a serious threat to survival of demersal or benthic
organisms.

The embryonic and larval stages of the life cycle of marine animals are generally considered to be the most sensitive
to oil pollution stress (Hyland and Schneider, 1976). Recently,
McKim (1977) evaluated the use of bioassays with early life
stagesof fish for preducting chronic toxicity and for establishing the maximum acceptable toxicant concentration (MATC) water
quality criteria. He found that in 56 life-cycle toxicity

[1]*Present Address: Scientific Applications, Inc., 2760 29th
Street, Boulder, Colorado 80302.*

85

tests performed with 34 organic and inorganic chemicals and 4 species of fresh water fish, the embryo-larval and early juvenile stages were the most, or among the most, sensitive. He concluded that tests with these stages can be used to estimate the MATC in most cases.

The objectives of the present investigation were to determine the toxicity of the water-soluble fraction (WSF) of a #2 fuel oil to the embryo-larval stages of the estuarine killifish, *Fundulus heteroclitus,* to determine the relative sensitivity of different embryonic stages to the WSF and to provide evidence concerning the physiological basis for this differential sensitivity.

MATERIALS AND METHODS

Adult male and female *Fundulus heteroclitus* (Walbaum) were maintained separately in the laboratory in 500 liter Living Stream recirculating aquaria (Frigid Units Inc., Toledo, Ohio) under conditions described by Boyd and Simmonds (1974) in order that viable gametes could be obtained year round. Culture conditions were maintained constant at 22 $\pm$ 1°C, 20 $^{\circ}$/oo salinity and a light:dark photoperiod of 16:8 hours provided by broad spectrum fluorescent lights. Fish culture and all experiments with fish eggs were conducted in artificial sea water prepared by mixing Instant Ocean artificial sea salts (Aquarium Systems, Inc., Eastlake, Ohio) with distilled water. Sea water salinities were determined with an optical refractometer (American Optical Co., Inc.). For each experiment, eggs were stripped from several females and fertilized with the sperm obtained from several males. Exposures were conducted in 8 cm glass finger bowls containing 50 ml of exposure medium and 20 to 25 eggs.

Preparation and Analysis of Oil-Water Mixtures

The water-soluble fraction (WSF) of #2 fuel oil (API reference oil III) was prepared by mixing 1 part oil over 9 parts artificial sea water for 20 hours (Anderson *et al.,* 1974). The stock 100% WSF was diluted with artificial sea water to the appropriate concentration and used immediately. Fresh WSF's were prepared daily. The concentrations of naphthalenes (naphthalene, methylnaphthalenes and dimethylnapthalenes) in the WSF were determined periodically during the exposures by the ultraviolet spectrophotometric method of Neff and Anderson

(1975). Total aqueous hydrocarbons were determined by infrared spectrometry (American Petroleum Institute, 1959) using a Miran I infrared analyzer (Wilks Scientific, Inc., S. Norwalk, Ct.).

Hatching Success

Three experiments were conducted to assess the overall sensitivity of *F. heteroclitus* embryos to #2 fuel oil WSF's and to determine which embryonic stages are the most sensitive to oil. In the first experiment, groups of embryos were exposed to 0 (control), 10, 20 and 25% dilutions of the WSF from fertilization until hatching. The larvae that successfully hatched from the 10 and 20% WSF exposure groups were exposed to WSF's for an additional 20 days so that a comparison of the relative sensitivity of the embryos and larvae could be made. Daily, all exposure media were changed and the number of embryos that hatched or died was counted. Initially, there were between 200 and 300 embryos in each exposure group.

In the second experiment, all groups of eggs, except the controls, were exposed to a 25% WSF, which was renewed daily. Exposure of all groups was initiated within 4 hours of fertilization, and was continued for 4, 8 or 12 days or until all embryos had hatched or died. Approximately 65 embryos were present in each exposure group and in the controls. At the end of the respective exposure periods, the developing embryos and larvae were returned to oil-free sea water for the duration of the experiment. Hatching success of each group was determined as noted above. These 2 experiments were conducted at room temperature (20-23^{0}C) and at a salinity of 20o/oo.

The third experiment was conducted at a constant temperature of 24 ± 1^{0}C at 20o/oo. Eggs were fertilized and allowed to develop in hydrocarbon free sea water for different lengths of time between 0 and 8 days before exposure to oil was initiated. Exposure to 25% WSF was continued until the thirtieth day of development, at which time all embryos had either died or hatched. Four replicates of 20 embryos each were used for each exposure group. Daily, embryos were checked for hatching and mortalities and exposure media were changed.

Uptake and Release of Naphthalenes and Water

The accumulation of naphthalenes (naphthalene, methylnaphthalenes and dimethylnaphthalenes) by embryos during continuous exposure to a 25% WSF of #2 fuel oil, renewed daily, was determined. Immediately after fertilization, a group of embryos was placed in the WSF in 20o/oo S sea water at 22^{0}C. After 1, 3, 5, 7, 9 or 11 days exposure, duplicate samples of 10 embryos

were removed, weighed and homogenized in 5 ml of spectrophotometric grade n-hexane. The hexane extract, after treatment
with activated Florisil, was analyzed for naphthalenes by the
method of Neff and Anderson (1975). Naphthalenes concentrations in the exposure water were also analyzed periodically
by the same method.

The rates of influx and efflux of naphthalene and water in
embryos at different stages of development were determined
using radioisotope methodology. Eggs were fertilized and
placed in hydrocarbon-free sea water. After 2, 4, 6, 8 or 10
days of development, 50 embryos were removed and placed in
finger bowls containing 50 ml of 20o/oo S sea water containing
10 μl 1-^{14}C-naphthalene (specific activity: 25 μCi/ml, Amer-
sham/Searle) and 50 μl ^{3}H$_2$O (specific activity: 1μCi/ml, New
England Nuclear). After 2 hours incubation, the eggs were
removed and rinsed in isotope-free sea water. In groups of
10, embryos were placed in scintillation vials containing 15
ml of Aquasol liquid scintillation coctail and homogenized
with a glass probe (Manner and Muehlman, 1975). ^{14}C and ^{3}H
activity were measured with a refrigerated Packard Tri Carb
liquid scintillation counter. All counts were corrected for
background, quench and counting efficiency by standard methods
and results were recorded as disintegrations per minute (DPM)
per 10 embryos. In a second experiment, embryos at 2, 4, 6,
8, 10 and 12 days of development were exposed to isotopically
labelled naphthalene and water as above. After 2 hours exposure, they were transferred to isotope-free sea water for 2
hours before being analyzed for radioactivity. The depuration
water was also analyzed for radioactivity.

Metabolism of Aromatic Hydrocarbons

In separate experiments 2, 4, 6, 8 or 10 day embryos were
exposed to 10 μl in 50 ml sea water of either ^{14}C-naphthalene
as above or of 5, 6 (11, 12)- ^{14}C-chrysene (specific activity:
18.9 μCi/ml, Amersham/Searle) for 2 hours. Embryos were then
placed in isotope-free sea water for an additional 2 hours.
The embryos and the depuration water were then fractionated
into nonpolar compounds (unmetabolized aromatics) and polar
compounds (polar metabolites) by the method of Roubal *et al.*
(1977) and each fraction was analyzed for ^{14}C activity as
described above.

Heart Beat Rate

Eggs were fertilized and placed in hydrocarbon free sea
water. After 2, 4, 6 or 8 days of development, groups of embryos were place in 25% WSF in 20o/oo S sea water at 24^{0}C.
The heart beat counts were made on days 9, 10 and 11 on 10
randomly selected embryos from each group of 50 embryos.

Respiratory Rate

Groups of embryos were exposed to the 25% WSF of #2 fuel oil for different lengths of time according to the same regime as described above for heart beat rate determinations. On day 10 of development, the respiratory rates of 16 groups of 4 randomly selected embryos from each exposure group were determined. Embryos were placed in groups of 4 in 5 ml all glass syringes containing 2 ml of Millipore-filtered (0.45µ) air-saturated 20o/oo S sea water. At time 0 and 3 hours, 70 µl aliquots of the water in the syringes were injected into a Radiometer BMS3Mk2 micro blood gas analyzer thermostated at 24^0C, and the pO_2 of the water was determined. The difference represented the O_2 consumption of the embryos. The embryos were dried to constant weight at 65^0C and weighed to the nearest 0.01 mg. Respiratory rates are expressed as ml O_2/g dry wt x hr.

The data obtained in the different experiments were analyzed statistically using several computer programs from Barr *et al.* (1976).

RESULTS

As has been demonstrated previously (Anderson *et al.*, 1974), the concentration of total hydrocarbons and total naphthalenes in the WSF's decreased during the 24 hours between media changes. At 20^0C, the mean initial naphthalene concentration in the 25% WSF of #2 fuel oil was 0.54 ± 0.01 ppm and dropped to 0.31 ± 0.08 ppm in 24 hours. The 10 and 20% WSF's had mean initial naphthalene concentrations of 0.21 and 0.44 ppm, respectively and showed a 24 hour concentration drop similar to that of the 25% WSF. At 24^0C the initial mean naphthalene concentrations of the WSF were approximately 10% higher than those at 20^0C but showed an approximately 70% drop in concentration in 24 hours. At 20^0C, initial mean total hydrocarbon concentrations in the WSF's were 0.85, 1.7 and 2.1 ppm for the 10, 20 and 25% WSF, respectively.

Hatching Success

Cumulative % hatch of embryos continuously exposed from fertilization to 0, 10, 20 and 25% WSF of #2 fuel oil is sum-marized in Figure 1. The first hatches of embryos were noted on day 15 for the controls, day 14 for the 10 and 20% WSF-exposed embryos and on day 19 for the embryos exposed to 25% WSF. At 0 and 10% WSF, the final hatching success was 91 and 92%, respectively. At 20% WSF, survival to hatching was 74%

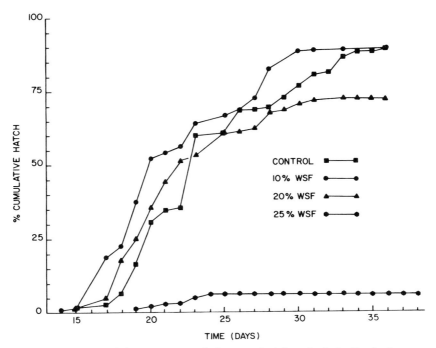

FIGURE 1.   *Hatching rate and % cumulative hatch* F. *hetero-clitus embryos continuously exposed to several concentrations of the WSF of #2 fuel oil.*

and at 25% WSF, it was 2% after 38 days of exposure. However at 25% WSF, embryos continued to hatch through day 52 with a total cumulative hatch of 22%. An analysis of variance showed that there was a significant difference in total cumulative hatch between controls and embryos exposed to 20% WSF ($P > F = 0.02$) and 25% WSF ($P > F = 0.0001$).

None of the larvae hatching from embryos continuously exposed to 25% WSF during development survived for more than 3 days, although these larvae were placed in hydrocarbon-free sea water immediately after hatching. Exposure to WSF's was continued for larvae hatching from embryos exposed to 10 and 20% WSF. Among these, 88% of the 10% WSF-exposed larvae and 61% of the 20% WSF-exposed larvae survived for 2 weeks. By

comparison 96% of the control larvae survived. Thus, cumulative survival of the embryo-larva life stages in the control,
10, 20 and 25% WSF exposure groups was 88, 82, 45 and 0%,
respectively. The median lethal concentration of the #2 fuel
oil WSF for the embryo-larva life stages can be estimated from
these data at about 1.5 ppm total hydrocarbons.

The duration of exposure to the 25% WSF had a profound
effect on the hatching rate and hatching success of the fish
embryos. The data for embryos continuously exposed to the WSF
from fertilization for different lengths of time are summarized
in Figure 2. The first hatches among control embryos and those
exposed for 4 days occurred on day 13. The first hatch occurred on days 14, 17 and 25 among the embryos exposed to the
WSF for 8, 12 days or continuously, respectively. Cumulative
hatching success of control and 4-day-exposed embryos was similar at 92 and 87%, respectively. Hatching success was much
lower in the other groups with 40, 6 and 1.5% of the embryos
eventually hatching in the 8-, 12-day and continuously exposed

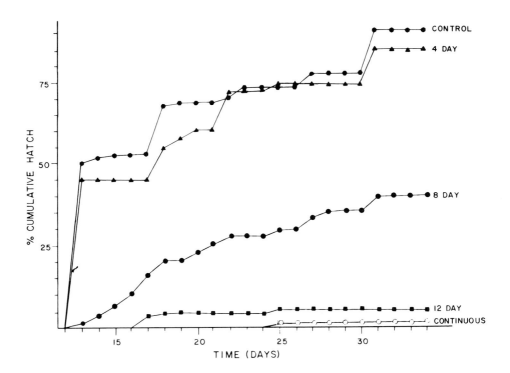

FIGURE 2. Hatching rate and % cumulative hatch of F.
heteroclitus embryos continuously exposed from fertilization
for different lengths of time to the 25% WSF of #2 fuel oil.

groups, respectively. Analysis of variance revealed that
duration of exposure had a highly significant effect on hatching success of the embryos (P > F = 0.0001). The Waller-
Duncan K-ratio T test for variable hatch revealed that there
were three exposure groups showing significantly different
hatching rates and success. These were the controls and 4-
day-exposed embryos (group 1), the 8-day-exposed embryos
(group 2) and the 12-day and continuously exposed embryos
(group 3).

When exposure to the 25% WSF was delayed for different
lengths of time after fertilization of the eggs, a similar
graded hatching success response was noted (Figure 3). Four
distinct subgroupings of exposure groups were evident. Hatching success of the controls and of embryos first exposed to
the 25% WSF after 7 or 8 days of development in hydrocarbon
free sea water was similar, 72.5, 76.3 and 78.8% respectively.
Hatching rates of the day 7 and 8 exposure groups appeared
to be somewhat higher than that of the controls. Embryos
first exposed on day 5 and 6 of development had a similar

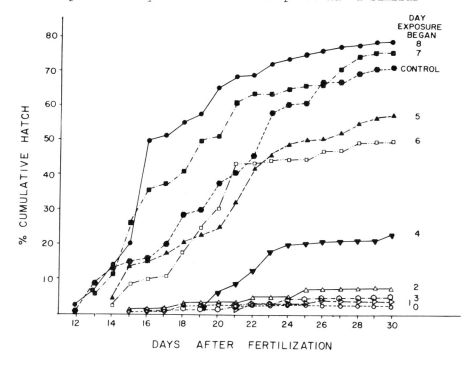

FIGURE 3. Hatching success and % cumulative hatch of F.
heteroclitus *embryos when continuous exposure to the 25% WSF
of #2 fuel oil was initiated after different periods of development in hydrocarbon-free sea water.*

hatching success of 57.5 and 50%, respectively and also had similar hatching rates. Hatching rate and success (22.5%) were greatly reduced in embryos first exposed to the WSF on day 4 of development. The fourth subgroup consisted of embryos first exposed on days 3, 2, 1 and 0 of development. Hatching success of these groups was 3.75, 7.5, 3.75 and 2.5% respectively. An analysis of variance of the hatching success data revealed that the day on which exposure to the WSF was initiated had a highly significant effect on hatching success (P > F = 0.0001). Duncan's multiple range test for variable hatch indicated 4 subgroups with significantly different hatching rates. These were the controls, 5, 7, and 8 day embryos (group 1), the 5 and 6 day embryos (group 2), the 4 day embryos (group 3), and the 0, 1, 2 and 3 day embryos (group 4).

The onset of hatching also was related to the time at which exposure to the WSF was initiated. However, there was not a clear cut dose x time/response relationship for this developmental parameter. The first hatch of controls and 8-day embryos occurred on day 12 of development. The 7 day embryos began hatching on day 13, the 5 and 6 day embryos on day 14, and 0 and 2 day embryos on day 15, the 3 day embryos on day 16, the 4 day embryos on day 19 and the 1 day embryos on day 21.

Uptake of Naphthalenes and Water

When embryos were exposed continuously from fertilization to a 25% WSF (0.57 ± 0.17 ppm mean aqueous naphthalenes concentration), they showed a nearly linear increase in mean tissue naphthalenes concentrations from 17 ppm on day 1 to 78 ppm on day 9 (Figure 4). The mean tissue naphthalenes concentration then dropped to 63 ppm on day 11 of exposure. The uptake experiment was repeated 2 additional times with essentially similar results. The 9-day biomagnification factor for naphthalenes was approximately 137.

Pulse labeling experiments revealed different temporal patterns of change in ^{14}C-naphthalene and ^{3}H-water uptake rates (Figure 5). The instantaneous naphthalene uptake rate decreased markedly from 2530 DPM ^{14}C/embryo x 2 hrs on day 2 of development in hydrocarbon-free sea water to 390 DPM ^{14}C/ embryo x 2 hrs on day 10 of development. The sharpest drops in the apparent permeability of the embryos to naphthalene occurred between day 2 and 4 and day 8 and 10. In contrast, instantaneous water uptake rates increased slightly in a nearly linear fashion from 220 DPM ^{3}H/embryo x 2 hrs on day 2 to 460 DPM ^{3}H/embryo x 2 hrs on day 10. Standard deviations were never more than 10% of the means. Thus, the apparent permeability of the embryos to naphthalene and water changed in different directions during the time course of development.

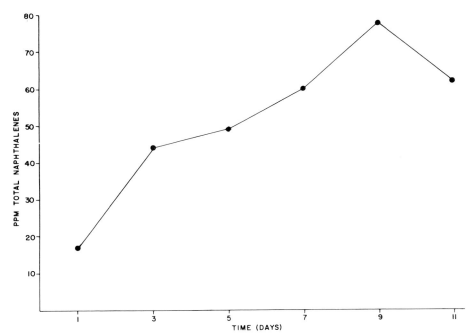

FIGURE 4. Accumulation of total naphthalenes (naphthalene, methylnaphthalenes and dimethylnaphthalenes) by intact F. heteroclitus *embryos during continuous exposure to the 25% WSF of #2 fuel oil (mean aqueous naphthalenes concentration, 0.57 ppm). Each point represents the mean naphthalenes concentration (ppm = µg naphthalenes/g wet wt. tissue) in 2 pooled samples of 10 embryos.*

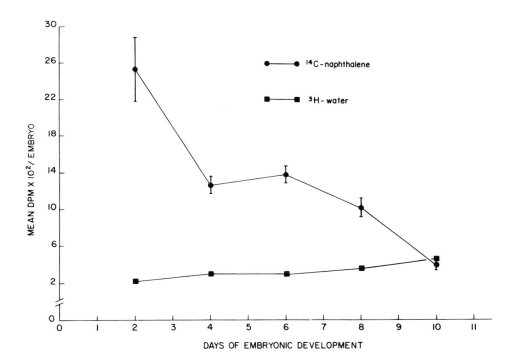

FIGURE 5. "Instantaneous" uptake rates of ^{14}C-naphthalene and ^{3}H-water by F. heteroclitus embryos at different times during development in hydrocarbon free seawater. Points represent radioactivity accumulated during 2-hours incubation in sea water containing the appropriate radioisotopes. Standard deviations for the ^{14}C-naphthalene values are included. Those for ^{3}H-water were always less than 10% of the mean value.

The rate at which accumulated naphthalene and water were
released from the embryos following 2 hours of pulse labeling
also changed during the time course of development (Figure 6).
In this case, the efflux rate of both naphthalene and water
decreased with increasing age of the embryo.   At all times,
the efflux rate of water was greater than that of naphthalene.
Water efflux rate remained between 23.2 and 24%/2 hrs between
day 2 and day 6 of development and then dropped in a nearly
linear fashion to 14.9%/2 hrs at day 12 of development, the
approximate time at which the embryos began hatching.   Naphthalene efflux rate varied from 17.3 to 19.9%/2 hrs between
day 2 and day 6 development and then dropped in a nearly
linear fashion to 11.7%/2 hrs on day 12 of development.

Metabolism of Aromatic Hydrocarbons

No evidence could be found that embryos at any stage of
development were able to metabolize either naphthalene or
chrysene.   In all cases the polar fractions of the embryo

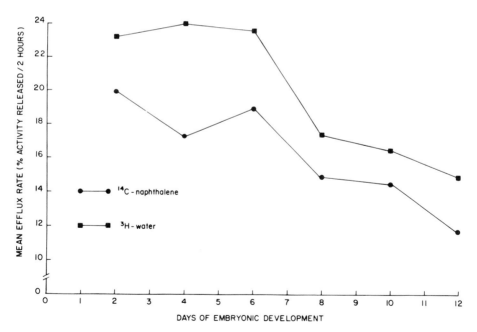

FIGURE 6.   Mean rates of release of ^{14}C-naphthalene and
^3H-water by F. heteroclitus embryos following 2-hours exposure
to the isotopically labeled compounds at different times of
embryonic development.

tissues and depuration water contained radioactive counts
per minute which were not significantly different from background. The radioactivity of the nonpolar fractions of naph-
thalene-exposed embryos was similar to that reported above
for the pulse uptake experiment indicating that the embryos
in this experiment were accumulating naphthalene. Chrysene
uptake rates were only about 10% of the naphthalene uptake
rates.

Heart Beat Rate

The heart beat rate of control animals increased slightly
but not significantly from 127 ± 10 beats per minute (bpm)
on day 9 of development to 135 ± 7.8 bpm on day 11 (Figure 7).
Embryos first exposed to the WSF on day 8 of development had
heart beat rates essentially identical to those of controls
at all sampling times. Embryos first exposed to the WSF on
day 6 had slightly depressed heart beat rates ranging from
113 ± 5.7 bpm on day 9 to 122 ± 9 6 bpm on day 11. Embryos
first exposed to the WSF on day 2 or 4 of development had
heart beat rates that were significantly depressed below those
of controls on days 10 and 11. In addition, heart beat rates

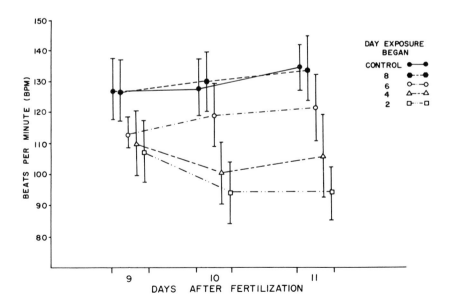

FIGURE 7. Mean heart beat rates of F. heteroclitus *embryos
on days 9, 10 and 11 of development during continuous exposure
to the 25% WSF of #2 fuel oil initiated at different times
after fertilization. Standard deviations are given.*

showed a decreasing trend between day 9 and 11. The heart
beat rates of day-2 embryos dropped from 107 ± 11 bpm on day
9 to 94.4 ± 7 bpm on day 11. The relatively large standard
deviations in the heart beat rates measured at any given time
of development seem to be typical for embryos of this species
(Glaser, 1929).

Respiration Rate

The mean respiratory rates of 10 day old embryos showed a
definite decreasing trend with increasing duration of exposure
to the 25% WSF during development (Figure 8). Control embryos
(the complete embryo complex, including embryo, yolk, perivette-
line fluid and chorion) had a mean respiratory rate of 0.170
± 0.012 ml O_2/g dry wt x hr. Embryos first exposed to the WSF
on day 2 (total length of exposure, 8 days) had a mean respiratory rate of 0.131 ± 0.031 ml O_2/g dry wt x hr, significantly
below the control rate. Embryos first exposed to the WSF on

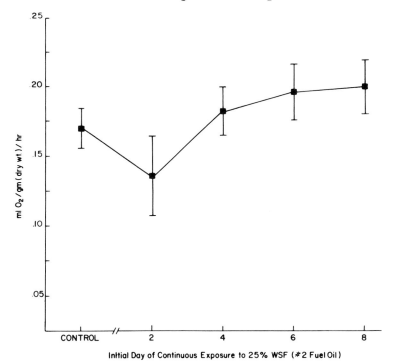

FIGURE 8. Mean respiratory rates, including standard
deviations, of 10-day old F. heteroclitus embryos during continuous exposure to the 25% WSF of #2 fuel oil initiated at
different times after fertilization.

days 4, 6 or 8 of development had mean respiratory rates between 0.183 ± 0.014 and 0.200 ± 0.017 ml O_2/g dry wt. x hr., slightly higher, but not significantly so, than that of the controls. The mean dry weight of embryos (complete embryo complex) in each exposure group was similar and ranged from 0.916 ± 0.042 mg/embryo for controls to 0.967 ± 0.416 mg/ embryo for day-2 embryos (Figure 9).

Behavior and Anomalies

Control embryos and those exposed to the WSF only late in development showed considerable activity (caudal and pectoral fin swimming motions, spiralling within the egg capsule, etc.) during development, and the newly hatched larvae were highly active. In contrast, surviving embryos which were first exposed to the WSF earlier in development, showed little activity within the egg capsule. After hatching the resulting larvae remained nearly motionless on the bottom of the culture

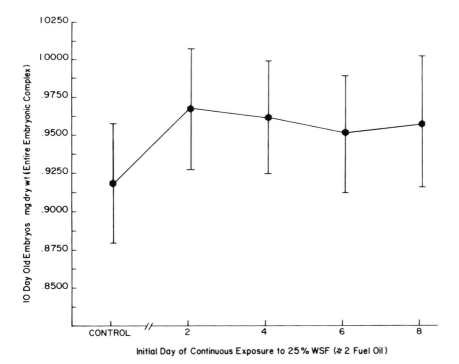

Initial Day of Continuous Exposure to 25% WSF (# 2 Fuel Oil)

FIGURE 9. *Mean dry weights, including standard deviations, of 10-day old* F. heteroclitus *embryos continuously exposed to the 25% WSF of #2 fuel oil initiated at different times after fertilization.*

dishes and made only periodic weak swimming motions. Very few gross developmental anomalies were noted in the emergent larvae from the control or any exposure group. Qualitative evaluation did not reveal any tendency toward an increase in the incidence of anomalies with increasing exposure concentration or increasing duration of exposure during embryonic development. However, larvae hatching from embryos that had been exposed to the WSF were shorter in length and had larger yold sacs than did unexposed controls. The deviations from control fry allometry seemed to increase with increasing duration of exposure to the 25% WSF.

DISCUSSION

The embryos of *Fundulus heteroclitus* were found to be relatively more tolerant to the WSF of a #2 fuel oil than the embryos and larvae of several other species of marine animals. In a review of the toxicity of oil to marine animals, Hyland and Schneider (1976) concluded that the embryos and larvae of most marine animals studied had LC_{50} values in the range of 0.1 to 1 ppm, while LC_{50} values of the adults usually fell in the 1 to 100 ppm range. In the present investigation, the LC_{50} of the embryo-larva life stages of *F. heteroclitus* was estimated to be about 1.5 ppm total aqueous hydrocarbons. A #2 fuel oil WSF of this concentration contains approximately 0.4 ppm total naphthalenes. The naphthalenes were earlier judged to be the major toxicants in the particular #2 fuel oil used in these experiments (Anderson *et al.,* 1974).

The 2 experiments in which the time and duration of exposure of embryos to the 25% WSF was varied reveal that the sensitivity of the embryos to oil changed during the time course of development. When exposure to the WSF was initiated immediately after fertilization, 4 days exposure had no effect on hatching rate or success, but 8 days exposure decreased hatching success to 40% and retarded hatching rate. If the initiation of continuous exposure to the WSF was delayed to several days after fertilization, longer exposure times were required to produce a similar decrease in hatching success and hatching rate. When exposure was initiated on day 5 and 6, hatching success was 57.5 and 50%, respectively. The mean hatching time (time at which 50% of the embryos eventually hatching had hatched) was at 19-20 days of development in these 2 exposure groups. Thus the mean duration of exposure of these embryos was 14 days. The first hatches of these embryos was on day 14 after 8-9 days of exposure. If it is assumed that the magnitude of toxic response to the WSF is a function of the exposure concentration and the duration of exposure, as is true for most slightly toxic pollutants that are bio-concentrated, then the earlier embryonic stages of *F. hetero-clitus* **were** more sensitive than later embryonic stages. Other investigators using somewhat different experimental designs have come to a similar conclusion. Linden (1974) reported that the embryos of Baltic herring, *Clupea harengus membras,* were mo**re** sensitive to oil spill dispersants if exposure was started in connection with fertilization than 6 hours after fertilization. Kuhnhold (1972, 1974) observed that embryos of cod, *Gadus morhua,* were more sensitive to the WSF of 3 crude oils during the gastrula stage than at later stages of embryonic development. This differential sensitivity of embryonic stages to oil is not surprising. Hinrichs (1925) and Rosenthal and Alderdice (1976) reported that a wide variety of environmental and chemical stressors produce greater dele-terious effects if exposure is initiated early in embryonic development than if it is started after embryonic differentia-tion and morphogenesis are well advanced.

In this investigation, newly hatched larvae of *F. hetero-clitus* appeared to be slightly less sensitive to the WSF of #2 fuel oil than were the embryos. Larvae hatching from embryos continuously exposed to 10 and 20% WSFs during development were exposed to the WSFs for an additional 2 weeks. The lar-vae experienced 88 and 61% survival, respectively compared to 92 and 74% survival for the corresponding groups of embryos. However, some of the mortality experienced by larvae was probably due to damage produced during embryonic exposure. Circumstantial evidence for this is provided by the observation that, of the 22% of embryos eventually surviving to hatching in 25% WSF, none survived for more than 3 days although the emergent larvae were immediately transferred to hydrocarbon-free sea water. Preliminary experiments in which newly hatched larvae, not previously exposed as embryos to oil, were exposed to WSF's revealed that larvae were not seriously affected by the 25% WSF. Similar results were obtained by Mirnov (1972) and Anderson *et al*. (1977). However, Kuhnhold (1972) reported that young larvae of cod, *Gadus morhua*; herring, *Clupea harengus*; and plaice, *Pleuronectes platessa,* were much more sensitive to dissolved oil components than embryos were. Similarly, Rice *et al*. (1975) reported that fry and alevins of pink salmon, *Oncorhynchus gorbuscha*, were more sensitive to Prudhoe Bay crude oil than were the embryos.

Embryos, continuously exposed to the 25% WSF of #2 fuel oil from fertilization, continued to accumulate naphthalenes in their tissues for 9 days to a maximum concentration of 79 ppm, 137 times the mean naphthalenes concentration in the exposure medium. The naphthalenes concentration in embryos exposed to the WSF for 11 days dropped to 63 ppm, suggesting that this was the tissue saturation concentration of naphthalenes in "equilibrium" with an ambient medium containing 0.57 ppm naphthalenes. Since a few embryos were able to survive to hatching following 12 days exposure to the 25% WSF, embryos of *F. heteroclitus* are unusually tolerant to high tissue naphthalene concentrations. Anderson *et al.* (1977) suggested that accumulation of hydrocarbons to lethal levels is the cause of death of fish embryos. Their data suggest that a concentration of approximately 5 ppm total naphthalenes in the embryos of *Fundulus similis* is the upper limit for embryonic survival. Our results indicate that embryos of some species or at least some individuals can tolerate tissue naphthalene concentrations substantially higher than this. The whole body burden of naphthalenes or other hydrocarbons may be a poor index of incipient lethality dye to hydrocarbon exposure. Different organs and tissues undoubtedly show substantial variation in their sensitivity to hydrocarbon insult. The hydrocarbon concentration in the most sensitive organs may be a better index of hydrocarbon-mediated stress. Since the distribution of aromatic hydrocarbons in the tissues of adult fish varies substantially during the time course of exposure to oil (Neff *et al.*, 1976; Dixit and Anderson, 1977; Roubal *et al.*, 1977), it is not possible to extrapolate from the whole body burden to the concentration of hydrocarbons in a particular organ. Although the precise distribution of accumulated hydrocarbons within the complete embryo complex is not known, it is likely that a majority of the accumulated hydrocarbons were sequestered in the lipid-rich yolk. Ernst *et al.* (1977) reported dramatic changes in the histological structure and appearance of the yolk of *Fundulus grandis* embryos during exposure to a WSF of #2 fuel oil. The hydrocarbons sequestered in the yolk would be mobilized during the periods of rapid yolk utilization toward the end of embryonic development and in the newly emergent yolk sac fry (Smith, 1957), partially explaining the delayed mortality often observed after the termination of exposure to hydrocarbons.

The more than six-fold decrease in the apparent permeability of embryos to ^{14}C-naphthalene between day 2 and day 10 of development may partially explain the greater sensitivity of early than late embryonic stages to oil. The chorion of the *Fundulus heteroclitus* egg, like that of most other teleosts, is highly permeable to water and low molecular weight solutes like sodium, sucrose and glycerol (Kao *et al.*, 1954; Shankalin, 1959). The chorion of this species is composed primarily of protein and a small amount of carbohydrate (about 4%) (Kaighn, 1964) and is therefore probably not an important site of hydrocarbon accumulation. It therefore seems likely that the temporal changes observed in the instantaneous uptake rate of naphthalene by embryos are not due to changes in the permeability of the chorion. This is supported by the observation of

Anderson *et al.* (1977) that surgical removal of the chorion from embryos of the sheepshead minnow *Cyprinodon variegatus* had little effect on the sensitivity of the embryos to the WSF of #2 fuel oil. Within the chorion are the perivitelline space, representing approximately 25-35% of the intrachorionic volume in fertilized *F. heteroclitus* eggs (Shankaline, 1959) and the embryo itself. The embryo, at least during early development, is covered by the vitelline membrane which is thought to represent the main barrier to diffusion of water and solutes between the medium and the embryonic tissues (Potts and Eddy, 1973). Thus, the change in the net uptake rate of naphthalene by embryos during development may reflect changes in the permeability properties to hydrophobic substances of the vitelline membrane or to changes in the pool size or turnover rate of hydrophobic materials (primarily lipids) within the embryo. The change in uptake rate might also be explained by an increase in the rate of efflux of accumulated hydrocarbons, either passively or by the development of hydrocarbon metabolizing enzyme systems as embryonic development progresses. However, data presented above indicate that the rate of efflux of accumulated naphthalene-derived ^{14}C activity also decreased during development. Furthermore, attempts to detect metabolites of ^{14}C-naphthalene and ^{14}C-chrysene in the embryos and the depuration water failed. The high permeability of the early embryo to naphthalene as reflected by the high influx-efflux rates may partially explain the greater sensitivity of early than late embryos to oil pollution.

The gradual increase in the permeability of the embryo to water from day 2 to day 10 of development has previously been documented in marine teleosts (Potts and Eddy, 1973), and reflects an increase in the permeability of the vitelline membrane to water. The apparent decrease in water efflux rate can probably be best explained by the increase in volume of the perivitelline fluid space as the volume of the embryo decreases during development. Tritium is thus effluxing from a larger pool late in development and so the apparent efflux rate seems to decrease.

Heart beat rate and respiratory rate were used as indices of sublethal hydrocarbon stress in the embryos. Heart beat rate was significantly depressed below control rates only under exposure regimes that were found to substantially reduce hatching success (initiation of exposure to the 25% WSF after 2 or 4 days of development). Similar results were reported by Anderson *et al.* (1977) in several species of cyprinodont embryos exposed to the oil WSF. Thus, heart beat rate is not a particularly sensitive index of sublethal hydrocarbon stress in fish embryos, although it has been shown to be sensitive to other types of chemical and natural stress (Linden, 1974;

Rosenthal and Alderdice, 1976). Oxygen consumption was depressed slightly by highly stressful exposure regimes and
stimulated slightly by mildly stressful exposure regimes.
The metabolic scope of fish embryos is undoubtedly limited
and energy reserves are small, accounting for the small respiratory response to stress.

In studies reported here, few gross developmental anomalies
that could be attributed to hydrocarbon exposure were observed.
Mirnov (1972), Kuhnhold (1974) and Linden (1975) reported a
high incidence of anomalies in marine teleost embryos exposed
to certain crude oils. Crude oils generally contain higher
concentrations of polycyclic aromatic hydrocarbons, some of
which are known to be teratogenic, than do #2 fuel oils. This
may explain the difference between our results and those of
Mirnov, Kuhnhold and Linden. Emergent larvae from embryos
that had been exposed to the WSF were generally shorter in
length and had larger yolk sacs than did control embryos.
These differences indicate that the oil-exposed embryos hatched
somewhat prematurely. As Rosenthal and Alderdice (1976),
noted, such premature larvae generally require a longer period
of time to complete the extremely vulnerable period between
hatching and the beginning of active feeding.

The data presented here all lend support to the conclusion
that hydrocarbon pollutants act on fish embryos as general
stressors and do not have a specific effect on any single
enzyme or physiological process. Hydrocarbon pollutants may
shunt limited metabolic energy away from critical differentia-
tion and morphogenic processes to maintenance functions. The
depressant effects of hydrocarbon stress early in development
may be reflected in later embryonic stages by depressed development rate, hatching success and survival of emergent fry
and increased incidence of developmental anomalies. The etiology of this generalized stress response in fish embryos has
been elegantly documented by Rosenthal and Alderdice (1976).

The aqueous hydrocarbon concentrations required in the
present investigation to produce significant mortality or sublethal effects in *F. heteroclitus* embryos are higher than those
usually encountered in the natural environment even after a
major oil spill. McAuliffe (1976) recently reviewed the available literature on the concentrations of hydrocarbons from all
sources in sea water and marine sediments. Low molecular
weight C_2-C_4 hydrocarbons are found at all depths in the oceans
at a mean concentration of about 45 ng/l (parts per trillion)
with higher levels near submarine gas and oil seeps. In the
surface waters near the natural oil seep at Coal Oil Point on
the Santa Barbara Channel, California, C_3-C_8 hydrocarbons were
present at a mean concentration of 290 ng/l (Koons and Brandon,
1975). Total dissolved C_1-C_{10} hydrocarbons under an oil slick
produced by a major oil spill in the Gulf of Mexico reached

maximum concentrations of 1-200 µg/l (parts per billion) while total dispersed oil in the water column ranged from 1-70 mg/l (parts per million) (McAuliffe *et al.*, 1975).

Hydrocarbon concentrations, and especially the medium and high molecular weight hydrocarbons, are generally much higher in marine sediments and may reach levels greater than 500 ppm in coastal areas receiving petroleum and industrial waste inputs (McAuliffe, 1976). Armstrong *et al.* (1977) reported that hydrocarbon concentrations in the water column near an oil separator platform in the shallow waters of Trinity Bay, Texas were low (10.5 ppb total hydrocarbons and 1.9 ppb total naphthalenes). However, naphthalenes levels in the sediments adjacent to the platform were quite high with mean naphthalenes concentrations of approximately 20 ppm being obtained over the 21 month period of the investigation. Our results suggest that demersal eggs, such as those of *F. heteroclitus,* coming in contact with such heavily contaminated sediments might be severely stressed.

CONCLUSIONS

The embryo-larval stages of the estuarine killifish *Fundulus heteroclitus* were moderately tolerant to the water-soluble fraction of a #2 fuel oil, showing an LC_{50} of approximately 1.5 ppm total hydrocarbons. The early embryonic stages were somewhat more sensitive to the 25% WSF than were later embryonic and larval stages. Embryos exposed continuously to the 25% WSF from fertilization accumulated naphthalenes to a maximum concentration of 78 ppm on day 9, representing a bioaccumulation factor of approximately 137. However, the instantaneous uptake rate of ^{14}C-naphthalene was highest in the 2-day old embryos and decreased in a nearly linear fashion as development progressed. The rate of ^3H-water uptake increased slightly during the same developmental interval. The rate of both naphthalene and water efflux decreased during the course of embryonic development. Embryos between 2 and 10 days post-fertilization were unable to metabolize either naphthalene or chrysene. These results indicate that the permeability to naphthalene of the vitelline membrane of the *Fundulus* embryo decreases during embryonic development, partially explaining the greater sensitivity of the early embryo to WSF. The heart beat rate of 9-11 day old embryos was significantly depressed only by WSF exposure regimes which also resulted in significant decreases in hatching success. Oxygen consumption of 10-day old embryos was slightly depressed by highly stressful exposure regimes and stimulated slightly by mildly stressful exposure regimes. Larvae hatching from embryos previously exposed to

the WSF were generally shorter in length and had larger yolk
sacs than controls. Metabolic scope and energy reserves of
fish embryos are small. When stressed by exposure to oil,
limited metabolic energy is shunted away from critical differ-
entiation and morphogenic processes to maintenance functions.
The depressant effects of hydrocarbon stress early in development may be reflected in later embryonic stages by depressed
development rate, hatching success and survival of emergent
larvae and increased incidence of development anomalies.

ACKNOWLEDGMENTS

     This research was supported by grants #206-76 (170) from
the American Petroleum Institute and #IDO75-04890 from NSF-
International Decade of Ocean Exploration.

LITERATURE CITED

American Petroleum Institute. 1958. Determination of volatile
     and nonvolatile oily material. Infrared spectrometric
     method, No. 733-48.
Anderson, J. W., D. B. Dixit, G. S. Ward and R. S. Foster.
     1977. Effects of petroleum hydrocarbons on the rate of
     heart beat and hatching success of estuarine fish embryos.
     In: Physiological Responses of Marine Biota to Pollutants,
     pp. 241-258, ed. by F. J. Vernberg, A. Calabrese, F. P.
     Thurberg, and W. B. Vernberg. New York: Academic Press.
Anderson, J. W., J. M. Neff, B. A. Cox, H. E. Tatem and G. M.
     Hightower. 1974. Characteristics of dispersions and
     water-soluble extracts of crude and refined oils and their
     toxicity to estuarine crustaceans and fish. Mar. Biol.
     75-78.
Armstrong, H. W., K. Fucik, J. W. Anderson and J. M. Neff.
     1977. Effects of oilfield brine effluent on benthic
     organisms from Trinity Bay, Texas. API Publ. No. 4291.
     81 pp. Washington, D. C.: American Petroleum Institute.
Barr, A. J., J. H. Goodnight, J. P. Sall and J. T. Helwig.
     1976. A users guide to SAS 76. 329 pp. Raleigh, N. C.:
     SAS Institute, Inc.
Blumer, M., G. Sauza and J. Sass. 1970. Hydrocarbon pollution
     of edible shellfish by an oil spill. Mar. Biol. 5: 195-202.
Boyd, J. F. and R. C. Simmonds. 1974. Continuous laboratory
     production of fertile *Fundulus heteroclitus* (Walbaum) eggs
     lacking chorionic fibrils. J. Fish. Biol. 6: 389-394.

Dixit, D. B. and J. W. Anderson. 1977. Distribution of naphthalenes within exposed *Fundulus similus* and correlations
with stress behavior. Proceedings of 1977 Oil Spill Conference (Prevention, Behavior, Control, Cleanup). pp. 633-
636. Washington, D. C.: American Petroleum Institute.

Ernst, V. V., J. M. Neff and J. W. Anderson. 1977. The
effects of water-soluble fractions of No. 2 fuel oil on
the early development of the estuarine fish, *Fundulus grandis* Baird and Girard. Environ. Pollut. 14: 25-35.

Glaser, O. 1929. Temperature and heart rate in *Fundulus*
embryos. Brit. J. Exp. Biol. 6: 325-339.

Hinrichs, M. A. 1925. Modification of development on the
basis of differential susceptibility to radiation. I.
Fundulus heteroclitus and ultraviolet radiation. J.
Morphol. Physiol. 41: 329-265.

Hyland, J. L. and E. D. Schneider. 1976. Petroleum hydrocarbons and their effects on marine organisms, populations,
communities and ecosystems. In: Sources, effects and sinks
of hydrocarbons in the aquatic environment. pp. 463- 506.
Washington, D. C.: American Institute of Biological Sciences.

Kaighn, M. E. 1964. A biochemical study of the hatching process in *Fundulus heteroclitus*. Develop. Biochem. 9: 56-80.

Kao, C. Y., R. Chambers and E. L. Chambers. 1954. Internal
hydrostatic pressure of the *Fundulus* egg. II. Permeability
of the chorion. J. Exp. Biol. 31: 447-461.

Koons, C. B. and D. E. Brandon. 1975. Hydrocarbons in water
and sediment samples from Coal Oil Point area, offshore
California. Proc. Offshore Technol. Conf. III: 513-521.

Kuhnhold, W. W. 1972. The influence of crude oils on fish
fry. In: Marine Pollution and Sea Life. pp. 315-318, ed.
by M. Ruivo. London: Fishery News (Books) Ltd.

Kuhnhold, W. W. 1974. Investigations on the toxicity of
seawater-extracts of three crude oils on eggs of cod
Gadus morhua L. Ber. dt. wiss. Komm. Meeresforsch. 23:
165-180.

Linden, O. 1974. Effects of oil spill dispersants on the
early development of Baltic herring. Ann. Zool. Fennici,
11: 141-148.

Linden, O. 1975. Acute effects of oil and oil/dispersant
mixture on larvae of Baltic herring. Ambio. 4: 130-133.

Manner, H. W. and C. Muehlman. 1975. Permeability and uptake
of ^{3}H-uridine during teleost development. Sci. Biol. J.
1: 81-82.

McAuliffe, C. D. 1976. Surveillance of the marine environment
for hydrocarbons. Mar. Sci. Commun. 2: 13-42.

McAuliffe, C. D., A. E. Smalley, R. D. Groover, W. M. Welsh, W. S. Pickle and G. E. Jones. 1975. Chevron main pass block 41 oil spill: chemical and biological investigations Proc. 1975 Conf. on Prevention and Control of Oil Pollution. pp. 555-566. Washington, D. C.: American Petroleum Institute.

McKim, J. M. 1977. Evaluation of tests with early life stages of fish for predicting long-term toxicity. J. Fish. Res. Bd. Can. 34: 1148-1154.

Mirnov, O. G. 1972. Effect of oil pollution on flora and fauna of the Black Sea. IN: Marine Pollution and Sea Life. pp. 222-224, ed. by M. Ruivo. London: Fishery News (Books) Ltd.

Neff, J. M. and J. W. Anderson. 1975. An ultraviolet spectrophotometric method for the determination of naphthalene and alkylnaphthalenes in the tissues of oil-contaminated marine animals. Bull. Environ. Contam. Toxicol. 14: 122-128.

Neff, J. M., B. A. Cox, D. Dixit and J. W. Anderson. 1976. Accumulation and release of petroleum-derived aromatic hydrocarbons by four species of marine animals. Mar. Biol. 38: 279-289.

Potts, W. T. W. and F. B. Eddy. 1973. The permeability to water of the eggs of certain marine teleosts. J. Comp. Physiol. 82: 305-315.

Rice, S. D., D. A. Moles and J. W. Short. 1975. The effect of Prudoe Bay crude oil on survival and growth of eggs, alevins and fry of pink salmon, *Oncorhynchus gorbuscha*. Proc. 1975 Conf. on Prevention and Control of Oil Pollution. pp. 503-507. Washington, D. C.: American Petroleum Institute.

Rosenthal, H. and D. F. Alderdice. 1976. Sublethal effects of environmental stressors, natural and pollutional, on marine eggs and larvae. J. Fish. Res. Bd. Can. 33: 2047-2065.

Roubal, W. T., T. K. Collier and D. C. Malins. 1977. Accumulation and metabolism of carbon-14 labeled benzene, naphthalene, and anthracene by young coho salmon (*Oncorhynchus kisutch*). Arch. Environ. Contam. Toxicol. 5: 513-529.

Shankalin, D. R. 1959. Studies on the *Fundulus* chorion. J. Cell Comp. Physiol. 53: 1-11.

Smith, S. 1957. Early development and hatching. In: The Physiology of Fishes, Vol. 1, pp. 323-355, ed. by M. E. Brown. New York: Academic Press.

PART II

Metals

TOXIC CATIONS AND MARINE BIOTA: ANALYSIS
OF RESEARCH EFFORT DURING THE THREE-
YEAR PERIOD 1974-1976[1]

Ronald Eisler

U. S. Environmental Protection Agency
Environmental Research Laboratory
Narragansett, Rhode Island

INTRODUCTION

There is a growing body of literature on biological and
toxicological effects of heavy metals and other cations to
marine biota (Waldichuk, 1974; Bernhard and Zattera, 1975;
Eisler and Wapner, 1975). This increase, which shows no
sign of abatement, is probably attributable to recent technological advances in atomic absorption spectrophotometry, neutron
activation analyses, and anodic stripping voltometry, as well
as to the greater availability of instrumentation to marine
scientists. However, the research approaches employed to-
date on evaluation of effects of toxic cations on estuarine
and oceanic flora and fauna are both numerous and disparate.
Awareness of the many factors involved in assessment of
biological properties of chemical pollutants in general, and
cations in particular, is essential for the establishment and
implementation of meaningful marine water quality standards.
This account categorizes available technical articles on
effects of toxic cations to marine organisms that were published during the three-year period 1974-1976. Specifically,
233 individual contributions are identified by effort generated
by cation, by major indicator organism assemblage and by
response parameter. Research effort is also identified by

[1]Note: Due to the abundance of information presented in
Appendix I, the numerical system of referencing will be used
in that section.

laboratory study or by field investigation. Major trends
are tabulated, research gaps are identified, and suggestions
are offered on direction and emphasis of future research in
this subject area.

ANALYSIS OF RESEARCH EFFORT

By Cation

During 1974-1976, marine biological effects studies were
conducted with 41 different heavy metals and cations (Table 1).
Data on sodium, potassium, and calcium were especially abundant, but these were arbitrarily excluded from Table 1. About
62% of the total effort was contributed to studies on five
metals: zinc, mercury, copper, cadmium, and lead. Another
25% was confined to the group consisting of iron, manganese,
chromium, nickel, silver, cobalt, and arsenic. The remaining
29 elements constituted about 13% of all effort.

Of the total cation effort, laboratory studies conprised
about 48% and field investigations 52% (Table 1). The average
laboratory study contained 2.09 cations/reference; this value
was substantially higher (3.16) for field investigation reports.
For all references, the average number of cations per reference
was 2.54. When compared to laboratory studies a slightly
higher percentage of the field effort was devoted to Zn, Pb,
and Ni, and a lower percentage to Co; otherwise, effort
for individual cations was within 1% for both categories.

It is remarkable that no research effort was recorded during the 1974-1976 interval on cations of known high toxicity to
aquatic biota, *viz* gold, platinum, tin, and tungsten. Elements
of potential environmental concern from the actinide series,
especially californium and neptunium, and practically the entire lanthanide group were also neglected.

By Indicator Assemblage

Organisms from 22 major marine taxonomic groups, representing more than 200 different species, were used to evaluate
toxic cation effects during the period 1974-1976 (Table 2).
About 76% of the total effort was restricted to four groups:
crustacea, mollusca, algae, and fish. An additional 11% was
contributed by annelids, bacteria, coelenterates, and echinoderms. The remaining 14 groups comprised about 13% of the
total indicator organism effort.

TABLE 1.   *Toxic cations and marine biota: cation effort from 233 papers published during the interval 1974-1976 inclusive.   N=number of papers referring to specific cation; %=percent of total responses contributed by individual cations*

		Total		Laboratory Studies		Field Studies	
		N	%	N	%	N	%
Zinc (Zn)		84	14.2	38	6.4	46	7.8
Mercury (Hg)		78	13.2	42	7.1	36	6.1
Copper (Cu)		75	12.7	37	6.2	38	6.4
Cadmium (Cd)		75	12.7	35	5.9	40	6.8
Lead (Pb)		54	9.1	22	3.7	32	5.4
Iron (Fe)		25	4.2	10	1.7	15	2.5
Manganese (Mn)		23	3.9	10	1.7	13	2.2
Chromium (Cr)		22	3.7	11	1.9	11	1.9
Nickel (Ni)		22	3.7	6	1.0	16	2.7
Silver (Ag)		20	3.4	9	1.5	11	1.9
Cobalt (Co)		17	2.9	13	2.2	4	0.7
Arsenic (As)		14	2.4	6	1.0	8	1.3
Magnesium (Mg)		10	1.7	7	1.2	3	0.5
Plutonium (Pu)		10	1.7	4	0.7	6	1.0
Strontium (Sr)		5	0.8	4	0.7	1	0.2
Selenium (Se)		5	0.8	3	0.5	2	0.3
Cesium (Cs)		5	0.8	3	0.5	2	0.3
Polonium (Po)		5	0.8	1	0.2	4	0.7
Ruthenium (Ru)		4	0.7	3	0.5	1	0.2
Vanadium (V)		4	0.7	2	0.3	2	0.3
Boron (B)		3	0.5	3	0.5	0	0.0
Americium (Am)		3	0.5	1	0.2	2	0.3
Antimony (Sb)		3	0.5	0	0.0	3	0.5
Technetium (Tc)		2	0.3	2	0.3	0	0.0
Thallium (Tl)		2	0.3	2	0.3	0	0.0
Aluminum (Al)		2	0.3	1	0.2	1	0.2
Cerium (Ce)		2	0.3	1	0.2	1	0.2
Molybdenum (Mo)		2	0.3	1	0.2	1	0.2
Yttrium (Y)		2	0.3	1	0.2	1	0.2
Bismuth (Bi)		2	0.3	0	0.0	2	0.3
Titanium (Ti)		2	0.3	0	0.0	2	0.3
Beryllium (Be)		1	0.2	1	0.2	0	0.0
Germanium (Ge)		1	0.2	1	0.2	0	0.0
Rubidium (Rb)		1	0.2	1	0.2	0	0.0
Silicon (Si)		1	0.2	1	0.2	0	0.0
Europium (Eu)		1	0.2	0	0.0	1	0.2
Niobium (Nb)		1	0.2	0	0.0	1	0.2
Radium (Ra)		1	0.2	0	0.0	1	0.2
Rhodium (Rh)		1	0.2	0	0.0	1	0.2
Thorium (Th)		1	0.2	0	0.0	1	0.2
Zirconium (Zr)		1	0.2	0	0.0	1	0.2
	TOTAL	592	100.0	282	47.8	310	52.6

TABLE 2. Research effort expended on biological effects of cations, by indicator organism groups, during the interval 1974-1976 inclusive. N=number of 233 published papers referring to a specific indicator group; %=percent of all responses (347) contributed by individual groups

	Total[a]		Laboratory Studies[b]		Field Studies[c]	
	N	%	N	%	N	%
Mollusca	78	22.4	38	11.0	40	11.4
Fish	69	19.8	27	7.8	42	12.0
Crustacea	66	19.0	41	11.8	25	7.2
Algae	51	14.7	33	9.5	18	5.2
Annelida	15	4.3	14	4.0	1	0.3
Mammals	11	3.2	0	0.0	11	3.2
Echinodermata	10	2.8	5	1.4	5	1.4
Bacteria	8.	2.3	5	1.4	3	0.9
Higher Plants	8	2.3	0	0.0	8	2.3
Tunicata	6	1.8	2	0.6	4	1.2
Coelenterata	6	1.7	5	1.4	1	0.3
Elasmobranchs	4	1.2	0	0.0	4	1.2
Insecta	3	0.9	0	0.0	3	0.9
Protozoa	2	0.6	2	0.6	0	0.0
Birds	2	0.6	0	0.0	2	0.6
Plankton	2	0.6	0	0.0	2	0.6
Turbellaria	1	0.3	1	0.3	0	0.0
Nematoda	1	0.3	1	0.3	0	0.0
Platyhelminthes	1	0.3	1	0.3	0	0.0
Bryazoa	1	0.3	1	0.3	0	0.0
Fungi	1	0.3	1	0.3	0	0.0
Chaetognaths	1	0.3	0	0.0	1	0.3
	347	100.0	177	51.0	170	49.0

[a] 233 references; 347 citations=1.49 indicators/reference
[b] 135 references; 177 citations=1.31 indicators/reference
[c] 98 references; 170 citations=1.74 indicators/reference

There were marked differences between laboratory and field studies in selection of indicators. For example, no laboratory studies were conducted on marine mammals, higher plants, elasmobranchs, insects, birds, plankton, and chaetognaths; whereas, field studies did not include protozoa, turbellarians, nematodes, platyhelminthids, bryozoans, or fungi as indicators. Only 9 of the 22 indicator groups were studied under both laboratory and field conditions: crustacea, mollusca, algae, fish, annelids, bacteria, coelenterates, echinoderms, and tunicates. Of these only crustacea, mollusca, algae and fish contributed significantly to the total research effort. Reports arising from field investigations cited a greater average number of indicator organism groups per reference than laboratory studies: 1.74 *vs.* 1.31; for all studies this was 1.49.

By Response Parameter

Criteria used by investigators during the 3-year period 1974-1976 to evaluate effects of heavy metals and other cations to selected indicator organism groups are listed by cation, by reference, and by laboratory or field effort (see Appendix 1). Factors modifying the major effect studied are also listed when appropriate.

More than 37% of the total response parameter effort was devoted to effects of modifiers (see Appendix 1); in many cases these modifiers could alter measured responses by an order of magnitude, or more. Thus, biotic modifiers which affected cation residue content in field collections of organisms included: age, size, body weight, sex and sexual condition of the organism; feeding frequency, diet composition, and feeding niche in water column; metabolism, organism complexity, and selective tissue specificity of cations; general condition of the animal, and whole body elemental composition. Abiotic modifiers affecting residues in field biota included: geographic and spatial factors such as distance from point source, latitudinal, seasonal, monthly, and yearly variations; physical factors such as temperature, salinity, turbidity, water depth, sediment particle size, and geothermal and tectonic activity; inorganic and organic composition of water column and sediment substrate; selective accumulation of different chemical species; and errors introduced in sampling *i.e.* Cu and Zn contamination of molluscs from a water pump. As was true for field studies, many factors strongly modified uptake, retention, and translocation mechanisms of cations under controlled laboratory conditions. These included temperature, salinity, light intensity, turbidity, size of particulates, season of year, toxicant concentration and mode of administration; age, size, sexual condition, molting state (in crustaceans), tissue specificity and bacterial

interaction effects; physical and chemical composition of
sediment substrate; and levels of inorganics and organics in
water column. Differential uptake of chemical species was
also important, with marked variations evident for chemical
species of Cd, Cr, Cu, Hg, Ni, Ru, Se, Zn; variations were
partly attributable to differences in anion constituent, partly
to organic vs inorganic forms, and partly to discrimination
among isotopes. Similar cases, on a more limited basis, are
made for factors modifying survival, growth, and development
(including photosynthesis), and reproduction.

Response parameter data are summarized in Table 3. It is
clear from these data that the major emphasis in field studies
is on collection of baseline residue data; for laboratory
studies it is on bioaccumulation, survival, growth, reproduction, food chain dynamics, behavior, respiration, biochemistry
and hematology, histology, and others, in that order. Publications resulting from field studies, when compared to laboratory investigations contained about twice the number of response
parameters per reference: 7.15 vs 3.71 (Table 3). But this
statistic may reflect the extensive elemental analyses conducted on field collections comprised of numerous species.

Analysis of cations as a function of response parameter
indicates that greatest effort for all studies was devoted
to Hg, Cd, Cu, Zn, Pb, Cr, Fe, and Mn, in that sequence. These
data are not shown, but they do not differ significantly in
ranking from those shown previously in Table 1.

For all studies, indicator groups used most frequently for
response parameters were molluscs (30.7%), fishes (20.9%),
algae (14.8%), crustaceans (10.1%), mammals (5.2%), annelids
(3.8%), and echinoderms (3.5%). Molluscs accounted for 30.7%
of the total response parameter effort compared to only 22.4%
when listed by citation (Table 2). For crustaceans the response
parameter effort was 10.2%, a drop from 19.0% when listed by
citation (Table 2); none of the other indicator assemblages
showed major directional trends when compared to Table 2.

DISCUSSION AND RECOMMENDATIONS

Continuing research effort appears warranted on cations
with known deleterious effects to human health, especially
mercury, cadmium, lead, arsenic, and selenium, and on metals
of proven high toxicity at comparatively low concentrations
to marine biota, viz copper, chromium, silver, and to a lesser
extent, zinc and nickel. Increased emphasis should be directed
to studies on environmental aspects of transuranics in marine
ecosystems, especially plutonium, americium, curium, neptunium,
and other alpha emitters. These will assume increasing

TABLE 3. Summary of laboratory and field research effort on effects of toxic cations to marine biota, by response parameters. Values are in % of total effort (N=1202)

	Laboratory Studies[a]	Field Studies[b]	Total[c]
Baseline residue data	0.0	35.4	35.4
-modifiers	0.0	19.5	19.5
Uptake, retention, translocation	8.7	0.0	8.7
-modifiers	8.8	0.0	8.8
Survival	4.9	0.0	4.9
-modifiers	2.3	0.0	2.3
Growth, development, photosynthesis	3.7	0.6	4.3
-modifiers	2.0	0.0	2.0
Reproduction	2.1	0.4	2.5
-modifiers	0.9	0.0	0.9
Food chain transfer	2.3	0.6	2.9
Behavior	1.1	0.0	1.1
Respiration	1.1	0.0	1.1
Role in cycling	0.0	1.1	1.1
Enzymes, proteins, blood chemistry	0.9	0.0	0.9
Histology, pathology	0.9	0.0	0.9
Seasonal dynamics	0.8	0.0	0.8
Osmoregulation	0.4	0.0	0.4
Health Hazard	0.2	0.2	0.4
Morphology	0.3	0.0	0.3
Others	0.3	0.5	0.8
	41.7	58.3	100.0

[a] 135 references; 501 citations=3.71 response parameters/reference
[b] 98 references; 701 citations=7.15 response parameters/reference
[c] 233 references; 1202 citations=5.16 response parameters/reference

importance if proposed construction of coastal nuclear steam
electric stations is implemented. Some effort should be
devoted to platinum in the marine environment, since this
metal could become an important contaminant owing to its use
in automobile emission control devices. Other potentially
harmful cations about which little is known regarding sources,
distribution, fate, and effects in marine ecosystems include
tungsten, thallium, tin, and beryllium; research on these
cations and their salts clearly merit additional effort. For
all studies the chemical species of the cation should be
identified together with the methylation potential and relation between organic and inorganic forms, if possible. Results
from field investigations would be substantially enhanced with
inclusion of data on anthropogenic loadings of cations and
other potential toxicants to the study area; amounts of these
same materials normally contributed from geological and other
sources should also be referenced.

Greater use of marine mammals, fish-eating birds, higher
plants, and bacteria in laboratory studies on effects of toxic
cations is suggested, as is continued use of these groups in
field investigations. Ship-board scientists should strive to
integrate their efforts with ongoing comprehensive monitoring
programs typical of those established by International Com-
missions on whaling, salmon, tuna, and seabirds. All studies
should continue to utilize molluscs, crustaceans, fishes, and
algae as indicators of environmental stress and to explore
possibilities of other groups, especially coelenterates,
annelids, and echinoderms, which seem to show promise of
becoming suitable indicators. Since conservation and maintenance of marine resources entails knowledge of toxicant fluxes,
including effects on several life stages of economically
important species and their food organisms, it behooves the
investigator to select appropriate indicators and developmental stages for purposes of testing and analyses. Where
human health is a prime factor, it appears reasonable to
prioritize data collections of metals in edible flesh of marine
products of commerce, especially teleosts, bivalve molluscs,
and decapod crustaceans. Metal concentrations in viscera and
other "scrap" marine biota now processed into meal supplements
for domestic poultry, swine, and cattle may indirectly affect
health; accordingly, residues in these products should also
be determined and the efficiency of food chain transfer
established. A word of caution: the number of species now
utilized or recommended as indicators of marine water quality
is proliferating rapidly, yet essential information on baseline
physiology, metabolism, behavior, and other indices of well-
being is either limited or missing. This is in sharp contrast
to the health sciences, which annually expends substantial
funds on normal physiology and behavior of small mammals used

in research--especially rodents, felines, canines, and primates. It is apparent that additional research is required on "unstressed" marine indicator groups, with emphasis on only a few selected species.

As recently as 20 years ago there were comparatively few researchers evaluating biological effects of metals in marine waters and most of this early effort was restricted to effects on survival and growth. Substantial advances have been made since, but, in perspective, response parameters currently used to measure cation-induced stress are still somewhat limited. For example, effects at the cellular and sub-cellular levels are scarce or absent while those on residues and bio-magnification receive a disproportionate share of research budgets. Towards this end, greater effort could be devoted to histopathological, biochemical, and behavioral parameters as early indicators of growth impairment, reproductive success, survival, metabolic imbalance, and alterations in community structure. These, however, must of necessity be tied to data on survival, growth, residue levels, metabolism, and general well-being.

Finally, it is strongly recommended that interaction effects of the cation under study should be evaluated with due consideration of other constituents in the waste stream. It is well-established that coastal waters adjacent to large population centers receive a variety of wastes as a result of domestic, agricultural, municipal, and industrial operations. Some of these components are harmful to marine biota, but their toxic action is not necessarily predictable on a simple additive basis; some may act synergistically, others antagonistically. Toxicological evaluation of complex wastes discharged into marine environments is still an infant science. However, a multiparametric approach is now routinely employed by industrial toxicologists, physicians, and others concerned with the health of persons who work with dangerous or hazardous materials. This approach is based on the acceptable hypothesis that metabolic disturbances have a far greater probability of detection if a number of physiological and biochemical parameters are simultaneously determined and plotted. This technique appears to hold substantial promise as a predictive tool for diagnosis of metabolic derangement in marine indicator species long before obvious morphological or behavioral effects become evident. With the multiparametric approach, laboratory-oriented scientists may find it profitable to devote an increasing proportion of their work load to assessment of real world concentrations and combinations of actual or potential toxicants, whereas field workers may find similar rewards in expansion of their baseline residue data effort to include physiological indications of environmental perturbation.

SUMMARY

Available technical articles published during 1974, 1975, and 1976 on effects of heavy metals and toxic cations upon estuarine and marine flora and fauna were reviewed. For all articles, research effort generated by toxicant, by response parameter, and by major indicator organism assemblage was tabulated for field investigations and for laboratory studies. Most of the effort during this period was restricted to 5 metals (Zn, Hg, Cu, Cd, Pb) and 4 indicator groups (molluscs, teleosts, crustaceans, algae). Laboratory investigators devoted a major portion of their workload to uptake, retention, and translocation studies, followed by studies on survival, on growth and development, reproduction, food chain transfer, behavior, respiration, body fluid chemistry, histology, and others, in that sequence. Results of field studies were confined almost exclusively to acquisition of data on baseline elemental composition and accompanying biotic and abiotic modifiers.
Apparent limitations and deficiencies of current projects on biological effects assessment of metals and toxic cations in marine environments were presented and discussed. Suggestions were offered on direction and emphasis of future research in this subject area.

ACKNOWLEDGMENTS

I am obligated to my colleagues at the Environmental Research Laboratory, especially Drs. J. H. Gentile, C. S. Hegre, E. H. Jackim, F. G. Lowman, D. Miller, and G. E. Zaroogian, for constructively reviewing early drafts of this manuscript, and to Ms. G. Gaboury for secretarial assistance.

LITERATURE CITED

1. Ahsanullah, M. 1976. Acute toxicity of cadmium and zinc to seven invertebrate species from Western Port, Victoria. Austral. J. Mar. Fresh. Res. 27: 187-196.

2. Alexander, G. V. and D. R. Young. 1976. Trace metals in southern californian mussels. Mar. Poll. Bull. 7: 7-9.

3. Alley, W. P., H. R. Brown, and L. Y. Kawasaki. 1974. Lead concentrations in the wooly sculpin, *Clinocottus analis*, collected from tidepools of California. Calif. Fish Game 60: 50-51.

4. Amiard, J. C. 1975. Interprétation d'une étude experimentale du metabolisme du radiostrontium chez la plie *(Pleuronectes platessa)* a l'aide des analyses factorielles. Rev. Int. Océan. Méd. 39-40: 177-212.

5. Amiard, J. C. 1976. Phototactic variations in crustacean larvae due to diverse metallic pollutants: demonstrated by a sublethal toxicity test. Mar. Biol. 34: 239-245.

6. Amiard-Triquet, C., and J. C. Amiard. 1974. Contamination de chaines trophiques marines par le cobalt 60. Rev. Int. Océan. Méd. 33: 49-59.

7. Amiard-Triquet, C. and J. C. Amiard. 1975. Etude experimentale du transfert du cobalt 60 dans une chaine trophique marine benthique. Helgol. Meeresunters 27: 283-297.

8. Amiard-Triquet, C., and J. C. Amiard. 1976. L'organotropisme du ^{60}Co chez *Scrobicularia plana* et *Carcinus maenas* en fonction du vecteur de contamination. Oikos: 27: 122-126.

9. Anas, R. E. 1974. Heavy metals in northern fur seals, *Callorhinus ursinus,* and harbor seals, *Phoca vitulina richardi*. U. S. Dept. Comm. Fish. Bull. 72: 133-137.

10. Anderlini, V. 1974. The distribution of heavy metals in the red abalone, *Haliotis refescens,* on the California coast. Arch. Envir. Cont. Tox. 2: 253-265.

11. Andryushchenko, V. V. and G. G. Polikarpov. 1974. An experimental study of uptake of Zn^{65} and DDT by *Ulva rigida* from seawater polluted with both agents. Hydrobiol. J. 10(4): 41-46.

12. Antia, N. J., and J. Y. Cheng. 1975. Culture studies on the effects from borate pollution on the growth of marine phytoplankters. J. Fish. Res. Bd. Canada 32: 2487-2494.

13. Arima, S. and S. Umemoto. 1976. Mercury in aquatic organisms. II. mercury distribution in muscles of tunas and swordfish. Bull. Jap. Soc. Sci. Fish. 42: 931-937.

14. Aubert, M., R. Bittel, F. Laumond, M. Roméo, B. Donnier, and M. Barelli. 1974. Utilisation d'une chaine trophodynamique de type néritique a mollusque pour l'étude des transferts des pollutants metalliques. Rev. Int. Océan. Médic. 33: 7-29.

15. Banus, M., I. Valiela, and J. M. Teal. 1974. Export of lead from salt marshes. Mar. Poll. Bull., 5: 6-9.

16. Beasley, T. M. and S. W. Fowler. 1976. Plutonium and americium: uptake from contaminated sediments by the polychaete *Nereis diversicolor*. Mar. Biol. 38: 95-100.

17. Benayoun, G., S. W. Fowler, and B. Oregioni. 1975. Flux of cadmium through euphausiids. Mar. Biol. 27: 205-212.

18. Bender, J. A. 1975. Trace metal levels in beach dipterans and amphipods. Bull. Envir. Cont. Tox. 14: 187-192.

19. Benijts-Claus, C. and F. Benijts. 1975. The effect of low-lead and zinc concentrations on the larval development of the mud-crab, *Rhithropanopeus harrisii* Gould. In: Koeman, J. H. and J. J. T. W. A. Strik (eds.). Sublethal effects of toxic chemicals on aquatic animals. Elsevier Sci. Publ. Co., Amsterdam: 43-52.

20. Berland, B. R., D. J. Bonin, V. I. Kapkov, S. Y. Maestrini and D. P. Arlhac. 1976. Action toxique de quatre métaux lourds sur la croissance d'algues unicellulaires marines. C. R. Acad. Sci. Paris, t. 282, Ser. D: 633-636.

21. Bernhard, M. and A. Zattera. 1975. Major pollutants in the marine environment. In: Pearson and Frangipane (eds.). Marine pollution and marine waste disposal. Pergamon Press, N. Y.: 195-300.

22. Betzer, S. B. and M. E. Q. Pilson. 1974. The seasonal cycle of copper concentration in *Busycon canaliculatum* L. Biol. Bull. 146: 165-175.

23. Betzer, S. B. and M. E. Q. Pilson. 1975. Copper uptake and excretion by *Busycon canaliculatum* L. Biol. Bull. 148: 1-15.

24. Betzer, S. B. and P. P. Yevich. 1975. Copper toxicity in *Busycon canaliculatum*. Biol. Bull. 148: 16-25.

25. Blake, N. J. and D. L. Johnson. 1976. Oxygen production consumption of the pelagic *Sargassum* community in a flow-through system with arsenic additions. Deep Sea Res. 23: 773-778.

26. Blankenship, M. L. and K. M. Wilbur. 1975. Cobalt effects of cell division and calcium uptake in the coccolitho-phorid *Cricosphaera carterae* (Haptophyceae). J. Phycol. 11: 211-219.

27. Bohn, A. 1975. Arsenic in marine organisms from West Greenland. Mar. Poll. Bull. 6:87-89.

28. Bohn, A. and R. O. McElroy. 1976. Trace metals (As, Cd, Cu, Fe, and Zn) in Arctic cod, *Boreogadus saida*, and selected zooplankton from Strathcona Sound, Northern Baffin Island. J. Fish. Res. Bd. Canada 33: 2836-2840.

29. Boyden, C. R. 1974. Trace element content and body size in molluscs. Nature 251: 311-314.

30. Boyden, C. R. and M. G. Romeril. 1974. A trace metal problem in pond oyster culture. Mar. Poll. Bull. 5: 74-78.

31. Boyden, C. R., H. Watling and I. Thornton. 1975. Effect of zinc on the settlement of the oyster *Crassostrea gigas*. Mar. Biol. 31: 227-234.

32. Braek, G. S., A. Jensen, and A. Mohus. 1976. Heavy metal tolerance of marine phytoplankton. III. Combined effects of copper and zinc ions on cultures of four common species. J. Exp. Mar. Biol. Ecol. 25: 37-50.

33. Brooks, R. R. and D. Rumsey. 1974. Heavy metals in some New Zealand commercial sea fishes. New Zealand J. Mar. Fresh. Res. 8: 155-166.

34. Bryan, G. W. 1974. Adaptation of an estuarine polychaete to sediments containing high concentrations of heavy metals. In: Vernberg, F. J. and W. B. Vernberg (eds.). Pollution and physiology of marine organisms. Academic Press, N. Y.: 123-135.

35. Buhler, D. R., R. R. Claeys, and B. R. Mate. 1975. Heavy metal and chlorinated hydrocarbon residues in California sea lions (*Zalophus californianus californianus*). J. Fish. Res. Bd. Canada 32: 2391-2397.

36. Calabrese, A. and D. A. Nelson. 1974. Inhibition of embryonic development of the hard clam, *Mercenaria mercenaria*, by heavy metals. Bull. Envir. Cont. Tox. 11: 92-97.

37. Calabrese, A., F. P. Thurberg, M. A. Dawson and D. R. Wenzloff. 1975. Sublethal physiological stress induced by cadmium and mercury in winter flounder, *Pseudopleuronectes americanus*. In: Koeman, J. H. and J. J. T. W. A. Strik (eds.). Sublethal effects of toxic chemicals on aquatic animals. Elsevier Sci. Pub. Co., Amsterdam: 15-21.

38. Cherry, R. D., S. W. Fowler, T. M. Beasley, and M. Heyraud. 1975. Polonium-210: its vertical oceanic transport by zooplankton-metabolic activity. Mar. Chemistry 3: 105-110.

39. Chow, T. J., C. C. Patterson and D. Settle. 1974. Occurrence of lead in tuna. Nature 251: 159-161.

40. Chow, T. J., H. G. Snyder, and C. B. Snyder. 1976. Mussels (*Mytilus* sp.) as an indicator of lead pollution. Science Total Envir. 6: 55-63.

41. Colwell, R. R. and J. D. Nelson, Jr. 1974. Bacterial mobilization of mercury in Chesapeake Bay. Proc. Int. Conf. Transport Persistent Chem. in Aquatic Ecosys., Ottawa, Canada: III 1 - III 10.

42. Colwell, R. R. and J. D. Nelson, Jr. 1975. Metabolism of mercury compounds in microorganisms. USEPA Rept. 600/3-73-007. USEPA, Off. Res. Dev., Envir. Res. Lab., Narragansett, RI:84 pp.

43. Colwell, R. R., G. S. Sayler, J. D. Nelson, Jr. and A. Justice. 1976. Microbial mobilization of mercury in the aquatic environment. In: Nriagu, J. D. (ed.). Environmental biogeochemistry, Vol. 2. Metals transfer and ecological mass balances. Ann Arbor Sci. Publ., Ann Arbor, Mich.: 437-487.

44. Cossa, D. 1976. Sorption du cadmium par une population de la diatomee *Phaeodactylum tricornutum* en culture. Mar. Biol. 34: 163-167.

45. Cowen, J. P., V. F. Hodge and T. R. Folsom. 1976. *In vivo* accumulation of radioactive polonium by the giant kelp, *Macrocystis pyrifera*. Mar. Biol. 37: 239-248.

46.   Cugurra, F. and G. Maura. 1976. Mercury content in several species of marine fish. Bull. Envir. Cont. Tox. 15: 568-573.

47.   Cunningham, P. A. and M. R. Tripp. 1975. Factors affecting the accumulation and removal of mercury from tissues of the American oyster *Crassostrea virginica*. Mar. Biol. 31: 311-319.

48.   Cunningham, P. A. and M. R. Tripp. 1975. Accumulation, tissue distribution and elimination of $^{203}HgCl_2$ and CH_3 $^{203}HgCl$ in the tissues of the American oyster *Crassostrea virginica*. Mar. Biol. 31: 321-334.

49.   Cutshall, N. 1974. Turnover of zinc-65 in oysters. Health Physics 26: 327-331.

50.   D'Agostino, A. and C. Finney. 1974. The effect of copper and cadmium on the development of *Tigriopus japonicus*. In: Vernberg, F. J. and W. B. Vernberg (eds.). Pollution and physiology of marine organisms. Academic Press, N. Y.: 445-463.

51.   D'Amelio, V., G. Russo, and D. Ferraro. 1974. The effect of heavy metal on protein synthesis in crustaceans and fish. Rev. Int. Océan. Médic. 33:111-118.

52.   Davies, A. G. 1974. The growth kinetics of *Isochrysis galbana* in cultures containing sublethal concentrations of mercuric chloride. J. Mar. Biol. Assn. U. K. 54: 157-169.

53.   Davies, A. G. 1976. An assessment of the basis of mercury tolerance in *Dunaliella tertiolecta*. J. Mar. Biol. Assn. U. K. 56: 39-57.

54.   DeClerck, R., R. Vanderstoppen, and W. Vyncke. 1974. Mercury content of fish and shrimps caught off the Belgian coast. Ocean Manage. 2: 117-126.

55.   Dorn, P. 1976. The feeding behavior of *Mytilus edulis* in the presence of methylmercury acetate. Bull. Envir. Cont. Tox. 15: 714-719.

56.   Drifmeyer, J. E. 1974. Zn and Cu levels in the eastern oyster, *Crassostrea virginica,* from the lower James River. J. Wash. Acad. Sci. 64: 292-294.

57.   Dunstan, W. M., H. L. Windom, and G. L. McIntire. 1975. The role of *Spartina alterniflora* in the flow of lead, cadmium and copper through the salt marsh ecosystem. In: Howell, F. G., J. B. Gentry and M. H. Smith (eds.). Mineral cycling in southeastern ecosystems. U. S. Energy Res. Dev. Admin.: 250-256. Available as CONF-740513 from NTIS, U.S. Dept. Comm., Springfield, VA 22161.

58.   Eiben, R. 1976. Influence of wetting tension and ions upon settlement and beginning of metamorphosis in bryozoan larvae (*Bowerbankia gracilis*). Mar. Biol. 37: 249-254.

59.   Eisler, R. 1974. Radiocadmium exchange with seawater by *Fundulus heteroclitus* (L.) (Pisces: Cyprinodontidae). J. Fish Biol. 6: 601-612.

60. Eisler, R. and M. Wapner. 1975. Second annotated bibliography on biological effects of metals in aquatic environments. U. S. Environ. Protect. Agen. Rept. EPA-600/3-75-008: 400 pp. Available from Nat. Tech. Inf. Serv., Springfield, VA 22161.

61. Establier, R. 1975. Contenido en mercurio de las anguilas (*Anguilla anguilla*) de la desembocadura del rio Guadalquiver y esteros de las salinas de la zona de Cádiz. Invest. Pesq. 39: 249-255

62. Establier, R. 1975. Concentracion de mercurio en los cabellos de la poblacion de Cádiz y pescadores de altura. Invest. Pesq. 39: 509-516.

63. Eustace, I. J. 1974. Zinc, cadmium, copper, and manganese in species of finfish and shellfish caught in the Derwent Estuary, Tasmania. Austral. J. Mar. Fresh. Res. 25: 209-220.

64. Fletcher, G. L., E. G. Watts, and M. J. King. 1975. Copper, zinc, and total protein levels in the plasma of sockeye salmon (*Oncorhynchus nerka*) during their spawning migration. J. Fish. Res. Bd. Canada 31: 78-82.

65. Foster, P. 1976. Concentrations and concentration factors of heavy metals in brown algae. Envir. Poll. 10: 45-53.

66. Fowler, B. A., D. A. Wolfe, and W. F. Hettler. 1975. Mercury and iron uptake by cytosomes in mantle epithelial cells of quahog clams (*Mercenaria mercenaria*) exposed to mercury. J. Fish. Res. Bd. Canada 32: 1767-1775.

67. Fowler, S., M. Heyraud, and T. M. Beasley. 1975. Experimental studies on plutonium kinetics in marine biota. In: Impacts of nuclear releases into the aquatic environments. Int. Atom. Ener. Agen., Vienna. Paper SM-198/23: 157-177.

68. Fowler, S. W. 1974. The effect of organism size on the content of certain trace metals in marine zooplankton. Rapp. Comm. Int. Mer Medit. 22 (9): 145-146.

69. Fowler, S. W. and G. Benayoun. 1974. Experimental studies on cadmium flux through marine biota. In: Comparative studies of food and environmental contamination, Inter. Atom. Ener. Agen., Vienna: 159-178.

70. Fowler, S. W. and G. Benayoun. 1976. Influence of environmental factors on selenium flux in two marine invertebrates. Mar. Biol. 37: 59-68.

71. Fowler, S. W. and G. Benayoun. 1976. Accumulation and distribution of selenium in mussel and shrimp tissues. Bull. Envir. Cont. Tox. 16: 339-346.

72. Fowler, S. W. and G. Benayoun. 1976. Selenium kinetics in marine zooplankton. Mar. Sci. Communications. 2:43-67.

73. Fowler, S. W., J. LaRosa, M. Heyraud, and W. C. Renfro. 1975. Effect of different radiotracer labelling techniques on radionuclide excretion from marine organisms. Mar. Biol. 30: 297-304.

74. Fowler, S. W. and B. Oregioni. 1976. Trace metals in mussels from the N. W. Mediterranean. Mar. Poll. Bull. 7: 26-29.

75. Frazier, J. M. 1975. The dynamics of metals in the American oyster, *Crassostrea virginica*. I. seasonal effects. Chesapeake Sci. 16: 162-171.

76. Frazier, J. M. 1976. The dynamics of metals in the American oyster, *Crassostrea virginica*. II. environmental effects. Chesapeake Sci. 17: 188-197.

77. Freeman, H. C., D. A. Horne, B. McTague, and M. McMenemy. 1974. Mercury in some Canadian Atlantic coast fish and shellfish. J. Fish. Res. Bd. Canada 31: 369-372.

78. Gardner, G. R. 1975. Chemically induced lesions in estuarine or marine teleosts. In: Ribelin, W. E. and G. Migaki (eds.). The pathology of fishes. Univ. Wisconsin Press, Madison, Wisc.: 657-693.

79. Gaskin, D. E., G. J. D. Smith, P. W. Arnold, M. V. Louisy, R. Frank, M. Holdrinet, and J. W. McWade. 1974. Mercury, DDT, dieldrin, and PCB in two species of Odontoceti (Cetacea) from St. Lucia, Lesser Antilles. J. Fish. Res. Bd. Canada 31: 1235-1239.

80. George, S. G., B. J. S. Pirie and T. L. Coombs. 1976. The kinetics of accumulation and excretion of ferric hydroxide in *Mytilus edulis* (L.) and its distribution in the tissues. J. Exp. Mar. Biol. Ecol. 23: 71-84.

81. Gould, E., R. S. Collier, J. J. Karolus, and S. Givens. 1976. Heart transaminase in the rock crab, *Cancer irroratus*, exposed to cadmium salts. Bull. Envir. Cont. Tox. 15: 635-643.

82. Gray, J. S. 1974. Synergistic effects of three heavy metals on growth rates of a marine ciliate protozoan. In: Vernberg, F. J. and W. B. Vernberg (eds.), Pollution and physiology of marine organisms. Academic Press, N. Y.: 465-485.

83. Green, F. A., Jr., J. W. Anderson, S. R. Petrocelli, B. J. Presley, and R. Sims. 1976. Effect of mercury on the survival, respiration, and growth of postlarval white shrimp, *Penaeus setiferus*. Mar. Biol. 37: 75-81.

84. Greig, R. A., B. A. Nelson, and D. A. Nelson. 1975. Trace metal content in the American oyster. Mar. Poll. Bull. 6: 72-73.

85. Greig, R. A., D. R. Wenzloff, and J. B. Pearce. 1976. Distribution and abundance of heavy metals in finfish, invertebrates, and sediments collected at a deepwater disposal site. Mar. Poll. Bull. 7: 185-187.

86. Greig, R. A., D. R. Wenzloff, and C. Shelpuk. 1975. Mercury concentrations in fish, North Atlantic offshore waters 1971. Pest. Monit. J. 9: 15-20.

87. Gromov, V. V. and V. I. Spitsyn. 1974. Uptake of plutonium, ruthenium, and technetium by phytoplankton. Hydrobiology 215: 147-150.

88. Gromov, V. V. and V. I. Spitsyn. 1974. Influence of phytoplankton on the physicochemical state of Pu^{239}, Ru^{106}, Tc^{99}, and Co^{60} in seawater. Doklady Biol. Sci. 215: 151-153.

89. Guary, J. C., M. Masson, and A. Fraizier. 1976. Etude preliminaire, *in situ*, de la distribution du plutonium dans differents tissus et organes de *Cancer pagurus* (Crustacea: Decapoda) et de *Pleuronectes platessa* (Pisces: Pleuronectidae). Mar. Biol. 36: 13-17.

90. Hardisty, M. W., S. Kartar, and M. Sainsbury. 1974. Dietary habits and heavy metal concentrations in fish from the Severn Estuary and Bristol Channel. Mar. Poll. Bull. 5: 61-63.

91. Haug, A., S. Melsom, and S. Omang. 1974. Estimation of heavy metal pollution in two Norwegian fjord areas by analysis of the brown alga *Ascophyllum nodosum*. Envir. Poll. 7: 179-192.

92. Heslinga, G. A. 1976. Effects of copper on the coral-reef echinoid *Echinometra mathaei*. Mar. Biology 35: 155-160.

93. Hessler, A. 1974. The effects of lead on algae. I. effects of Pb on viability and motility of *Platymonas subcordiformis* (Chlorophyta: Volvocales). Water, Air, Soil Poll. 3: 371-385.

94. Hessler, A. 1975. The effects of lead on algae. II. mutagenesis experiments on *Platymonas subcordiformis* (Chlorophyta: Volvocales). Mutation Res. 31: 43-47.

95. Heyraud, M., S. W. Fowler, T. M. Beasley and R. D. Cherry. 1976. Polonium-210 in euphausiids: a detailed study. Mar. Biol. 34: 127-136.

96. Hoffman, F. L., V. F. Hodge, and T. R. Folsom. 1974. Polonium radioactivity in certain mid-water fish of the eastern temporal Pacific. Health Physics 26: 65-70.

97. Holden, A. V. 1975. The accumulation of oceanic contaminants in marine mammals. Rapp. P. v. Réun. Const. int. Explor. Mer 169: 353-361.

98. Hutcheson, M. S. 1974. The effect of temperature and salinity on cadmium uptake by the blue crab, *Callinectes sapidus*. Chesapeake Sci. 15: 237-241.

99. Ireland, M. P. 1974. Variations in the zinc, copper, manganese and lead content of *Balanus balanoides* in Cardigan Bay, Wales. Envir. Poll. 7:65-75.

100. Ishikawa, M., T. Koyanagi, and M. Saiki. 1976. Studies on the chemical behavior of ^{106}Ru in seawater and its uptake by marine organisms. I. accumulation and excretion of ^{106}Ru by clam. Bull. Jap. Soc. Sci. Fish. 42: 287-297.

101. Jensen, A., B. Rystad, and S. Melsom. 1974. Heavy metal tolerance of marine phytoplankton. I. the tolerance of three algal species to zinc in coastal seawater. J. Exp. Mar. Biol. Ecol. 15: 145-157.

102. Jensen, A., B. Rystad, and S. Melsom. 1976. Heavy metal tolerance of marine phytoplankton. II. copper tolerance of three species in dialysis and batch cultures. J. Exp. Mar. Biol. Ecol. 22: 249-256.

103. Jones, D., K. Ronald, D. M. Lavigne, R. Frank, M. Holdrinet, and J. F. Uthe. 1976. Organochlorine and mercury residues in the harp seal (*Pagophilus groenlandicus*). Science Total Environ. 5: 181-195.

104. Jones, L. H., N. V. Jones and A. J. Radlett. 1976. Some effects of salinity on the toxicity of copper to the polychaete *Nereis diversicolor*. Estuar. Coast. Mar. Sci. 4: 107-111.

105. Jones, M. B. 1975. Effects of copper on survival and osmoregulation in marine and brackish water isopods (Crustacea). Proc. 9th Europ. Mar. Biol. Symp.: 419-431.

106. Jones, M. B. 1975. Synergistic effects of salinity, temperature and heavy metals on mortality and osmoregulation in marine and estuarine isopods (Crustacea). Mar. Biol. 30: 13-20.

107. Kayser, H. 1976. Waste-water assay with continuous algal cultures: the effect of mercuric acetate on the growth of some marine dinoflagellates. Mar. Biol. 36: 61-72.

108. Kennedy, V. S. 1976. Arsenic concentrations in some coexisting marine organisms from Newfoundland and Labrador. J. Fish. Res. Bd. Canada 33: 1388-1393.

109. Klaunig, J., S. Koepp, and M. McCormick. 1975. Acute toxicity of a native mummichog population (*Fundulus heteroclitus*) to mercury. Bull. Envir. Cont. Tox. 14: 534-536.

110. Klemmer, H. W., C. S. Unninayer, and W. I. Ukubo. 1976. Mercury content of biota in coastal waters in Hawaii. Bull. Envir. Cont. Tox. 15: 454-457.

111. Koeman, J. H., W. S. M. van de Ven, J. J. M. de Goeij, P. S. Tjioe and J. L. van Haaften. 1975. Mercury and selenium in marine mammals and birds. Science Total Envir. 3: 279-287.

112. Kopfler, F. C. 1974. The accumulation of organic and inorganic mercury compounds by the eastern oyster (*Crassostrea virginica*). Bull. Envir. Cont. Tox. 11: 275-280.

113. Kustin, K., K. V. Ladd, and G. C. McLeod. 1975. Site and rate of vanadium assimilation in the tunicate *Ciona intestinalis*. J. Gen. Physiol. 65: 315-328.

114. Lake, P. S. and V. J. Thorp. 1974. The gill lamellae of the shrimp *Paratya tasmaniensis* (Atyidae: crustacea). Normal ultra structure and changes with low levels of cadmium. Eighth Int. Cong. Electron Microsc., Canberra, Australia, Vol. II: 448-449.

115. Larsson, A. 1975. Some biochemical effects of cadmium on fish. In: Koeman, J. H. and J. J. T. W. A. Strik (eds.). Sublethal effects of toxic chemicals on aquatic animals. Elsevier Sci. Publ. Co., Amsterdam: 3-13.

116. Leatherland, T. M. and J. D. Burton. 1974. The occurrence of some trace metals in coastal organisms with particular reference to The Solent region. J. Mar. Biol. Assn. U. K. 54: 457-468.

117. Lewin, J. and C. H. Chen. 1976. Effects of boron deficiency on the chemical composition of a marine diatom. J. Exp. Botany 27: 916-921.

118. Lindberg, E. and C. Harriss. 1974. Mercury enrichment in estuarine plant detritus. Mar. Poll. Bull. 5: 93-95.

119. Lockwood, A. P. M. and C. B. E. Inman. 1975. Diuresis in the amphipod, *Gammarus duebeni* induced by methylmercury, D. D. T., lindane and fenithrothion. Comp. Biochem. Physiol. 52(2C): 75-80.

120. Lorz, H. W. and B. P. McPherson. 1976. Effects of copper or zinc in fresh water on the adaptation to seawater and ATPase activity and the effects of copper on migratory disposition of coho salmon *(Oncorhynchus kisutch)*. J. Fish. Res. Bd. Canada 33: 2023-2030.

121. Mackay, N. J., R. J. Williams, J. L. Kacprzac, M. N. Kazacos, A. J. Collins, and E. H. Auty. 1975. Heavy metals in cultivated oysters (*Crassostrea commericalis = Saccostrea cucullata)* from the estuaries of New South Wales. Austral. J. Mar. Fresh. Res. 26: 31-46.

122. Mandelli, E. F. 1975. The effects of desalination brines on *Crassostrea virginica* (Gmelin). Water Research 9: 287-295.

123. Marchand, M. 1974. Considérations sur les formes physico-chimiques du cobalt, manganèse, zinc, chrome et fer dans une eau de mêr enriche ou non de matière organique. J. Cons. int. Explor. Mer 35: 130-142.

124. Martin, J. H. and W. W. Broenkow. 1975. Cadmium in plankton: elevated concentrations off Baja California. Science 190: 884-885.

125. Martin, J. H., P. D. Elliot, V. C. Anderlini, D. Girvin, S. A. Jacobs, R. W. Risebrough, R. L. Delong, and W. G. Gilmartin. 1976. Mercury-selenium-bromine imbalance in premature parturient California sea lions. Mar. Biol. 35: 91-104.

126. Martin, J. H. and A. R. Flegal. 1975. High copper concentrations in squid livers in association with elevated levels of silver, cadmium, and zinc. Mar. Biol. 30: 51-55.

127. McDermott, D. J., G. V. Alexander, D. R. Young, and A. J. Mearns. 1976. Metal contamination of flatfish around a large submarine outfall. J. Water Poll. Contr. Fed. 48: 1913-1918.

128. McLeese, D. W. 1974. Toxicity of copper at two temperatures and three salinities to the American lobster (*Homarus americanus*). J. Fish. Res. Bd. Canada 31: 1949-1952.

129. McLeod, G. C., K. V. Ladd, K. Kustin, and D. L. Toppen. 1975. Extraction of vanadium (V) from seawater by tunicates: a revision of concepts. Limnol. Ocean. 20: 491-493.

130. McLerran, C. J. and C. W. Holmes. 1974. Deposition of zinc and cadmium by marine bacteria in estuarine sediments. Limnol. Ocean. 19: 998-1001.

131. McLusky, D. S. and C. N. K. Phillips. 1975. Some effects of copper on the polychaete *Phyllodoce maculata*. Estuar. Coast. Mar. Sci. 3:103-108.

132. McNaughton, S. J., T. C. Folsom, T. Lee, F. Park, C. Price, D. Roeder, J. Schmitz, and C. Stockwell. 1974. Heavy metal tolerance in *Typha latifolia* without the evolution of tolerant races. Ecology 55: 1163-1165.

133. Middaugh, D. P. and C. L. Rose. 1974. Retention of two mercuricals by striped mullet, *Mugil cephalus*. Water Res. 8: 173-177.

134. Milanovich, F. P., R. Spies, M. S. Guram, and E. E. Sykes. 1976. Uptake of copper by the polychaete *Cirriformia spirabrancha* in the presence of dissolved yellow organic matter of natural origin. Estuar. Coast. Mar. Sci. 4: 585-588.

135. Mitchell, R. and I. Chet. 1975. Bacterial attack of corals in polluted seawater. Microb. Ecol. 2: 227-233.

136. Moore, M. N. and A. R. D. Stebbing. 1976. The quantitative cytochemical effects of three metal ions on a lysosomal hydrolase of a hydroid. J. Mar. Biol. Assn. U. K. 56: 995-1005.

137. Morisawa, M. and H. Mohri. 1974. Heavy metals and spermatozoan motility. II. turbidity changes induced by divalent cations and adenosinetriphosphate in sea urchin sperm flagella. Exper. Cell. Res. 83: 87-94.

138. Morris, O. P. and G. Russell. 1974. Inter-specific differences in responses to copper by natural populations of *Ectocarpus*. Brit. Phycol. J. 9: 269-272.

139. Negilski, D. S. 1976. Acute toxicity of zinc, cadmium, and chromium to the marine fishes, yellow-eye mullet (*Aldrichetta forsteri* C. & V.) and small-mouthed hardyhead (*Atherinasoma microstoma* Whitley). Austral. J. Mar. Fresh. Res. 27: 137-149.

140. Nelson, D. A., A. Calabrese, B. A. Nelson, J. R. MacInnes and D. R. Wenzloff. 1976. Biological effects of heavy metals on juvenile bay scallops, *Argopeeten irradians*, in short-term exposures. Bull. Envir. Cont. Tox. 16: 275-282.

141. Nelson, J. D., Jr. and R. R. Colwell. 1975. The ecology of mercury-resistent bacteria in Chesapeake Bay. Microb. Ecol. 1: 191-218.

142. Noshkin, V. E., R. J. Eagle and K. M. Wong. 1976. Plutonium levels in Kwajalein lagoon. Nature 262: 745-748.

143. Noshkin, V. E., K. M. Wong, R. J. Eagle, and C. Gatrousis. 1975. Transuranics and other radionuclides in Bikini lagoon: concentration data retrieved from aged coral sections. Limnol. Ocean. 20: 729-742.

144. Noshkin, V. E., Jr. and C. Gatrousis. 1974. Fallout 240_{Pu} and ^{239}Pu in Atlantic marine samples. Earth Planet. Sci. Lett. 22: 111-117.

145. Nuorteva, P. and E. Hasanen. 1975. Bioaccumulation of mercury in *Myoxocephalus quadricornis* (L.) (Teleostei, Cottidae) in an unpolluted area of the Baltic. Ann. Zool. Fennici 12: 247-254.

146. Nuorteva, P., E. Hasanen, and S. L. Nuorteva. 1975. The effectiveness of the Finnish anti-mercury measurements in the moderately polluted area of Hameenkyro. Ymparisto ja Terveys 6(8): 611-635.

147. Olafson, R. W. and J. A. J. Thompson. 1974. Isolation of heavy metal binding proteins from marine vertebrates. Mar. Biol. 28: 83-86.

148. Oshida, P. S., A. J. Mearns, D. J. Reish, and C. S. Word. 1976. The effects of hexavalent and trivalent chromium on *Neanthes arenaceodentata* (Polychaeta: Annelida). S. Calif. Coast. Water Res. Proj., 1500 E. Imperial Hwy., El Segundo, Calif., TM 225: 58 pp.

149. Oshida, P. and D. J. Reish. 1975. Effects of chromium on reproduction in polychaetes. S. Calif. Coast. Water Res. Proj., 1500 E. Imperial Hwy., El Segundo, Calif. Ann. Report: 55-60.

150. Overnell, J. 1975. The effect of heavy metals on photosynthesis and loss of cell potassium in two species of marine algae, *Dunaliella tertiolecta* and *Phaeodactylum tricornutum*. Mar. Biol. 29: 99-103.

151. Parrish, K. M. and R. A. Carr. 1976. Transport of mercury through a laboratory two-level marine food chain. Mar. Poll. Bull. 7: 90-91.

152. Patel, B., C. D. Mulay, and A. K. Ganguly. 1975. Radioecology of Bombay Harbour--a tidal estuary. Estuar. Coast. Mar. Sci. 3: 13-42.

153. Penot, M. and C. Videau. 1975. Absorption du 86_{Rb} et du 99_{Mo} par deux algues marines: le *Laminaria digitata* et le *Fucus serratus*. Z. Pflanzenphysiol. Bd. 76 (Suppl): 285-293.

154. Penrose, W. R., R. Black, and M. J. Hayward. 1975. Limited arsenic dispersion in seawater, sediments, and biota near a continuous source. J. Fish. Res. Bd. Canada 32: 1275-1281.

155. Pentreath, R. J. 1976. Some further studies on the
accumulation and retention of Zn-65 and Mn-54 by the plaice,
Pleuronectes platessa L. J. Exp. Mar. Biol. Ecol. 21: 179-189.

156. Pentreath, R. J. 1976. The accumulation of mercury
from food by the plaice, *Pleuronectes platessa* L. J. Exp. Mar.
Biol. Ecol. 25: 51-65.

157. Pentreath, R. J. 1976. The accumulation of organic
mercury from seawater by the plaice, *Pleuronectes platessa* L.
J. Exp. Mar. Biol. Ecol. 24: 121-132.

158. Pentreath, R. J. and M. B. Lovett. 1976. Occurrence
of plutonium and americium in plaice from the northeastern
Irish Sea. Nature 262: 814-816.

159. Persoone, G. and G. Uyttersprot. 1975. The influence
of inorganic and organic pollutants on the rate of reproduction of a marine hypotrichous ciliate: *Euplotes vannus* Muller.
Rev. Int. Ocean. Med. 37-38: 125-151.

160. Phelps, D. K., G. Telek, and R. L. Lapan, Jr. 1975.
Assessment of heavy metal distribution within the food web.
In: Pearson and Frangipane (eds.). Marine pollution and
marine waste disposal. Pergamon Press, N. Y.: 341-348.

161. Phillips, J. H. 1976. The common mussel *Mytilus
edulis* as an indicator of pollution by zinc, cadmium, lead
and copper. I. effects of environmental variables on uptake
of metals. Mar. Biol. 38: 59-69.

162. Phillips, J. H. 1976. The common mussel *Mytilus
edulis* as an indicator of pollution by zinc, cadmium, lead
and copper. II. relationship of metals in the mussel to
those discharged by industry. Mar. Biol. 38: 71-80.

163. Pilson, M. E. Q. 1974. Arsenate uptake and reduction
by *Pocillopora verrucosa*. Limnol. Ocean. 19: 339-341.

164. Pouvreau, B. and J. C. Amiard. 1974. Etude experimentale de l'accumulation de l'argent 110 m chez divers
organismes marins. Comm. a l'Energie Atomique France, Rept.
CEA-R-4571:19 pp.

165. Ratkowsky, D. A., T. G. Dix, and K. C. Wilson. 1975.
Mercury in fish in the Derwent Estuary, Tasmania and its rela-
tion to the position of the fish in the food chain. Austral.
J. Mar. Fresh. Res. 26: 223-231.

166. Ratkowsky, D. A., S. J. Thrower, I. J. Eustace, and J.
Olley. 1974. A numerical study of the concentration of some
heavy metals in Tasmanian oysters. J. Fish. Res. Bd. Canada
31: 1165-1171.

167. Reinhart, K. and T. D. Myers. 1975. Eye and tentacle
abnormalities in embryos of the Atlantic oyster drill,
Urosalpinx cinera. Chesapeake Sci. 16: 286-288.

168. Reish, D. J., J. M. Martin, F. M. Piltz, and J. Q. Word.
1976. The effect of heavy metals on laboratory populations of
two polychaetes with comparisons to the water quality conditions
and standards in southern California marine waters. Water
Res. 10: 299-302.

169. Renfro, J. L., B. Schmidt-Neilson, D. Miller, D. Benos,
and J. Allen. 1974. Methyl mercury and inorganic mercury:
uptake, distribution, and effect on osmoregulatory mechanisms
in fishes. In: Vernberg, F. J. and W. B. Vernberg (eds).
Pollution and physiology of marine organisms. Academic Press,
N. Y.: 101-122.

170. Renfro, W. C., S. W. Fowler, M. Heyraud, and J. LaRosa.
1975. Relative importance of food and water in long-term zinc-
65 accumulation by marine biota. J. Fish. Res. Bd. Canada 32:
1339-1345.

171. Roberts, D. and C. Maguire. 1976. Interactions of
lead with sediment and meiofauna. Mar. Poll. Bull. 7: 211-213.

172. Roberts, T. M., P. B. Heppleston and R. D. Roberts.
1976. Distribution of heavy metals in tissues of the common
seal. Mar. Poll. Bull. 7: 194-196.

173. Roesijadi, G., S. R. Petrocelli, J. W. Anderson, B. J.
Presley, and R. Sims. 1974. Survival and chloride ion regulation of the porcelain crab *Petrolisthes armatus* exposed to
mercury. Mar. Biol. 27: 213-217.

174. Romeril, M. G. 1974. Trace metals in sediments and
bivalve mollusca in Southampton Water and the Solent. Rev.
Int. Ocean. Medic. 33: 31-47.

175. Rosenthal, H. and K. R. Sperling. 1974. Effects of
cadmium on development and survival of herring eggs. In:
Blaxter, J. H. S. (ed.). The early life history of fish.
Springer-Verlag: 383-396.

176. Ryndina, D. D. and G. G. Polikarpov. 1974. Role of
Cystoseira barbata (Good et Wood) AG. polysaccharides in the
extraction of some radionuclides from seawater. Hydrobiol. J.
10(5): 61-65.

177. Saenko, G. N., M. D. Koryakova, V. F. Makienko, and
I. G. Dobrosmyslova. 1976. Concentration of polyvalent metals
by seaweeds in Vostok Bay, Sea of Japan. Mar. Biol. 34: 169-176.

178. Saliba, L. J. and R. M. Krzyz. 1976. Effects of
heavy metals on hatching of brine-shrimp eggs. Mar. Poll. Bull.
7: 181-182.

179. Saliba, L. J. and R. M. Krzyz. 1976. Acclimation and
tolerance of *Artemia salina* to copper salts. Mar. Biol. 38:
231-238.

180. Sano, K. and H. Mohri. 1976. Fertilization of sea
urchins needs magnesium ions in seawater. Science 192: 1339-
1340.

181. Saward, D., A. Stirling, and G. Topping. 1975. Experimental studies on the effects of copper on a marine food chain. Mar. Biol. 29: 351-356.

182. Sayler, G. S., J. D. Nelson, Jr., and R. R. Colwell. 1975. Role of bacteria in bioaccumulation of mercury in the oyster Crassostrea virginica. Appl. Microbiol. 30: 91-96.

183. Schell, W. R. and R. L. Watters. 1975. Plutonium in aqueous systems. Health Physics 29: 589-597.

184. Schulz-Baldes, M. 1974. Lead uptake from sea water and food, and lead loss in the common mussel Mytilus edulis. Mar. Biol. 25: 177-193.

185. Shealy, M. H. and P. A. Sandifer. 1975. Effects of mercury on survival and development of the larval grass shrimp, Paleomonetes vulgaris. Mar. Biol. 33: 7-16.

186. Sherwood, M. J. 1975. Toxicity of chromium to fish. Ann. Rep. S. Calif. Coast. Water Res. Proj., 1500 E. Imperial Hwy.; El Segundo, Calif.,: 61-62.

187. Shultz, C. D., D. Crear, J. E. Pearson, J. B. Rivers, and J. W. Hylin. 1976. Total and organic mercury in Pacific blue marlin. Bull. Envir. Cont. Tox. 15: 230-234.

188. Skei, J. M., M. Saunders, and N. B. Price. 1976. Mercury in plankton from a polluted Norwegian fjord. Mar. Poll. Bull. 7: 34-35.

189. Small, L. F., S. Keckes, and S. W. Fowler. 1974. Excretion of different forms of zinc by the prawn, Palaemon serratus (Pennant). Limnol. Ocean. 19: 789-793.

190. Spinelli, J. and C. Mahnken. 1976. Effect of diets containing dogfish (Squalus acanthias) meal on the mercury content and growth of pen-reared coho salmon (Oncorhynchus kisutch). J. Fish. Res. Bd. Canada 33: 1771-1778.

191. Stebbing, A. R. D. 1976. The effects of low metal levels on a clonal hydroid. J. Mar. Biol. Assn. U. K. 56: 977-994.

192. Stenner, R. D. and G. Nickless. 1974. Absorption of cadmium, copper and zinc by dog whelks in the Bristol Channel. Nature 247: 198-199.

193. Stenner, R. D. and G. Nickless. 1974. Distribution of some heavy metals in organisms in Hardangerfjord and Skjerstadfjord, Norway. Water, Air, Soil Poll. 3: 279-291.

194. Stenner, R. D. and G. Nickless. 1975. Heavy metals in organisms of the Atlantic coast of S. W. Spain and Portugal. Mar. Poll. Bull. 6: 89-92.

195. Stephenson, R. R. and D. Taylor. 1975. The influence of EDTA on the mortality and burrowing activity of the clam (Venerupis decussata) exposed to sublethal concentrations of copper. Bull. Envir. Cont. Tox. 14: 304-308.

196. Stewart, J. and M. Schulz-Baldes. 1976. Long term
lead accumulation in abalone *(Haliotis* spp.) fed on lead-
treated brown algae *(Egregia laevigata).* Mar. Biol. 36: 19-24.

197. Stickney, R. R., H. L. Windom, D. B. White, and F. E.
Taylor. 1975. Heavy metal concentrations in selected Georgia
estuarine organisms with comparative food habit data. In:
Howell, F. G., J. B. Gentry and M. H. Smith (eds.). Mineral
cycling in southeastern ecosystems. U. S. Energy Res. Dev.
Admin.: 257-267. Available as CONF-740513 from NTIS, U. S.
Dept. Comm., Springfield, VA 22161.

198. Strohal, P., D. Huljev, S. Lulic, and M. Picer. 1975.
Antimony in the coastal marine environment, North Adriatic.
Estuar. Coast. Mar. Sci. 3: 119-123.

199. Styron, C. E., T. M. Hagen, D. R. Campbell, J. Harvin,
N. K. Whittenberg, G. A. Baughman, M. E. Bransford, W. H.
Saunders, D. C. Williams, C. Woodle, N. K. Dixon, and C. R.
McNeill. 1976. Effects of temperature and salinity on growth
and uptake of Zn-65 and Cs-137 for six marine algae. J. Mar.
Biol. Assn. U. K. 56: 13-20.

200. Sunda, W. and R. R. L. Guillard. 1976. The relationship between cupric ion activity and the toxicity of copper
to phytoplankton. J. Mar. Res. 34: 511-529.

201. Svansson, A. 1975. Physical and chemical oceanography
of the Skagerrak and the Kattegat. I. Open sea conditions.
Fish. Bd. Sweden, Inst. Mar. Res., Goteborg, Sweden, Rept.
No. 1:88 pp.

202. Swinehart, J. H., W. R. Biggs, D. J. Halko, and N. C.
Schroeder. 1974. The vanadium and selected metal contents
of some ascidians. Biol. Bull. 146: 302-312.

203. Talbot, V. W., R. J. Magee and M. Hussain. 1976.
Cadmium in Port Phillip Bay mussels. Mar. Poll. Bull. 7:
84-86.

204. Thomas, W. H. and A. N. Dodson. 1974. Inhibition of
diatom photosynthesis by germanic acid: separation of diatom
productivity from total marine primary productivity. Mar.
Biol. 27: 11-19.

205. Thompson, J. A. J., J. C. Davis, and R. E. Drew. 1976.
Toxicity, uptake amd survey studies of boron in the marine
environment. Water Res. 10: 869-875.

206. Thurberg, F. P., W. D. Cable, M. A. Dawson, J. R.
MacInnes, and D. R. Wenzloff. 1975. Respiratory response of
larval, juvenile and adult surf clams, *Spisula solidissima,*
to silver. In: Cech, J. J., Jr. D. W. Bridges, and D. B.
Horton (eds.). Respiration of marine organisms. TRIGOM Publ.
S. Portland, ME:41-52.

207. Thurberg, F. P., A. Calabrese, and M. A. Dawson. 1974.
Effects of silver on oxygen consumption of bivalves at various
salinities. In: Vernberg, F. J. and W. B. Vernberg (eds.).
Pollution and physiology of marine organisms. Academic Press,
N. Y.: 67-78.

208. Tolkach, V. V., V. V. Gromov and V. I. Spitsyn. 1975.
An investigation of absorption of Cs-137 by krill. Doklady
Biol. Sci. Proc. Acad. Sci. USSR 220:11-13.

209. Tripp, M. and R. C. Harris. 1976. Role of mangrove
vegetation in mercury cycling in the Florida everglades. In:
Nriagu, J. D. (ed.). Environmental Biogeochemistry. Vol 2.
Metals transfer and ecological mass balances. Ann Arbor Sci.
Publ., Ann Arbor, Mich:489-497.

210. Turekian, K. K., J. K. Cochran, D. P. Kharkar, R. M.
Cerrato, J. R. Vaisnys, H. L. Sanders, J. F. Grassle, and J.
A. Allen. 1975. Slow growth rate of a deep-sea clam determined by Ra-228 chronology. Proc. Nat. Acad. Sci., USA 72:
2829-2832.

211. Ueda, T., R. Nakamura, and Y. Suzuki. 1976. Comparison of [115 m] Cd accumulation from sediments and seawater by
polychaete worms. Bull. Japan. Soc. Sci. Fish. 42: 299-306.

212. Valiela, I., M. D. Banus, and J. M. Teal. 1974.
Response of salt marsh bivalves to enrichment with metal-
containing sewage sludge and retention of lead, zinc, and
cadmium by marsh sediments. Environ. Poll. 7: 149-157.

213. Vernberg, W. B., P. J. DeCoursey, and J. O'Hara. 1974.
Multiple environmental factor effects on physiology and behavior of the fiddler crab, *Uca pugilator*. In: Vernberg, F. J.
and W. B. Vernberg (eds.). Pollution and physiology of
marine organisms. Academic Press, Inc. N. Y.: 381-425.

214. Voyer, R. A. 1975. Effect of dissolved oxygen con-
centration on the acute toxicity of cadmium to the mummichog,
Fundulus heteroclitus (L.) at various salinities. Trans. Amer.
Fish. Soc. 104:129-134.

215. Vucetic, T., W. B. Vernberg, and G. Anderson. 1974.
Long-term annual fluctuations of mercury in the zooplankton
of the east central Adriatic. Rev. Int. Ocean. Medic. 33:
75-81.

216. Waldichuk, M. 1974. Some biological concerns in
heavy metal pollution. In: Vernberg, F. J. and W. B. Vernberg
(eds.). Pollution and physiology of marine organisms. Academic
Press, N. Y.: 1-57.

217. Walker, G., P. S. Rainbow, P. Foster, and D. J. Crisp.
1975. Barnacles: possible indicators of zinc pollution? Mar.
Biol. 30: 57-65.

218. Walker, G., P. S. Rainbow, P. Foster, and D. L. Holland.
1975. Zinc phosphate granules in tissue surrounding the midgut
of the barnacle *Balanus balanoides*. Mar. Biol. 33: 161-166.

219. Watling, H. R. and R. J. Watling. 1976. Trace metals in oysters from Knysna Estuary. Mar. Poll. Bull. 7: 45-48.

220. Watling, H. R. and R. J. Watling. 1976. Trace metals in *Choromytilus meridionalis*. Mar. Poll. Bull. 7: 91-94.

221. Weis, J. S. 1976. Effects of mercury, cadmium, and lead salts on regeneration and ecdysis in the fiddler crab, *Uca pugilator*. U. S. Dept. Comm., Fish. Bull. 74: 464-467.

222. Weis, P. and J. S. Weis. 1976. Effects of heavy metals on fin regeneration in the killifish, *Fundulus heteroclitus*. Bull. Envir. Cont. Tox. 16: 197-201.

223. Westernhagen, H. V. and V. Dethlefsen. 1975. Combined effects of cadmium and salinity on development and survival of flounder eggs. J. Mar. Biol. Assn. U. K. 55: 945-957.

224. Westernhagen, H. V., V. Dethlefsen, and H. Rosenthal 1975. Combined effects of cadmium and salinity on development and survival of garpike eggs. Helgol. wiss. Meeres. 27: 268-282.

225. Westernhagen, H. V., H. Rosenthal, K. R. Sperling. 1974. Combined effects of cadmium and salinity on development and survival of herring eggs. Helgol. wiss. Meeres. 26: 416-433.

226. Whitfield, P. H. and A. G. Lewis. 1976. Control of the biological availability of trace metals to a calanoid copepod in a coastal fjord. Estuar. Coast. Mar. Sci. 4: 255-266.

227. Windom, H., W. Gardner, J. Stephens, and F. Taylor. 1976. The role of methylmercury production in the transfer of mercury in a salt marsh ecosystem. Estuar. Coast. Mar. Sci. 4: 579-583.

228. Wolfe, D. A. 1974. The cycling of zinc in the Newport River estuary, North Carolina. In: Vernberg, F. J. and W. B. Vernberg (eds.). Pollution and physiology of marine organisms. Academic Press, N. Y.: 79-99.

229. Wright, D. A. 1976. Heavy metals in animals from the North East coast. Mar. Poll. Bull. 7: 36-38.

230. Yoshinari, T. and V. Subramanian. 1976. Adsorption of metals by chitin. In: Nriagu, J. D. (ed.). Environmental biogeochemistry. Vol. 2. Metals transfer and ecological mass balances. Ann Arbor Sci. Publ., Ann Arbor, Mich.: 541-555.

231. Young, D. 1974. Cadmium and mercury in the southern California Bight. Summary of findings, 1971 to 1973. S. Calif. Coast. Water Res. Proj., 1500 E. Imperial Highway, El Segundo, Calif.: 16 pp.

232. Young, D. R. and D. J. McDermott. 1975. Trace metals in harbor mussels. Ann Rep. S. Calif. Coast. Water Res. Proj., 1500 E. Imperial Highway, El Segundo, Calif.: 139-142.

233. Young, L. G. and L. Nelson. 1974. The effects of heavy metal ions on the motility of sea urchin spermatozoa. Biol. Bull. 147: 236-246.

234.   Young, M. L.   1975.   The transfer of ^{65}Zn and ^{59}Fe along a *Fucus serratus* (L.) *Littorina obtusata* (L.) food chain. J. Mar. Biol. Assn. U. K. 55: 583-610.

235.   Zaroogian, G. E. and S. Cheer.   1976.   Cadmium accumulation by the American oyster, *Crassostrea virginica*.   Nature 261: 408-409.

236.   Zitko, V. and W. V. Carson.   1975.   Accumulation of thallium in clams and mussels.   Bull. Envir. Cont. Tox. 14: 530-533.

APPENDIX I

Response parameters used to evaluate effects of toxic cations
to marine biota. Groupings are by indicator organisms, by
laboratory studies, and by field investigations.

ALGAE

Laboratory Studies

Survival: Cd (20); Cu (20); Hg (20); Pb (20, 93, 94)
-Modifier:
--resistance of lab strains *vs* wild cells: Pb (20)
Food chain transfer: Cu (181)
Respiration: As (25); Hg (150)
Photosynthesis and Growth: B (12, 117); Cd (20); Co (26);
    Cs (199); Cu (20, 32, 102, 138, 150, 181); Ge (204); Hg
    (20, 52, 53, 107, 150); Pb (20, 93, 94); Tl (150);
    Zn (32, 102, 199)
-Modifiers:
--chemical species: Cu (200)
--prior exposure affects growth: B (12)
--inorganics: Cu (32); Ge (204); Hg (150); Zn (32)
--essential trace: B (117); Co (26)
--developmental stage: Pb (93)
--temperature: Cs (199); Zn (199)
--salinity: Cs (199); Zn (199)
--organics: Cu (200)
--pH: Cu (200)
--species specific: Ge (204)
Uptake, retention, translocation: Cd (44, 211); Cs (199);
    Ce (176); Co (7, 123); Cr (123); Cu (102, 200); Fe (123,
    234); Pb (196); Mn (123, 176); Hg (52); Mo (153); Pu (87,
    88); Rb (153); Ru (87, 88); Ag (164); Tc (87, 88); Zn
    (11, 123, 199, 234)
-Modifiers:
--chemical species: Cd (44)
--particle size: Pu (88); Ru (88)
--growth stage: Cd (44); Fe (234); Zn (234)
--temperature: Mo (153); Rb (153); Zn (199)
--season of year: Zn (234)
--light: Rb (153)
--inorganics: Cd (44)
--organics: Cd (44); Ce (176); Co (123); Cr (123); Fe (123);
    Mn (123); Mo (153); Rb (153); Zn (11, 123).

Biochemical, morphological, and other sublethal effects:
-proteins: Co (26)
-cell volume: Co (26)
-Ca metabolism: Co (26)
-cell structure: Co (26); Hg (107)
-accelerates K release: Cu (150)
-cell permeability: Cu (150)
-non-mutagenesis: Pb (94)
-detoxification mechanisms: Hg (52, 53)
-motility: Pb (93)

Field Studies

Residues: As (27, 116, 154); Cd (65, 91, 116, 124, 193, 194);
      Co (177); Cr (65, 160, 177); Cu (65, 91, 177, 193, 194);
      Fe (65, 177); Hg (86, 91, 116, 188, 193); Mn (65, 177);
      Mo (177); Ni (65, 177); Pb (65, 91, 193, 194); Po (45);
      Pu (144, 183); Sb (116, 198); Ti (177); V (177); Zn (65,
      91, 116, 177, 193, 194, 228)
-Modifiers:
--distance from point source: Cd (91, 116); Hg (91, 188);
      Zn (91)
--relation to water levels: Cd (65, 124); Mn (65); Ni (65);
      Zn (65)
--inorganics: Cr (177)
--age of organism: Cu (91); Zn (91)
--season of year: Cu (91); Zn (91)
--differential uptake of chemical species: Pu (144)
--species differences: Ti (177)
Food chain amplification: Hg (188)
Role in cycling: Zn (228)

ANNELIDA

Laboratory Studies

Survival: Cd (1, 168); Cr (148, 149, 168); Cu (34, 104, 131,
      134, 168); Hg (168); Pb (168); Zn (1, 34, 168)
-Modifiers:
--age of organism: Cd (168); Cr (168); Cu (168); Hg (168);
      Pb (168); Zn (168)
--chemical species: Cr (148, 149)
--salinity: Cu (104)
Uptake, retention, translocation: Ag (34); Am (16); Cd (211);
      Co (6); Cr (148); Cu (134); Pu (16,67)
-Modifiers:
--reflection of substrate: Ag (34); Am (16); Cu (34); Pb
      (34); Zn (34)
--tissue specificity: Co (6)

--organics: Cu (134)
--chemical species: Pu (67)
Food chain transfer: Co (6)
Reproduction: Cr (148, 149)
Behavior:
-tube building: Cr (149)

Field Studies

Residues: Cr (160)
-Modifiers:
--diet: Cr (160)
--reflection of substrate: Cr (160)

BACTERIA

Field Studies

Isolation of resistant strains: Hg (41, 141)
Seasonal abundance: Hg (41)
-Modifier:
--correlated with oxygen: Hg (141)
Role in cycling: Cd (130); Zn (130)

BIRDS

Field Studies

Residues: Hg (111, 145); Se (111)
-Modifiers:
--age: Hg (111)
--inorganics: Hg (111); Se (111)
--diet: Hg (145)

BRYAZOA

Laboratory Studies

Growth and Development: Cd (58); Mg (58)

CHAETOGNATHA

Field Studies

Residues: As (28); Cd (28); Cu (28); Fe (28); Zn (28)

COELENTERATA

Laboratory Studies

Survival: Cu (135)
-Modifier:
--pretreatment sensitizes: Cu (136)
Growth: Cd (136, 191); Cu (136, 191); Hg (136, 191)
Uptake, retention, translocation: As (163)
Respiration: As (25)
Mucous production: Cu (135)
Interference with enzyme staining response: Cd (136); Cu
    (136); Hg (136)
Bacterial interaction: Cu (135)

Field Studies

Residues: Am (143); Bi (143); Co (143); Cs (143); Eu (143);
    Pb (143); Po (143); Pu (143); Rh (143); Sb (143); Sr (143);
    Y (143)
Discrimination among chemical species: Pu (143)

CRUSTACEANS

Laboratory Studies

Survival: B (205); Cd (1); Pb (171); Zn (1)
-Modifiers:
--salinity: B (205); Cd (98)
--temperature: Cd (98)
Uptake, retention, translocation: Ag (164); B (205); Cd (69,
    98); Co (7, 8); Pb (171); Pu (67); Se (70, 71)
-Modifiers:
--tissue specificity: Ag (164); Cd (98); Co (8); Se (71)
--temperature: Cd (69, 98)
--salinity: Cd (98)
--inorganics: Cd (69)
--molting: Pu (67); Se (70)
--chemical species: Pu (67)
Respiration: As (25)
Food chain transfer: Co (6, 7, 8); Pu (67); Se (70, 71)
Protein synthesis: Pb (51)

Field Studies

Residues: Ag (18, 85); As (27, 28, 85, 154); Cd (18, 28, 85,
    124, 193, 194, 197, 229); Ce (152); Cr (85); Cs (152);
    Cu (18, 28, 85, 99, 193, 194, 197, 229); Fe (18, 28);
    Hg (54, 84, 85, 110, 145, 188, 194, 197, 215, 227);

Mn (18, 99); Nb (152); Ni (18, 85, 229); Pb (18, 85, 99,
    193, 194, 197); Po (95); Pu (89, 183); Ru (152); Sb (198);
    Zn (18, 28, 85, 99, 193, 194, 197, 218, 229); Zr (152)
-Modifiers:
--metabolism: As (27)
--diet: As (27)
--reflect substrate levels: As (154); Cd (124); Zn (218)
--age of organism: Cd (18); Po (95)
--tissue specificity: Cs (152); Hg (54); Po (95)
--monthly variation: Cs (152)
--seasonal and latitudinal variations: Cu (99); Hg (54);
    Mn (99); Pb (99); Zn (99)
--yearly fluctuations: Hg (215)
--niche in water column: Hg (110)
--organism complexity: Pu (183)
Food chain transfer: Hg (188)
Role in cycling: Po (95)
Non-hazard to man: Pu (89)

ECHINODERMATA

Laboratory Studies

Survival: Cu (92)
Growth: Cu (92)
Reproduction:
-Essential for successful reproduction: Mg (180)
-Fertilization success: Cu (92)
-Sperm motility: Cd (137); Co (137); Cu (137, 233); Fe (137);
    Hg (137, 233); Mn (137, 233); Ni (137); Sr (137); Zn (137,
    233)
-Sperm motility modifiers:
--chelators: Cu (92); Hg (233); Mn (233); Zn (233)
--nucleoside phosphates: Cd (137); Cu (137); Fe (137); Hg
    (137); Mn (137); Ni (137); Zn (137)

Field Studies

Residues: As (154); Cd (63, 193); Cu (63, 193); Hg (110);
    Mn (63); Pb (193); Pu (183); Zn (63, 193)
-Modifiers:
--distance from point source: As (154)
--feeding niche: Hg (110)
--tissue specificity: As (154)

FISH

Laboratory Studies

Survival: B (205); Cd (139, 214, 224, 225); Cr (139, 186);
    Cu (120); Hg (109, 133, 222); Zn (120, 139)
-Modifiers:
--salinity: B (205); Cd (224, 225)
--chemical species: Cr (139); Hg (133, 156, 157)
--age of organism: Cu (120)
--temperature: Zn (139)
Uptake, retention, translocation: Ag (164); B (205); Be (59);
    Cd (37, 59, 175, 224, 225); Co (59); Cu (59); Hg (37, 133,
    156, 157, 169); Mn (155); Ni (59); Sr (4); Zn (59, 155)
-Modifiers:
--sediment type: Cu (181); Sr (4)
--salinity: Hg (169); Sr (4)
--temperature: Sr (4)
--inorganics: Hg (169); Sr (4)
--mode of administration: Zn (170)
Growth and Development: Cd (175, 224, 225); Cu (181)
-Modifiers:
--inorganics: Cd (175)
--organics: Cd (175)
--chemical species: Cd (175)
--salinity: Cd (224, 225)
Behavior:
-sluggishness: Hg (109)
-uncoordinated movements: Hg (109); Zn (139)
-negative phototaxy: Hg (109)
-feeding response: Cr (186)
-migration: Cu (120)
-pigmentation change: Zn (139)
Histopathology: Ag (78); Cd (78); Cu (78); Hg (78); Zn (78)
Respiration: As (25); Cd (37, 214); Hg (37, 109)
-Modifier:
--salinity: Cd (214)
Heartbeat rate: Cd (224)
Fin erosion and regeneration: Cd (222); Pb (222); Zn (139)
Food chain transfer: Cu (181); Mn (155); Hg (156); Zn (155,
    170)
Blood chemistry: Cd (37, 115); Hg (37)
Enzymes: Cu (120); Hg (169)
Hepatic binding proteins: Cd (147)

Field Studies

Residues: Ag (85, 127); Al (127); Am (158); As (27, 28, 85,
    108, 116, 154); Cd (28, 33, 63, 85, 90, 116, 127, 193,

194, 197, 201, 229, 231); Ce (152); Cr (33, 85, 127, 160); Cs (152); Cu (28, 33, 63, 85, 127, 193, 194, 197, 201, 229); Fe (28, 33, 127); Hg (13, 46, 54, 61, 77, 84, 85, 110, 116, 145, 146, 165, 187, 190, 193, 194, 197, 201, 231); Mn (33, 63); Nb (152); Ni (33, 85, 127, 229); Pb (3, 33, 39, 85, 90, 127, 194, 197, 201); Po (96); Pu (89, 142, 158, 183); Ru (152); Sb (116, 198); Ti (127); Zn (28, 33, 63, 85, 90, 116, 127, 193, 194, 197, 201, 228, 229); Zr (152)

-Modifiers:

--tissue specificity: Al (127); Cd (33); Cr (127); Cu (63); Fe (127); Hg (13); Ti (127)

--associated with fin damage: Al (127); Cd (127); Cr (127); Fe (127); Ti (127)

--associated with changes in body elemental composition: Cd (127); Cu (33); Fe (33); Mn (33); Zn (33)

--seasonal variations: Am (158); Hg (85); Pu (158)

--age or size of fish: As (27, 28); Cd (28); Cu (28, 33); Fe (28, 33); Hg (54, 77, 145, 201); Mn (33); Zn (28, 33, 90)

--metabolism of fish: As (27)

--diet: Cd (90); Cu (63); Hg (110, 145, 165); Pb (90); Po (96); Pu (89); Zn (63, 90)

--reproductive state: Cu (64); Zn (64)

--sediment substrate: Hg (54, 110)

--depth of water column: Hg (85)

--geothermal activity: Hg (187)

--sex: Hg (54)

--distance from point source: Hg (146)

--geographic variations: Hg (85); Pb (3)

--chemical species: Pu (183)

--salinity: Cu (64); Zn (64)

Food chain amplification: As (108); Hg (190)

Contamination during canning process: Pb (39)

Role in cycling: Zn (228)

HIGHER PLANTS

Laboratory Studies

Survival: Cu (57); Pb (57)
Growth: Cu (57); Pb (57)

Field Studies

Residues: Cd (57); Co (177); Cr (177); Cu (57, 177, 194); Fe (177); Hg (118, 209, 227); Mn (177); Mo (177); Ni (177); Pb (15, 57); Ti (177); V (177); Zn (177, 194)

-Modifier:
--reflection of substrate: Cr (177); Fe (177); Mn (177); Ti
    (177)
Growth: Cd (132); Pb (132); Zn (132)
Food chain: Hg (118)
Role in cycling: Cd (57); Cu (57); Hg (209, 227); Pb (15,
    57); Zn (228)

INSECTA

Field Studies

Residues: Ag (18); Cd (18); Cu (18); Fe (18); Hg (146); Mn
    (18); Ni (18); Pb (18); Zn (18)
-Modifiers:
--time: Fe (18)
--feeding substrate: Hg (146)
--diet composition affects retention time: Hg (146)
--growth stage: Ni (18)

MAMMALS

Field Studies

Residues: Ag (125); As (9); Cd (9, 35, 97, 125, 147, 172);
    Cu (97, 125); Fe (125); Hg (9, 35, 62, 79, 97, 103, 111,
    125, 146, 172); Mg (125); Mn (125); Pb (9, 97, 172); Se
    (111, 125); Zn (97, 125)
-Modifiers:
--tissue specificity: Cd (9, 125, 147, 172); Cu (125); Hg (9,
    35, 97, 103, 125, 172); Mn (125); Pb (9, 172); Zn (97)
--age of animal: Cd (172); Cu (125); Hg (9, 97); Mn (125)
--sex: Hg (103)
--sexual condition: Cd (125)
--tectonic activity: Hg (79)
--starvation: Hg (103)
--chemical species: Hg (79)
Health hazard: Hg (62, 146)
Reproduction: Cd (125); Cu (125); Hg (125); Mn (125); Se
    (125)
Hg detoxification aid: Se (111)

MOLLUSCA

Laboratory Studies

Survival: Ag (36, 140, 207); As (140); B (105); Cd (1, 140);
    Cu (24, 122, 195); Hg (36, 112, 140); Ni (36); Pb (36);
    Zn (1, 31, 36)

-Modifiers:
--salinity: Ag (207); Cu (122)
--age of organism: Cu (122)
--temperature: Cu (122)
--season: Cu (122)
Respiration: Ag (206, 207); Cd (140)
-Modifiers:
--salinity: Ag (207)
Uptake, retention, accumulation: Ag (164, 206, 207); As (140);
    B (205); Cd (140, 161, 235); Co (7, 8); Cu (23, 24, 161,
    181); Hg (42, 43, 47, 48, 66, 112, 140, 182); Fe (80); Pb
    (161, 184, 196); Pu (67); Ru (100); Se (70, 71); Tl (236);
    Zn (49, 161)
-Modifiers:
--temperature: Cd (69, 161); Cu (161); Hg (47, 112); Pb (161);
    Se (71); Zn (161)
--salinity: Cd (161); Cu (161); Pb (161); Zn (161)
--inorganics: Cd (69, 161); Cu (161); Hg (66, 112); Pb (161);
    Se (171); Zn (161)
--tissue specificity: Co (7, 8); Cu (23); Hg (47, 48); Fe
    (80); Pb (184, 196); Pu (67); Ru (100); Se (70, 71)
--organics: Cr (14); Cu (14); Hg (14); Pb (14); Zn (14)
--chemical species: Cr (14); Cu (14); Hg (14, 48, 112); Se
    (71); Zn (14)
--role of bacteria: Hg (42, 43, 182)
--age of organism: Hg (47); Pb (184); Se (71)
--reproductive stage: Hg (47)
--particulates: Fe (234)
--mode of administration: Pb (184)
--initial concentration: Pb (184)
Reproduction:
-gametogenesis: Cu (122)
-birth defects: Hg (167)
-larval settling: Zn (31)
Behavior:
-valve movements: Ag (206)
-burrowing ability: Cu (195)
--modified by chelators: Cu (195)
-feeding rate: Hg (55)
Histology: Cu (24); Hg (66); Fe (80)
Growth: Cu (122, 181); Pb (196); Zn (31)
Health hazards: Cd (235); Hg (112)
Food chain transfer: Cd (69); Co (7, 8); Cr (14); Cu (14,
    181); Fe (234); Hg (14, 48); Pb (14, 184, 196); Zn (14,
    234)
Seasonal dynamics: Cd (75); Co (75); Cr (75); Cu (75); Fe
    (75); Mn (75); Ni (75); Pb (75); Zn (75)

Field Studies

Residues: Ag (2, 10, 74, 86, 126, 219, 220); As (27, 116, 121,
     154); Bi (220); Cd (10, 29, 63, 74, 76, 86, 116, 121, 126,
     161, 162, 165, 192, 193, 194, 203, 212, 219, 220, 231); Ce
     (152); Co (74, 220); Cr (2, 10, 74, 160, 220, 232); Cs
     (152); Cu (2, 10, 22, 29, 30, 56, 63, 74, 76, 86, 121, 126,
     161, 162, 165, 174, 192, 193, 194, 219, 220, 232); Fe (29,
     74, 76, 126, 174, 219, 220); Hg (10, 77, 86, 110, 116, 193,
     194, 227, 231); Mn (63, 74, 76, 194, 219, 220); Ni (2, 10,
     29, 74, 194, 219, 220, 232); Pb (2, 10, 29, 40, 74, 86,
     121, 161, 162, 193, 194, 212, 220); Pu (183); Ru (152);
     Sb (116, 198); Zn (2, 10, 29, 30, 56, 63, 74, 76, 86, 116,
     121, 126, 161, 162, 165, 174, 192, 193, 194, 212, 219,
     220, 232)
-Modifiers:
--distance from point source: Ag (2); As (154); Cd (165); Cr
     (2); Cu (2, 165, 232); Hg (10); Pb (2,40); Zn (165, 232)
--latitude: Ag (10); Cu (10)
--seasonal variations: Ag (74); Cu (22, 74, 76)
--inorganics: Ag (126); Cd (126); Fe (126); Zn (126)
--age of organism: Ag (220); As (121); Bi (220); Cd (29, 121,
     220); Co (220); Cr (220); Cu (29, 121, 220); Fe (29, 220);
     Mn (220); Ni (29, 220); Pb (29, 121, 220); Zn (29, 121,
     220)
--body weight: As (121); Cd (121); Cu (29, 121)
--sex: Ag (220); Bi (220); Cd (220); Co (220); Cr (220); Cu
     (220); Fe (29, 220); Mn (220); Ni (220); Pb (220); Zn
     (220)
--tissue specificity: As (116); Cd (86, 116); Cu (86); Hg
     (116); Sb (116); Zn (116, 218)
--water depth: Cd (161); Cu (161); Pb (161); Zn (161)
--reflects substrate composition: Cd (76, 192); Cu (76, 192);
     Fe (76); Mn (76); Zn (76, 192)
--seasonal variations: Cd (74, 76); Co (74); Cr (74); Fe
     (74, 76); Mn (74, 76); Ni (74); Pb (74); Zn (74, 76)
--particulates: Ce (152); Cs (152); Ru (152)
--diet: Cr (160); Hg (110)
--contamination from test pump: Cu (30); Zn (30)
Growth: Cd (212); Pb (212); Ra (210); Zn (212)
Role in cycling: Hg (227)

NEMATODA

Laboratory Studies

Uptake: Pb (171)

PLATYHELMINTHES

Laboratory Studies

Uptake: Pb (171)

PROTOZOA

Laboratory Studies

Reproduction: Cd (159); Cu (159); Hg (159); Pb (159); Zn
    (159)
Growth: Hg (82); Pb (82); Zn (82)
-Modifier:
--inorganics: Hg (82); Pb (82); Zn (82)

TUNICATA

Laboratory Studies

Uptake, retention, accumulation: V (113, 129)
-Modifiers:
--tissue specificity: V (113)
--organics: V (129)
--inorganics: V (129)

Field Studies

Residues: As (116); Cd (63, 116); Cu (63); Hg (116); Mn (63);
    Sb (116); V (202); Zn (63, 116)
-Modifiers:
--chemical species: V (202)

EFFECTS OF MERCURY, CADMIUM, AND LEAD COMPOUNDS
ON REGENERATION IN ESTUARINE FISHES AND CRABS

Judith S. Weis[1]
Peddrick Weis [2]

INTRODUCTION

Pollution of the coastal marine environment has received
much attention and is of great concern. Biological research
efforts have been devoted both to the establishment of lethal
levels of various toxicants to different species of aquatic
organisms, and to the delineation of effects of sublethal
levels of pollutants on biochemical, physiological, developmental and behavioral functions. One aspect of our work has
been directd toward determination of effects of heavy metals
on developmental processes in estuarine fishes and crabs.
We have studied embryonic development of the commonly found
killifish, in the presence of inorganic mercury, and methylmercury and have reported on the teratogenic effects of these
metals (Weis and Weis, 1977a; 1977b). Another aspect of
development which we have investigated and on which this paper
will focus, is regeneration. Regeneration is a process whereby
animals can replace appendages or part of the body following
amputation (or autotomy). The removal of an appendage exposes
tissues to injury and initiates wound healing. Cells then
undergo dedifferentiation and proliferation, followed by
re-differentiation and morphogenesis appropriate to the structure being regenerated. The rate of growth and morphogenesis
can be altered by environmental conditions. Regeneration is
also dependent on physiological factors, distant from the
amputation site. These factors, necessary for regeneration,

[1]Department of Zoology and Physiology, Rutgers-The State
University of New Jersey, Newark, New Jersey
[2]Department of Anatomy, New Jersey Medical School, Newark,
New Jersey

may be neural or hormonal in nature, and can also be affected
by environmental conditions. Regeneration, therefore, can
also be used as a parameter of an animal's physiological
response to pollutants.

HEAVY METALS AND REGENERATION OF FISH FINS

Fish fins are composed rays, or lepidotrichia, separated
by connective tissue. The lepidotrichia, which are segmented,
are each composed of two demi-rays which, in cross section,
fit together like a pair of parentheses (). Through the
hollow center run blood vessels and nerves. At the distal
end of each ray are collagenous fibers known as actinotrichia,
and at the proximal end each ray is attached to a basal bone
embedded in the body musculature. When a fin is cut, cells
from the bony rays and connective tissue dedifferentiate to
form a blastema. The blastema then gives rise to new lepidotrichia which become attached to the ends remaining in the
stump (Kemp and Park, 1970). The rate of regeneration is
proportional to the amount of fin removed (Tassava and Goss,
1966) and is dependent on an adequate nerve supply (Goss and
Stagg, 1957). It is affected by such factors as the age and
size of the fish (Comfort and Doljanski, 1958) as well as
water quality of the environment.

In studying the effects of mercury, cadmium and lead salts
on fin regeneration (Weis and Weis, 1976; 1978), fish (killifish
or mullets) were collected by seining and the lower portion
of each caudal fin was amputated with a scalpel about 1-2 mm
from the base. This procedure did not severely impair the
swimming ability of the fish as would have happened had the
entire tail fin been amputated. Fish of comparable size
(4-5 cm) were kept in groups of 14 in 20-gallon all-glass
aquaria in seawater (30o/oo) at approximately 25^0C which
received either lead nitrate, anhydrous cadmium chloride
or mercuric chloride at 0.01, 0.1, or 1.0 mg/l, or methylmercuric chloride at 0.001 or 0.01 mg/l dissolved first in
0.2% $NaHCO_3$. Controls for the methylmercury experiments were
also dosed with $NaHCO_3$. Aquaria were washed, filled with new
water, and re-dosed twice a week. Fish were fed Tetramin[R]
prior to changing the water. Regenerates were measured
twice weekly with a calibrated ocular micrometer in a stereo-
microscope, and lengths of regenerates compared by Students
t-test for independent observations. Mercuric chloride at
0.1 mg/l caused severe mortality in that more than half the
fish were dead by the end of one and one-half weeks. However,
even at the time of death the amount of fin regenerated was
comparable to controls. Lead had no effect on regeneration

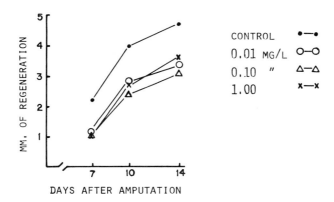

FIGURE 1. Effects of cadmium chloride on caudal fin
regeneration in Fundulus heteroclitus. Reprinted from Weis
and Weis (1976), courtesy of Springer-Verlag.

and was not toxic to the killifish at the concentrations used
in our studies. Cadmium, on the other hand, showed dramatic
effects on regeneration in *F. heteroclitus,* especially on
the early stages of wound healing and blastema formation,
which were greatly retarded. For one-half a week after the
other groups of fish had commenced blastema formation, the
fish in cadmium chloride at 0.01, 0.1, and 1.0 mg/1 remained
with bloody stumps. Once healing had occurred and delayed
regeneration had commenced, it proceded at a rate comparable
to controls (Figure 1). No dose-response relationship was
seen. In view of several recent reports of fin rot of unknown
etiology in benthic flatfishes living in polluted waters
(Ziskowski and Murchelano, 1975; Wellings *et al.,* 1976), we
suggest that cadmium be investigated as an agent delaying
healing of fins that are eroded by the benthic substrate. In
a recent report, Dixon and Compher (1977) have found similar
inhibition of regeneration in the newt by cadmium, and have
noted that cadmium-treated limbs were also erythematous due
to ruptured capillaries, a condition close to that observed
in the fish. Previous reports that cadmium reduces oxygen
consumption in fish (Thurberg and Dawson, 1974), damages the
gills (Gardner and Yevich, 1970) and reduces body growth
(Cearly and Coleman, 1974) may be related to the delay in
regenerative growth.
     Methylmercury produced a significant retardation of growth
in mullet *(Mugil cephalus)* at 0.01 and 0.001 mg/1 in diluted
seawater (8-9o/oo) (Figure 2) (Weis and Weis, 1978). Analysis
of mercury uptake showed it to be dose-dependent. Killifish
(F. confluentus) exposed to 0.01 mg/1 in seawater (36o/oo)
regenerated more slowly than controls, but at a rate comparable

to controls and methylmercury-exposed fish in diluted water
(8-9o/oo) (Figure 3). The more rapid growth in full-strength
seawater is consistent with other reports of more rapid growth

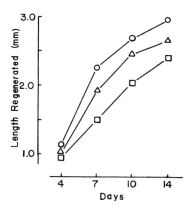

FIGURE 2. Length of caudal fin regenerated by Mugil
cephalus *treated with methylmercury in diluted seawater
(8-9o/oo salinity), Control = O-O; 0.001 mg/l = Δ-Δ;0.01 mg/l=
□-□). Reprinted from Weis and Weis (1978) courtesy Academic
Press.*

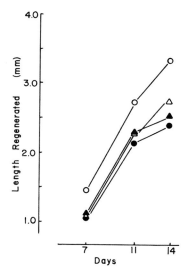

FIGURE 3. Length of caudal fin regenerated by Fundulus
confluentus *treated with methylmercury in diluted seawater
(8-9o/oo salinity) and in full-length seawater (36o/oo).
Seawater control = O-O; Seawater, treated = ●-●; Diluted
water control = Δ-Δ; Diluted water treated = ▲-▲. Reprinted
from Weis and Weis (1978) courtesy Academic Press.*

of fish in more saline environments (Gibson and Hirst, 1955;
Cangaratnam, 1959; Holliday and Blaxter, 1960; Alderdice and
Forrester, 1968). The reduction of regenerative growth by
methylmercury is consistent with some of its other effects on
organisms. It is a mitotic inhibitor (Ramel, 1969) and has
been shown to be teratogenic in a variety of organisms
(Gilani, 1975; Spyker and Smithberg, 1972b). Chang, Reuhl
and Dudley (1973) reported arrest of development in tadpoles,
and Christensen (1975) reported reduction of general body
growth of brook trout alevins treated with methylmercury.
However, Chang, Mak and Martin (1976) found no retardation
of regeneration rate in newts treated with methylmercury,
although there was some delay in initiation of regeneration
at some of the concentrations they used.

The observation that methylmercury but not organic mercury retarded fin growth is not surprising. Renfro *et al.*
(1974) have reported that while mercuric chloride is incorporated primarily into the gills, methylmercury is more mobile
and concentrates in the liver and kidney. Organic mercury,
furthermore, accumulates more rapidly than mercuric chloride
in fish (Middaugh and Rose, 1974). Fox *et al.* (1975) found
that methylmercury but not inorganic mercury depresses the
metabolic rate of nervous tissue. Methylmercury as a potent
nerve poison could further inhibit regenerative growth by
interference with the trophic influence of the nervous system
on regeneration.

The lack of effect of methylmercury on killifish in diluted
seawater is puzzling, especially since reports have indicated
increased toxicity of heavy metals to aquatic organisms at
lower salinity (Eisler, 1971; Vernberg and Vernberg, 1972).
We can only suggest that the effect of the methylmercury was
masked by the already slower growth at low salinity. It was
probably not due to differential uptake, since unpublished
observations on *F. heteroclitus* indicated equaivalent uptake
in 9 and 36o/oo salinities.

To study effects of combinations of methylmercury and
cadmium chloride, killifish (*F. confluentus*) were exposed to
0.025 mg/l of cadmium chloride, 0.025 mg/l of methylmercury,
0.025 mg/l of one plus 0.005 mg/l of the other, or to 0.025
mg/l of both simultaneously (Weis and Weis, 1978). The data
(analyzed by Duncan's multiple range test) indicated that
while cadmium or methylmercury alone retarded regeneration,
the effect was diminished when both were present (Table 1).
This antagonistic action in most cases reduced the retardation
to insignificance.

Other reports have indicated interference of heavy metals
with each other. For example, Roales and Perlmutter (1974)
have found that copper is antagonistic to the toxic action
of methylmercury in fish. Kim, Birks, and Heisinger (1977)

TABLE 1. Mean fin regeneration in mm at 14 days of Fundulus confluentus exposed to combinations of methylmercury and cadmium chloride in seawater. The condentrations of toxicants are in mg/l. F-values were derived by one-way analysis of variance. Reprinted from Weis and Weis (1978) courtesy Academic Press

Run	Controls	Hg 0.025	Cd 0.025	Hg 0.025 Cd 0.005	Hg 0.005 Cd 0.025	Hg 0.025 Cd 0.025	d.f.	F-ratio	F-probability
1	3.68	2.65[a]	3.41	3.50	3.43	2.80[a,b]	5,55	15.99	0.001
2	3.22	2.82[a]	2.88[a]	3.25	3.04	3.09	5,61	4.62	0.001
3	3.55	3.22[a]	3.28[a]	3.52	3.36	3.63	5,53	2.94	0.02

[a] Excluded by Duncan's multiple-range test from a homogeneous subset which includes all the other groups in the same run, at a significance level under 0.05.
[b] This figure may not be realistic due to a high variance ($S^2 = 0.21$).

have shown a protective action of selenium against methylmercury
in fish and Eaton (1973) has found a reduction in toxic effects
of cadmium to the fathead minnow when copper was present. Dixon
and Compher (1977) have observed a protective action of zinc
against the cadmium-caused inhibition of newt limb regeneration.
However, Eisler and Gardner (1973) have found synergistic
effects of cadmium, copper, and zinc in fish.

HEAVY METALS AND LIMB REGENERATION IN FIDDLER CRABS

    Brachyuran crabs can autotomize limbs at a pre-formed
breakage plane in the basi-ischium, from which a limb bud
subsequently regenerates. The regenerating limb buds grow in
a folded position within a layer of cuticle, and unfold when
the animal molts. The length of regenerating limb buds is
generally expressed in terms of "R" value" (Bliss, 1956),
which is:

length of limb bud X 100.
carapace width

Such a regeneration index is useful for comparisons of crabs
of different sizes. The regenerative period can be divided
into stages. After an initial lag period basal growth occurs
in which the basic limb structures are formed. This may be
followed by a plateau of anecdysis when the limb is approximately half grown. The second growth phase is one of rapid
growth, correlated with rising titers of ecdysones probably
produced by the Y-organ (Passano and Jyssum, 1963; Rao *et al.*,
1972; Chang and O'Connor, 1977) and culminating in ecdysis
and the unfolding of the limb. There is, sometimes, a terminal plateau just prior to ecdysis.
    Since regeneration always terminates with a molt, the pre-
sence of regenerating limbs can affect the timing of ecdysis,
and factors which influence ecdysis will also affect regeneration. For example, removal of eyestalks, a source of molt-
inhibiting hormones from the X-organ, is a standard way of
inducing precocious molting. Animals with missing eyestalks
regenerate missing limbs rapidly and go directly from basal
to proecdysial growth without a plateau, but will generally
die at ecdysis. Skinner and Graham (1972) have shown that
multiple autotomy, producing many regenerating limbs, can
cause accelerated regeneration also leading to precocious molt.
    The effects of inorganic salts of lead, cadmium, and
mercury, and of methylmercury on regeneration and molting in
the fiddler crab, *Uca pugilator,* have been performed by (Weis,
(1976a, 1977). Fiddler crabs, common estuarine intertidal

organisms, are subject to heavy metal pollution from industrial waste effluent and have been found to accumulate methylmercury from salt marsh ecosystems (Windom *et al.*, 1976). Multiple autotomy was induced by pinching the merus (of one chela and six walking legs) with a hemostat. Crabs were then placed in solutions of lead nitrate, mercuric chloride or anhydrous cadmium chloride at concentrations of 0.1 or 1.0 mg/l of the metal ion, or to 0.5 mg/l of methylmercury in shallow containers of filtered seawater at about 25^0C. The containers were washed twice weekly and re-dosed. Prior to changing the water crabs were fed Purina "Fly Chow". In all experiments, groups, which consisted of 10 or more crabs, were arranged to have the same mean carapace width and sex ratio. It was not necessary to "synchronize" each animal's stage in the molt cycle prior to limb removal since most adult crabs are in inter-molt and the multiple autotomy itself acts as a synchronizer. The very small percentage of crabs which failed to begin regeneration were removed from the experiment. Any crab which was in proecdysis at the time of limb removal would molt shortly thereafter but without forming any limb buds and was not considered in the data.

Limb buds were measured twice a week under a dissecting microscope with a calibrated ocular micrometer. In all cases the bud of the first walking leg was measured as a representative limb bud. Values were converted to R values, and the mean for each group compared by use of a t-test. Times of ecdysis were recorded for all animals. Limb buds reached R-values of about 20 just prior to ecdysis, and smaller crabs completed molting sooner than larger individuals. Cadmium at 0.1 mg/l had an inhibitory effect on regeneration and molting (Table 2), though most individuals did molt within one week after controls, while at 1.0 mg/l there was an even greater effect (Table 3). Mercuric chloride was toxic at 1.0 mg/l and those crabs (40%) which survived the experiment showed negligible growth. When returned to clean water they showed no evidence of recovery after one month. Lead had no effect on regeneration at either concentration, although residue analysis of crabs held at 0.1 mg/l revealed that the crabs had absorbed 2 ppm of lead, and only 0.5 ppm cadmium and 0.026 ppm mercury after two weeks of exposure. The greater effect of cadmium than mercury at 0.1 mg/l may have been due to the greater absorption, although lead, which accumulated to the greatest degree, had no effect.

Cadmium has similarly been found to inhibit molting in the grass shrimp, *Palaemonetes pugio* (Vernberg *et al.*, 1977), and to delay larval development of the blue crab, *Callinectes*

TABLE 2. R-values (mean ± standard error) of first walking legs of crabs after multiple autotomy and treatment with Pb, Hg and Cd at 0.1 mg/liter

Chemical	Days after limb removal				
	7	10	14	17	21
Carapace width 15 mm:					
Controls	1.8 ± 0.3	7.3 ± 0.6	12.0 ± 0.7	18.4 ± 0.8	60% molt
Pb	2.8 ± 0.5	10.2 ± 0.7	14.7 ± 1.1	20.1 ± 0.6	80% molt
Hg	2.3 ± 0.2	8.8 ± 1.2	13.8 ± 1.3	17.7 ± 0.8	70% molt
Cd	1.0 ± 3.3	3.3 ± 0.7*	8.6 ± 1.2	11.0 ± 1.3*	13.5 ± 1.5
Carapace width 13 mm:					
Controls	4.8 ± 0.4	10.6 ± 1.0	17.7 ± 1.0	20.2 ± 0.7 / 40% molt	70% molt
Pb	4.2 ± 0.4	9.2 ± 0.9	17.7 ± 0.8	18.2 ± 0.9 / 50% molt	90% molt
Hg	3.9 ± 0.7	9.4 ± 0.9	16.2 ± 1.1	17.9 ± 0.5 / 30% molt	60% molt
Cd	3.5 ± 0.6	8.0 ± 1.0	14.2 ± 1.1*	17.0 ± 0.8* / 0% molt	70% molt

*p = 0.05 or less. Reprinted courtesy Fish. Bull. U.S.

TABLE 3. R-values (mean ± standard error) of first walking legs after multiple autotomy and treatment with Pb, Hg, and Cd at 1.0 mg/liter   Reprinted courtesy Fishery Bull. U.S.

Chemical	Days after limb removal					
	7	10	14	17	21	24
Carapace width 15 mm:						
Controls	4.2 ± 0.4	8.0 ± 0.6	13.1 ± 1.0	15.9 ± 0.9	18.1 ± 0.3	70% molt
Pb	2.8 ± 0.6	6.2 ± 0.7	11.4 ± 1.0	14.5 ± 1.2	17.6 ± 0.7	70% molt
Hg	0*	0*	0*	0*	0.01 ± 0.01*	0.01±0.01*
Cd	0.3 ± 0.2*	2.2 ± 0.8*	4.3 ± 1.2*	5.6 ± 1.5*	8.3 ± 2.5*	7.6 ± 2.3*
						20% molt
Carapace width 13 mm:						
Controls	4.6 ± 0.5	10.2 ± 0.7	15.7 ± 0.9	18.0 ± 0.6	70% molt	90% molt
Hg	1.0 ± 0.6	1.5 ± 0.8	1.6 ± 0.8	[a]2.1 ± 1.0	2.7 ± 1.1*	2.9 ± 1.1*
Cd	3.5 ± 0.2	6.8 ± 0.6	11.5 ± 1.3	16.0 ± 2.0*	16.0 ± 2.0*	50% molt

[a]Returned to clean water.

*p = 0.05 or less.

sapidus, (Rosenberg and Costlow, 1976). Since cadmium reduces oxygen consumption in crustaceans (Collier *et al.,* 1973), reduced metabolic rate could be associated with the retardation of growth and molting.

Methylmercury retarded regeneration and molting in fiddler crabs at 0.5 mg/l (Weis, 1977). At 0.1 mg/l the regeneration rate was unaffected, but another effect was seen: absence of melanin pigment in the regenerates. The lack of pigmentation was believed to be due to inhibition of melanin synthesis rather than inhibition of cell migration, since when the crabs were returned to clean water after ecdysis, melanocytes appeared. They were initially very pale and gradually became darker. The cells were scattered evenly throughout the limb, showing no proximo-distal gradient, thereby indicating delayed pigment synthesis rather than delayed cell migration. Old limbs and carapace of methylmercury-exposed crabs also appeared paler than controls, reflecting inhibition of pigment synthesis in them as well as in regenerating structures. This effect of methylmercury was not seen in developing fish embryos nor in regenerating fish fins, and, to our knowledge, has not been reported in other organisms.

The growth retardation observed cannot be called a specific effect on regeneration, but one on regeneration and ecdysis, since most animals with retarded regeneration delayed ecdysis until regeneration was complete. A major effect of methylmercury as a nerve poison might be on the neuroendocrine system which controls the molt cycle. The retarding effect of methylmercury on growth and ecdysis is in keeping with the results of Shealy and Sandifer (1975) who found a delay of molting and extended development time in larval grass shrimp exposed to inorganic mercury. However, Green *et al.* 1976) found no effect of mercury at 0.5 and 1.0 ppb (much lower levels) on molting in *Penaeus setiferus.* As Calabrese *et al.* (1977) point out, retardation of larval growth can prolong the pelagic stages of organisms, thereby increasing the risk of predation, disease, and dispersal. Growth retardation also may be related to depressed metabolic rate of fiddler crabs exposed to mercury (Vernberg and Vernberg, 1972). Methylmercury is accumulated more rapidly and lost more slowly than inorganic mercury in shrimp and mussels (Fowler, Heyraud and LaRosa, 1976).

Methylmercury retarded regeneration to a greater extent in animals maintained at 8-9 o/oo salinity than those at 36 o/oo (Fig. 4). While controls at 36 o/oo seawater molted somewhat sooner than controls at the lower salinity, experimental crabs at 8-9 o/oo had taken up more mercury (2 ppm) than those at 36 o/oo (0.8 ppm). A similar increase in severity of effects at lower salinity was observed in *U. minax,* a species which normally inhabits water of low salinities (Weis, in prep.).

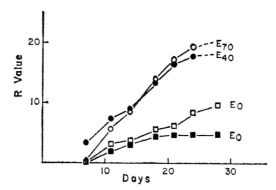

FIGURE 4. *Growth of regenerates of the first walking leg of 12 mm* Uca pugilator *in 36o/oo seawater (open circle), 9o/oo salinity (solid circle), 0.5 mg/l methylmercury in 36o/oo (solid square). "E" with a subscript indicates the time at which that percentage of crabs had undergone ecdysis. From Weis (1977). Reprinted courtesy of Biological Bulletin.*

The greater inhibition produced by methylmercury at lower salinity is consistent with the findings of Vernberg and Vernberg (1972) who demonstrated that fiddler crabs were more susceptible to mercury at lower salinities, and Vernberg and O'Hara (1972) who noted that the crabs take up more mercury at the gills in lower salinity. Low salinity similarly increased cadmium toxicity to larvae of the mud crab *Rhithropanopeus harrisi* (Rosenberg and Costlow, 1976), as well as cadmium uptake and toxicity to fiddler crabs (O'Hara 1973a, 1973b).

When crabs were subjected to cadmium and methylmercury combinations in seawater (30o/oo) the antagonistic effect seen in the fish did not occur. Instead, the retarding effects of the two metals were additive in that doubly exposed animals regenerated more slowly than those exposed to either toxicant alone (Figure 5). When these data were subjected to analysis of variance, highly significant effects of mercury and cadmium were revealed, but with no significant interaction of the two, thereby reflecting additivity. Since both metals reduced the metabolic rate, this common action may have been responsible for the additivity observed.

IMPLICATIONS

The organisms which have been used in most of this work, killifish and fiddler crabs, are important members of our coastal and estuarine environment. They are easy to collect, handle, and maintain in the laboratory. They are, for these reasons,

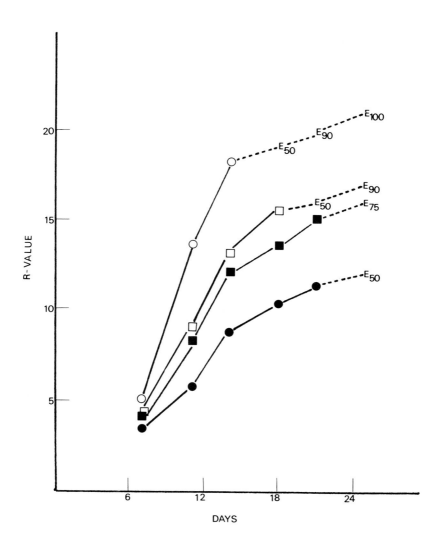

FIGURE 5. *Growth of regenerates of first walking legs of 15 mm Uca pugilator in 30o/oo salinity seawater (open circle), 1.0 mg/l cadmium (open square), 0.5 mg/l methylmercury (solid square), or 1.0 mg/l cadmium plus 0.5 mg/l methylmercury (solid circle).*

ideal experimental animals, and information gained from them
may indicate how other animals in our coastal food web may
react to change in their environment. Since they are hardy
animals, however, it is likely that lower levels of toxicants
would effect animals which are less tolerant.

The static systems used here exposed the organisms initially
to the proper concentration of the toxicant with a subsequent
decrease due to adsorption onto the container and absorption into the animals during the three or four days between water
changes. A constant-flow system might have shown more severe
effects at any given concentration, since the animals would have
been continuously exposed to the proper level of the toxicant.

Regenerative ability varies widely throughout the animal
kingdom. The ability to regenerate lost appendages clearly
conveys an advantage to an animal. This ability, however, is
not without drawbacks: a regenerating appendage remains relatively unprotected and vulnerable to attack or infection
for a prolonged period of time. Most terrestrial vertebrates
and adult terrestrial arthropods lack the ability to regenerate appendages, but, instead, have a rapid wound healing
process, thereby reducing the chances of abrasion, infection,
and attack. In an animal which does have the ability to
regenerate appendages, it is adaptive to replace the missing
part as rapidly as possible. In the case of fish with parts
of fins missing, the coordination of swimming could be adverse-
ly affected and it is therefore advantageous to replace the
lost fin rapidly. In the case of crabs, the autotomy reflex
is clearly of adaptive significance in avoiding predation.
The loss of one or two walking legs is probably of little
significance to a decapod, but should a crab lose several
legs (such crabs have been found, though rarely; McVean, 1976,
has noted cumulative limb losses in *Carcinus maenas*) the loco-
motor ability would be impaired. It has been found that the
rate of regeneration is proportional to the number of limbs
missing (Fingerman and Fingerman, 1974). The loss of feeding
claws would reduce the ability to eat, and the loss of a
male's major chela would reduce the ability to attract a mate.
For all these reasons, it is beneficial to replace lost appendages in a short period of time.

The presence of heavy metal pollutants, in addition to all
their other deleterious effects, places animals in a disadvan-
tageous position by slowing down their speed of regeneration.
If, as we have suggested, a prolonged period of regeneration
makes an animal more subject to predation because of impaired
locomotion, this not only places the animal at an ecological
disadvantage, but also serves as a means of passing on more
of the toxicant to higher levels in the food chain.

SUMMARY

    Fin regeneration in the killifish, *Fundulus heteroclitus*
was retarded by exposure to $CdCl_2$ at 0.01 and 0.1 mg/l. $HgCl_2$
had no effect at levels which were sublethal.   However,
regeneration in *F. confluentus* was retarded by exposure to
methylmercuric chloride at 0.01 and 0.05 mg/l, but the effect
was reduced to insignificance when the salinity of the water
was reduced or when $CdCl_2$ was also present in the water.
    Regeneration of limbs in the fiddler crab, *Uca pugilator*
was studied after exposure to heavy metals following multiple
autotomy. $CdCl_2$ at 0.1 and 1.0 mg/l retarded the regeneration
rate but $HgCl_2$ inhibited regeneration only at 1.0 mg/l; this
concentration proved lethal to many crabs.   Methylmercury,
however, severly inhibited regeneration at 0.5 mg/l and caused
inhibition of melanogenesis in regenerates at 0.1 mg/l.   Combinations of methylmercury and $CdCl_2$ were additive in sea
water.

ACKNOWLEDGMENTS

    This research was supported in part by NIH Biomedical
Grant #RR7059 and by a grant #04-7-158-44042 from NOAA Office
of Sea Grant, Department of Commerce.   Part of the work was
performed at the New York Ocean Science Laboratory, Montauk,
New York and part at the Department of Biological Sciences
at Florida Atlantic University, Boca Raton, Florida in addition to the investigator's home institutions.   Thanks are
extended to Ms. Rhett Virbickas for technical assistance and
to Jennifer and Eric Weis for assistance in collecting the
organisms.

LITERATURE CITED

Alderdice, D. F. and C. R. Forrester.   1968.   Some effects of
    salinity and temperature on early development and survival
    of the English sole (*Parophrys vetulus*).   J. Fish. Res.
    Bd. Can. 25: 495-521.
Bliss, D. E.   1956.   Neurosecretion and the control of growth
    in a decapod crustacean.   In: K. G. Wingstrand (ed.) Bertil
    Hanstrom, Zoological Papers in Honor of his Sixty-fifth
    Birthday, Nov. 20, 1956, pp. 56-75. Zool. Inst., Lund,
    Sweden.

Calabrese, A., J. R. MacInnes, D. A. Nelson, and J. E. Miller.
    1977. Survival and growth of bivalve larvae under heavy
    metal stress. Mar. Biol. 41: 179-184.
Cangaratnam, P. 1959. Growth of fishes in different salinities.
    J. Fish. Res. Bd. Can. 16: 121-130.
Cearly, J. E. and R. L. Coleman. 1974. Cadmium toxicity and
    bioconcentration in largemouth bass and bluegill. Bull.
    Environ. Contam. Toxicol. 11: 146-151.
Chang, E. and J. O'Connor. 1977. Secretion of α ecdysone by
    crab Y organs *in vitro*. Proc. Nat. Acad. Sci. 74: 615-618.
Chang, L. W., L. M. Mak and A. Martin. 1976. Dose-dependent
    effects of methylmercury on limb regeneration in newts
    (Tritutus viridescens). Environ. Res. II 305-309.
Christensen, G. M. 1975. Biochemical effects of methylmercuric
    chloride, cadmium chloride, and lead nitrate on embryos and
    alevins of the brook trout, *Salvelinus fontinalis*. Toxicol.
    Appl. Pharm. 32: 191-197.
Collier, R. S., J. E. Miller, M. A. Dawson and F. P. Thurberg.
    1973. Physiological response of the mud crab, *Eurypanopeus
    depressus,* to cadmium. Bull. Environ. Contam. Toxicol. 10:
    378-382.
Comfort, A. and F. Doljanski. 1958. The relation of size and
    age to the rate of tail regneration in *Lebistes reticulatus*.
    Gerontologia 2: 266-283.
Dixon, C. and K. Compher. 1977. Protective action of zinc
    against the deleterious effects of cadmium in the regen-
    erating forelimb of the adult newt, *Notophthalmus virides-
    cens*. Growth 41: 95-103.
Eaton, J. G. 1973. Chronic toxicity of a copper, cadmium,
    and zinc mixture to the fathead minnow *Pimephales promelas
    Rafinesque)*. Water Res. 7: 1723-1736.
Eisler, R. 1971. Cadmium poisoning in *Fundulus heteroclitus*
    (Pisces: Cyprinodontidae) and other marine organisms. J.
    Fish. Res. Bd. Can. 28: 1225-1234.
Eisler, R. and G. R. Gardner. 1973. Acute toxicity to an
    estuarine telecost of mixtures of cadmium, copper, and
    zinc salts. J. Fish. Biol. 5: 131-142.
Fingerman, M. and S. W. Fingerman. 1974. The effect of limb
    removal on the rates of ecdysis of eyed and eyestalkless
    fiddler crabs, *Uca pugilator*. Zool. Jahrb. Physiol. 78:
    301-309.
Fowler, S. W., M. Heyraud and J. LaRosa. 1976. Heavy metal
    cycling studies: the cycling of mercury in shrimp and
    mussels. In: Activities of the International Laboratory
    of Marine Radioactivity, 1976 Report. IAEA - 187, pp. 11-
    20, IAEA, Vienna.
Fox, J. H., K. Patel-Mandlik and M. M. Cohen. 1975. Compara-
    tive effects of organic and inorganic mercury on brain-
    slice respiration and metabolism. J. Neurochem. 24: 757-762.

Gardner, G. R. and P. P. Yevich. 1970. Histological and hematological response of an estuarine telecost to cadmium. J. Fish. Res. Bd. Can. 27: 2185-2196.

Gibson, M. B. and B. Hirst. 1955. Effects of salinity and temperature on the pre-adult growth of guppies. Copeia 1955 (3): 241-242.

Gilani, S. H. 1975. Congenital abnormalities in methylmercury poisoning. Environ. Res. 9: 128-134.

Goss. R. J. and M. Stagg. 1957. The regeneration of fins and fin rays in *Fundulus heteroclitus*. J. Exper. Zool. 136: 487-508.

Green, F. A., J. W. Anderson, S. R. Petrocelli, B. J. Presley and R. Sims. 1976. Effects of mercury on the survival, respiration, and growth of post-larval white shrimp, *Penaeus setiferus*. Mar. Biol. 37: 75-81.

Holliday, F. G. and J. H. Blaxter. 1960. The effect of salinity on the developing eggs and larvae of the herring *(Clupea harengus)*. J. Mar. Biol. Assoc. U. K. 39: 591-603.

Kemp, N. E. and J. H. Park. 1970. Regeneration of lepido-thrichia and actinotrichia in the tailfin of the teleost *Tilapia mossambica*. Devel. Biol. 22: 321-342.

Kim, J. H., E. Birks and J. F. Heisinger. 1977. Protective action of selenium against mercury in the Northern creek chub. Bull. Environ. Contam. Toxicol. 17: 132-136.

McVean, A. 1976. The incidence of autotomy in *Carcinus maenas* (L.) J. Exper. Mar. Biol. Ecol. 24: 177-187.

Middaugh, D. P. and C. L. Rose. 1974. Retention of two mercurials by striped mullet, *Mugil cephalus*. Water Res. 8: 173-177.

O'Hara, J. 1973a. Cadmium uptake by fiddler crabs exposed to temperature salinity stress. J. Fish. Res. Bd. Can. 30: 846-848.

O'Hara, J. 1973b. The influence of temperature and salinity on the toxicity of cadmium to the fiddler crab, *Uca pugilator*. Fish. Bull. 71: 149-153.

Passano, L. M. and S. Jyssum. 1963. The role of the Y organ in crab proecdysis and limb regeneration. Comp. Biochem. Physiol. 9A: 195-213.

Ramel, C. 1969. Genetic effects of organic mercury compounds. I. Cytological investigations on *Allium* roots. Hereditas 61: 208-230.

Rao, K. R., M. Fingerman and C. Hays. 1972. Comparison of the abilities of ecdysone and 20-hydroxyecdysone to induce precocious proecdysis and ecdysis in the fiddler crab, *Uca pugilator*. Z. Vergl. Physiol. 76: 270-284.

Renfro, J. L., B. Schmidt-Nielson, D. Benos and J. Allen. 1974. Methylmercury and inorganic mercury: uptake, distribution, and effect on osmoregulatory mechanisms in fishes. In: Pollution and Physiology of Marine Organisms ed. F. J. Vernberg and W. B. Vernberg. pp. 101-122. Acad. Press, New York.

Roales, R. R. and A. Perlmutter. 1974. Toxicity of methyl-mercury and copper applied singly and jointly to the blue gourami, *Trichogaster trichopterus*. Bull. Environ. Contam. Toxicol. 12: 633-639.

Rosenberg, R. and J. D. Costlow, Jr. 1976. Synergistic effects of cadmium and salinity combined with constant and cycling temperatures on the larval development of two estuarine crabs. Mar. Biol. 38: 291-303.

Shealy, M. H. and P. A. Sandifer. 1975. Effects of mercury on survival and development of the larval grass shrimp, *Palaemonetes vulgaris*. Mar. Biol. Mar. Biol. 33: 7-16.

Skinner, D. M. and D. E. Graham. 1972. Loss of limbs as a stimulus to ecdysis in Brachyura (true crabs). Biol. Bull. 143: 222-233.

Spyker, J. M. and M. Smithberg. 1972. Effects of methyl-mercury on prenatal development in mice. Teratology 5: 181-189.

Tassava, R. and R. J. Goss. 1966. Regeneration rate and am-tutation level in fish fins and lizard tails. Growth 30: 9-21.

Thurberg, F. P. and M. A. Dawson. 1974. Physiological response of the cunner, *Tautogolabrus adspersus*, to cadmium. III. Changes in osmoregulation and oxygen consumption. NOAA Tech. Rep. NMFS SSRF-681: 11-13.

Vernberg, W. B., P. J. DeCoursey, M. Kelly and D. M. Johns. 1977. Effects of sublethal concentrations of cadmium on adult *Palaemonetes pugio* under static and flow-through conditions. Bull. Environ. Contam. Toxicol. 17: 16-24.

Vernberg, W. B. and J. O'Hara. 1972. Temperature-salinity stress and mercury uptake in the fiddler crab, *Uca pugilator* J. Fish. Res. Bd. Can. 29: 1491-1494.

Vernberg, W. B. and F. J. Vernberg. 1972. The synergistic effects of temperature, salinity, and mercury on survival and metabolism of the adult fiddler crab, *Uca pugilator*. Fish. Bull. 70: 415-420.

Weis, J. S. 1976a. Effects of mercury, cadmium, and lead salts on regeneration and ecdysis in the fiddler crab, *Uca pugilator*. Fish. Bull. 74: 464-467.

Weis, J. S. 1976b. Effects of environmental factors on re-generation and molting in fiddler crabs. Biol. Bull. 150: 52-62.

Weis, J. S. 1977. Limb regeneration in fiddler crabs: species differences and effects of methylmercury. Biol. Bull. 152: 263-274.

Weis, J. S. and P. Weis, 1977a. Effects of heavy metals on embryonic development of the killifish, *Fundulus heteroclitus*. J. Fish. Biol. 11: 49-54.

Weis, P. and J. S. Weis. 1976. Effects of heavy metals on fin regeneration in the killifish, *Fundulus heteroclitus*. Bull. Environ. Contam. Toxicol. 16: 197-202.

Weis, P. and J. S. Weis. 1977b. Methylmercury teratogenesis in the killifish, *Fundulus heteroclitus*. Teratology 16: 317-326.

Weis, P. and J. S. Weis. 1978. Methylmercury inhibition of fin regeneration in fishes and its interaction with salinity and cadmium. Estuar. Coast. Mar. Sci. 6:327-334.

Wellings, S. R., C. E. Alpers, B. B. McCain and B. S. Miller. 1976. Fin erosion disease of starry flounder (*Platichthys stellatus*) and English sole *(Parophrys vetulus)* in the estuary of the Duwamish River, Seattle, Washington. J. Fish. Res. Bd. Can. 33: 2577-2586.

Windom, H., W. Gardner, J. Stephens and F. Taylor. 1976. The role of methylmercury production in the transfer of mercury in a salt marsh ecosystem. Estuar. Coast. Mar. Sci. 4: 579-583.

Ziskowski, J. and R. Murchelano. 1975. Fin erosion in winter flounder (*Pseudopleuronectes americanus*) from the New York Bight. Mar. Pollut. Bull. 6: 26-28.

HEMATOLOGICAL EFFECTS OF LONG-TERM MERCURY EXPOSURE
AND SUBSEQUENT PERIODS OF RECOVERY
ON THE WINTER FLOUNDER, *PSEUDOPLEURONECTES AMERICANUS*

Margaret A. Dawson

National Marine Fisheries Service
Northeast Fisheries Center
Milford Laboratory
Milford, Connecticut

INTRODUCTION

In recent years, hematological methods have become increasingly important in assessing the condition of fish and their
responses to the environment. Hesser (1960), Blaxhall (1972),
and Blaxhall and Daisley (1973) have attempted to standardize
and interpret hematological tests as applied to fish. It has
been suggested, based on mammalian work, that anemia may be
one of the earliest indications of metal toxicity (Bresnick,
1978). The work of Katz (1950) suggests that poor condition
in fish is more quickly reflected in hematological changes than
in other commonly-measured variables. The speed with which the
effects occur, as well as the possibility of testing an animal
without killing it, makes hematology a valuable tool, particularly for such institutions as fish hatcheries, which depend
upon immediate correction of detrimental conditions to maintain healthy animals. Although a number of investigators have
reported measuring only one or two variables, Hickey (1976)
emphasized the importance of using a variety of hematological
tests in obtaining an accurate estimate of the condition of
an animal.

The winter flounder, *Pseudopleuronectes americanus*, is
important to both the recreational and the commercial fisheries.
Because of the importance and availability of the species, the
normal hematology of the winter flounder and its variation with
season have been well studied (Umminger and Mahoney, 1972;
Bridges, Cech, and Pedro, 1975, 1976). Its presence along the
coast of the densely-populated areas of the eastern United

States exposes the flounder to considerable domestic and industrial pollution. Because many pollutants are concentrated
in the sediment, a benthic animal such as the flounder may be
a particularly valuable indicator species for a variety of
substances.

The effects of long-term mercury exposure upon several
hematological parameters in the winter flounder have already
been demonstrated at this laboratory (Calabrese *et al.*, 1975).
The present study concerns the rate and extent of recovery from
mercury exposure; in addition, it supplements the original
study by including information about mercury effects upon ion
balance.

MATERIALS AND METHODS

Collection and Exposure

Winter flounder were collected by otter trawl in Long
Island Sound near Milford, Connecticut, and held in flowing
sand-filtered seawater for at least two weeks prior to mercury
exposure. Throughout the acclimation and exposure periods the
fish were fed minced surf clams, *Spisula solidissima,* daily.
The exposure and recovery periods were conducted from November
1976 through March 1977 during which time the temperature
ranged from 0 to 8^0C and the salinity from 25.7 to 29.5 o/oo.
The fish were exposed to 285-liter fiberglass aquaria filled
to 225 liters with sand-filtered seawater by a proportional-
dilution apparatus (Mount and Brungs, 1967). The dilutor controlled the intermittent delivery of toxicant-containing or
control seawater at a flow rate of 1.5 liters every 2.5 min
throughout the test period. This provided a flow of 864 liters
per day and an estimated 90% replacement time of 9 hr (Sprague,
1969). Mercury, as mercuric chloride, was added at concentrations of 10 and 20 parts per billion (ppb). Mercury levels
refer to nominal concentrations of the mercury ion in solution,
not including the background level which was less than 0.7 ppb.

Each test group consisted of 20 fish per concentration.
The fish were exposed for 60 days, at which time one group was
sacrificed for testing. Following the exposure period, one
group of fish allowed to remain for 15 days and a second group
for 60 days in clean, running seawater prior to testing. Five
fish were placed in each tank. The fish had a mean length of
270 mm (range 219-342) and a mean weight of 208 g (range 118-
376).

Hematology

Blood was collected from each animal by cardiac puncture using a 3-ml plastic syringe and a 20- or 22-gauge needle. The sample was transferred gently into an 8-ml glass vial containing 150 units of dried ammonium heparinate as an anticoagulant. A portion of each blood sample was immediately centrifuged at 12,000 X g and the plasma frozen for later determination of osmolality, sodium, potassium, calcium, and protein. The remaining whole-blood sample was used for the determination of hemoglobin (Hb), hematocrit (Hct), and erythrocyte counts (RBC). Hemoglobin was determined by the cyanmethemoglobin method. Microhematocrits were determined following centrifugation for 5 min at 13,500 X g. Erythrocyte counts (RBC) were made in a hemacytometer; blood samples were diluted 1:200 with Natt-Herrick's solution (Natt and Herrick, 1952). Plasma osmolality was determined with an Advanced 3L osmometer; the effect of the added heparin upon the osmolality was negligible. Sodium, potassium, and calcium were determined with a Coleman 51 flame photometer. Protein was determined using the biuret method as modified by Layne (1957). Mean corpuscular volume (MCV), mean corpuscular hemoglobin (MCH), and mean corpuscular hemoglobin concentration (MCHC) were determined using the following equations:

$$\text{MCV in } \mu^3/\text{cell} = \text{Hct/RBC X 10} \tag{1}$$

$$\text{MCH in pg/cell} = \text{Hb/RBC X 10} \tag{2}$$

$$\text{MCHC in g/100 ml packed red cells} = \text{Hb/Hct X 100} \tag{3}$$

All data were analyzed by the Student's t-test.

RESULTS

Following the 60-day mercury-exposure period, the control fish had an average hematocrit of 37%, a hemoglobin of 6.6 g/100 ml, and red cell count of 3.01×10^6 cells/mm^3. These values were similar to those reported earlier for this species at the same season and location (Calabrese *et al.*, 1975) and slightly higher than those reported by Bridges *et al.* (1976) for winter flounder in Casco Bay, Maine. By the end of the 60-day recovery period, or a total of 120 days in the exposure system, the hematocrit, hemoglobin, and RBC had decreased slightly but significantly in control fish, and the MCV, MCH, and MCHC remained unchanged (Table 1).

The 60-day mercury exposure had a marked effect upon both the number and the characteristics of the red blood cells (Table 1). The hematocrit of fish exposed to either 10- or 20-ppb mercury decreased to 23%, or 62% of the control value. Greater decreases were noted in the hemoglobin and RBC of exposed fish. Hemoglobin in the 10- and 20-ppb exposed groups fell to 54 and 52%, and RBC to 56 and 66%, of the control values, respectively. The lack of change in the MCH was an indication that the lower total hemoglobin concentration in mercury-exposed fish reflected a decrease in the number of red cells, rather than a change in the amount of hemoglobin present in each cell. Both the MCV and the MCHC changed significantly in fish exposed to the higher test mercury concentration.

Virtually no recovery of the red cells was noted during the initial 15-day period in clean water (Table 1). The hematocrit for exposed animals remained at the same low value. Hemoglobin decreased further in the 10-ppb group, as did the RBC in the 20-ppb group. The MCV remained at the same elevated level in the 20-ppb group. In addition, the MCHC declined in both exposed groups.

The red cells showed considerable recovery in 60 days. In the fish exposed to 10-ppb mercury, all six of the variables measured returned to the control levels. In the 20-ppb exposure group, the hematocrit and MCHC returned to the control levels, whereas the MCV remained at its previous elevated level. Hemoglobin and RBC moved toward the control levels but did not recover completely.

Plasma chemistry was also altered significantly by mercury exposure and recovery (Table 2). Following the 60-day exposure period the plasma osmolalities of the 10- and 20-ppb exposed fish were 394 and 397 mOsm/kg water, respectively, each significantly higher than the control value of 367. The plasma osmolality of fish exposed to 20-ppb mercury remained elevated at the end of the 15-day recovery period, returning to the control value after 60 days. The plasma of fish exposed to 10 ppb returned to the control osmolality within 15 days.

There was no significant difference in plasma sodium concentration between controls and exposed animals immediately following sodium concentrations were 190 and 195 meq/l in the 10- and 20-ppb exposed groups, respectively, both significant increases over the control value of 182. Following the 60-day recovery period the sodium concentrations were highly variable and, despite the even higher mean values for both exposed groups, despite the even higher mean values for both exposed groups, were not significantly different from controls. The variability may simply represent individual variation in the speed of recovery; about half the values in the 20-ppb-exposed animals were typical of controls, while the others were elevated.

TABLE 1. Effects of Mercury Exposure with Subsequent Periods of Recovery on the Erythrocytes of the Winter Flounder, Pseudopleuronectes americanus

Test	Concentration	60-day exposure		15-day recovery		60-day recovery	
		Mean ± SE	N	Mean ± SE	N	Mean ± SE	N
Hematocrit (% packed red cells)	Control	37 ± 1.6	16	35 ± 2.5	9	31 ± 1.5	16
	10 ppb	23 ± 1.6[b]	16	23 ± 1.1[b]	14	30 ± 1.2	19
	20 ppb	23 ± 1.4[b]	18	25 ± 1.0[b]	19	28 ± 1.8	17
Hemoglobin (g/100 ml whole blood)	Control	6.6 ± 0.29	17	6.6 ± 0.10	9	5.4 ± 0.27	17
	10 ppb	3.6 ± 0.28[b]	16	2.9 ± 0.26[b]	18	5.0 ± 0.17	18
	20 ppb	3.4 ± 0.26[b]	18	3.4 ± 0.28[b]	19	4.3 ± 0.17[a]	17
RBC (10^6 cells/mm^3)	Control	3.01 ± 0.12	16	2.91 ± 0.18	9	2.52 ± 0.15	17
	10 ppb	1.68 ± 0.08[b]	16	1.74 ± 0.13[b]	16	2.23 ± 0.08	18
	20 ppb	2.00 ± 0.12[b]	18	1.67 ± 0.14[b]	17	1.84 ± 0.10[b]	17
Mean Corpuscular Volume (μ^3/cell)	Control	124 ± 3.9	16	119 ± 4.6	9	121 ± 4.0	13
	10 ppb	134 ± 6.0	16	140 ± 9.3	11	137 ± 7.7	11
	20 ppb	158 ± 7.7[b]	18	155 ± 6.4[b]	17	155 ± 8.6[a]	14
Mean Corpuscular Hemoglobin (pg/cell)	Control	22.4 ± 1.0	16	23.1 ± 0.7	9	21.4 ± 0.7	13
	10 ppb	21.1 ± 1.0	16	17.4 ± 1.1[b]	15	22.7 ± 0.6	11
	20 ppb	22.8 ± 0.7	18	20.9 ± 0.8	17	23.0 ± 0.9	14
Mean Corpuscular Hb Concentration (g/100 ml packed red cells)	Control	18.2 ± 0.75	16	19.5 ± 0.77	9	18.0 ± 0.43	16
	10 ppb	16.1 ± 0.82	16	12.8 ± 0.92[b]	10	16.7 ± 0.70	19
	20 ppb	14.9 ± 0.75[a]	18	13.4 ± 0.87[b]	17	16.0 ± 1.0	17

[a] significantly different from controls at the .01 level.
[b] Significantly different from controls at the .001 level.

TABLE 2. Effects of Mercury Exposure with Subsequent Periods of Recovery on the Plasma Chemistry of the Winter Flounder, Pseudopleuronectes americanus

Test	Concentration	60-day exposure		15-day recovery		60-day recovery	
		Mean ± SE	N	Mean ± SE	N	Mean ± SE	N
Osmolality (mOsm/kg H_2O)	Control	367 ± 5.5	9	375 ± 5.8	9	369 ± 6.1	14
	10 ppb	394 ± 9.8[a]	8	382 ± 8.1	10	373 ± 9.2	11
	20 ppb	397 ± 8.2[b]	10	402 ± 5.3[b]	7	362 ± 5.6	11
Sodium (meq/l)	Control	187 ± 1.7	13	182 ± 2.4	9	182 ± 7.1	11
	10 ppb	190 ± 4.1	8	190 ± 2.6[a]	12	194 ± 5.5	11
	20 ppb	191 ± 3.2	15	195 ± 1.9[c]	11	205 ± 10.7	11
Potassium (meq/l)	Control	5.00 ± 1.39	13	4.22 ± 0.35	9	6.71 ± 0.65	11
	10 ppb	4.82 ± 0.30	8	4.66 ± 0.34	12	6.53 ± 0.37	11
	20 ppb	5.12 ± 0.27	15	4.99 ± 0.29	11	7.52 ± 0.76	11
Calcium (meq/l)	Control	4.47 ± 0.13	13	4.14 ± 0.20	9	3.55 ± 1.04	11
	10 ppb	3.71 ± 0.31[a]	8	4.41 ± 0.18	11	4.56 ± 1.24	11
	20 ppb	4.05 ± 0.11[a]	14	4.23 ± 0.26	12	4.35 ± 1.33	11
Protein (g/100 ml)	Control	4.51 ± 0.25	5	4.00 ± 0.27	8	2.50 ± 0.27	15
	10 ppb	4.36 ± 0.29	8	4.00 ± 0.26	10	4.49 ± 0.37[c]	10
	20 ppb	4.88 ± 0.20	12	3.80 ± 0.25	10	4.59 ± 0.31[c]	9

[a] significantly different from controls at .05 level.
[b] significantly different from controls at .01 level.
[c] significantly different from controls at .001 level.

There was no significant difference in potassium concentrations between control and exposed animals on any test date. However, plasma levels rose in both control and exposed animals during the 60-day recovery period.

Plasma calcium levels dropped significantly in exposed fish, from the control level of 4.47 meq/l to 3.72 and 4.05 in flounders exposed to 10- and 20-ppb mercury, respectively. A 15-day recovery period was sufficient to return plasma calcium concentrations of exposed animals to the control lever.

Plasma protein concentrations in exposed animals were not significantly different from those of controls either immediately after the exposure period or after the 15-day recovery period. Following the 60-day recovery period, there was a significant difference between the controls and each of the exposed groups. This was the result of a drop in the control level compared to the earlier test dates, rather than an actual elevation in the exposed animals. Umminger and Mahoney (1972) noted a similar decrease in plasma protein of winter flounder during the winter months. It has been suggested that the drop in control levels in plasma protein may be the result of breakdown of the flounder's "antifreeze" as the water temperature increases and that mercury exposure may delay or prevent the breakdown (A. L. DeVries, personal communication).

DISCUSSION

The rate and extent of recovery by fish from heavy metal exposure have received little attention in the literature. Pentreath (1976), in his study of whole-body mercury uptake and retention in the plaice, *Pleuronectes platessa,* demonstrated a rapid initial rate of uptake followed by a gradual accumulation over the 90-day period of exposure. The loss of mercury was a similarly slow process; after 60 days in clean seawater, the point which corresponds to the longer recovery period in the present study, the plaice still retained over half of the accumulated mercury, and the loss was not much greater when the experiment was terminated at 81 days.

Information about physiological recovery of fish after heavy metal exposure is even more limited than that about retention of metals. Yet, this type of information is required for the adequate interpretation of existing toxicity data; an irreversible change might make survival difficult even though a short-lived change of similar magnitude would have little effect on the fish population. The mercury-induced changes demonstrated in the present study were long-lived but generally reversible. A 15-day period allowed only limited recovery. In 60 days the fish exposed to 10-ppb mercury

recovered completely in every variable measured, with the
exception of plasma protein. Animals exposed to 20-ppb mercury
recovered to control levels in five of the variables tested
and toward control values in two. The only exceptions were
MCV, which showed no recovery whatsoever, and plasma protein,
which was significantly different from controls only following
the 60-day recovery period.

The mercury-induced changes in plasma chemistry are probably
the result of damage to gill tissue. Meyer (1952) found
inhibited sodium uptake and increased sodium loss in the gills
of mercury-exposed goldfish in freshwater. Olson, Bergman,
and Fromm (1973) determined that, in trout exposed to $HgCl_2$,
the highest mercury concentration was found in the gills.
Olson, Fromm, and Frantz (1973) found ultrastructural damage
in trout gills following mercury exposure. Renfro et al. (1974)
demonstrated mercury uptake by teleost gill in freshwater and
concimitant inhibition of sodium uptake. A previous study
at this laboratory demonstrated uptake of mercury from seawater
into the gills of winter flounder (Calabrese et al., 1975).

Mercury-induced changes in the red-cell component of the
blood may be the result of mercury accumulation at any of a
number of sites. Pentreath (1976) reported that, following a
60-day exposure of the plaice, Pleuronectes platessa, to
labelled mercury, the kidney had the highest total mercury
concentration of the 14 tissues and organs tested and was
among the highest in ^{203}Hg concentration. After an 81-day
recovery period, the mercury concentration in the kidney
remained high. Olson, Bergman, and Fromm (1973) also reported
a high mercury concentration in the kidney of the rainbow
trout, Salmo gairdneri, following a 24-hr mercury exposure.
A high mercury concentration in the kidney would very likely
affect renal hemopoiesis and hence such variables as hematocrit,
hemoglobin, and red cell count. Both studies also reported
uptake of mercury into the blood which could lead to direct
cell damage. In addition, inhibition by mercuric chloride of
uroporphyrinogen I synthetase, a heme biosynthetic enzyme, has
been demonstrated in vitro; no reversal of inhibition was obtained with a liver factor which protects against similar
inhibition by lead and to some extent against inhibition by
cobalt and cadmium (Piper, Rios, and Tephly, 1977; Tephly et
al., 1978).

A slight drop in the hematocrit and hemoglobin of control
fish was noted following the 60-day recovery period. This may
have been a seasonal effect or it may have been caused by the
long period of captivity. Bridges et al. (1975) noted that
fish frequently become anemic if they are held too long in
the laboratory. Regardless of the reason for the drop, the

recovery of the exposed fish appears to be real. The hematocrit and hemoglobin levels for the exposed animals following the 60-day recovery period increased in absolute value, as well as in relation to the control values.

The results reported here for the initial mercury exposure are consistent with the earlier study at this laboratory using the same animal (Calabrese *et al.*, 1975). The greater magnitude of change in the present study reflects a higher test concentration of mercury. Similar effects have been demonstrated in the striped bass, *Morone saxatilis*, following mercury exposure (unpublished data).

Hematology is a valuable tool for assessing a variety of stresses in fish. Its usefulness is often impaired by a lack of information about the normal range of values in fish. Fish blood simply has not received the same critical study as human blood. In addition, fish are normally subjected to a wide range of temperature, salinity, and nutrient availability, all of which are likely to be reflected in their hematology. These facts compound the problem of interpreting hematological studies in fish. However, the growing body of literature on the subject is gradually providing the necessary background information. This already useful indicator of the general condition of fish should become even more valuable in the future.

SUMMARY

Winter flounder were exposed to 10- and 20-ppb mercury as the chloride for 60 days and subsequently allowed to recover for 15 or 60 days in clean seawater in order to determine mercury effects and the rate and extent of recovery as measured by a variety of hematological tests. The results were as follows:

1. Following the 60-day exposure period, the hemoglobin, hematocrit, RBC, and MCHC decreased in exposed fish. MCV increased in exposed animals. There was no change in MCH.

2. The 15-day recovery period brought about virtually no recovery of the red-cell component of the blood. After the 60-day recovery period, animals exposed to 20-ppb mercury returned to the control levels in hemoglobin and RBC; MCV remained at its previous elevated level. Fish exposed to 10-ppb mercury returned to control levels in all these variables.

3. Plasma osmolality increased in both exposure groups. Fish exposed to 10-ppb mercury recovered in 15 days and those exposed to 20-ppb mercury recovered within 60 days.

4. Plasma sodium concentration rose in both exposure
groups following the 15-day recovery period. After the 60-day
recovery, the values were highly variable and not significantly
different from controls.

5. Plasma calcium concentrations dropped in exposed fish,
returning to control values after 15 days of recovery.

6. There was no significant difference between controls
and exposed animals in plasma potassium concentration on any
test date.

7. Plasma protein concentrations in exposed animals were
not significantly different from controls following the exposure period or the 15-day recovery period. At the end of the
60-day recovery period the exposed animals had a significantly
higher plasma protein concentration, reflecting a drop in the
control values.

ACKNOWLEDGMENTS

The author thanks D. W. Bridges, D. N. Pedro, and J. J.
Cech, Jr. for their valuable advice and R. S. Riccio for her
critical reading and typing of this manuscript. Use of trade
names is to facilitate description and does not imply endorsement by the National Marine Fisheries Service.

LITERATURE CITED

Blaxhall, P. C. 1972. The hematological assessment of fresh-
     water fish. A review of selected literature. J. Fish
     Biol. 4: 593-604.
Blaxhall, P. C. and K. W. Daisley. 1973. Routine hematological
     methods for use with fish blood. J. Fish Biol. 5: 771-781.
Bresnick, E. 1978. Biological and pharmacological effects of
     metal contaminants. Introduction. Fed. Proc. 37: 15.
Bridges, D. W., J. J. Cech, Jr., and D. N. Pedro. 1975. Hema-
     tological responses of winter flounder to long and short
     term environmental change, pp. 171-182. In: J. J. Cech, Jr.,
     D. W. Bridges, and D. B. Horton (eds.), Respiration of
     Marine Organisms. TRIGOM Publications, South Portland,
     Maine.
Bridges, D. W., J. J. Cech, Jr., and D. N. Pedro. 1976. Sea-
     sonal hematological changes in winter flounder, *Pseudo-
     pleuronectes americanus*. Trans. Am. Fish. Soc. 105: 596-
     600.

Calabrese, A., F. P. Thurberg, M. A. Dawson, and D. R. Wenzloff. 1975. Sublethal physiological stress induced by cadmium and mercury in the flounder, *Pseudopleuronectes americanus,* pp. 15-21. In: J. H. Koeman and J. J. T. W. A. Strik (eds.), Sublethal Effects of Toxic Chemicals on Aquatic Animals. Elsevier Scientific Publishing Company, Amsterdam.

Hesser, E. F. 1960. Methods for routine fish hematology. Prog. Fish-Cult. 22: 164-171.

Hickey, C. R. 1976. Fish hematology, its uses and significance. N. Y. Fish Game J. 23: 170-175.

Katz, M. 1950. The number of erythrocytes in the blood of the silver salmon. Trans. Am. Fish. Soc. 80: 185-193.

Layne, E. 1957. Spectrophotometric and turbidometric methods for measuring proteins. III. Biuret method, pp. 456-461. In: S. P. Colowick and N. O. Kaplan (eds.), Methods in Enzymology. Vol. III. Academic Press, Inc., New York.

Meyer, D. K. 1952. Effects of mercuric ion on sodium movement through the gills of goldfish. Fed. Proc. 11: 107-108.

Mount, D. I. and W. A. Brungs. 1967. A simplified dosing apparatus for fish toxicology studies. Water Res. 1: 21-29.

Natt, M. P. and C. A. Herrick. 1952. A new blood diluent for counting the erythrocytes and leucocytes of the chicken. Poultry Sci. 31: 735-738.

Olson, K. R., H. L. Bergman, and P. O. Fromm. 1973. Uptake of methyl mercuric chloride and mercuric chloride by trout: a study of uptake pathways into the whole animal and up-take by erythrocytes *in vitro.* J. Fish. Res. Bd. Can. 30: 1293-1299.

Olson, K. R., P. O. Fromm, and W. L. Frantz. 1973. Ultrastruc-tural changes of rainbow trout gills exposed to methyl mer-cury or mercuric chloride. Fed. Proc. 32: 261.

Pentreath, R. J. 1976. The accumulation of inorganic mercury from seawater by the plaice, *Pleuronectes platessa* L. J. exp. mar. Biol. Ecol. 24: 103-119.

Piper, W. N., G. Rios, and T. R. Tephly. 1977. Studies on the role of a factor regulating lead inhibition of erythro-cytic and hepatic uroporphyrinogen I synthetase activity, pp. 98-109. In: Biological Implications of Metals in the Environment. Technical Information Center, Energy Research and Development Administration, U. S. Government Printing Office, Washington, D. C.

Renfro, J. L., B. Schmidt-Nielson, D. Miller, D. Benos, and J. Allen. 1974. Methyl mercury and inorganic mercury: uptake, distribution, and effect on osmoregulatory mechanisms in fishes, pp. 101-122. In: F. J. Vernberg and W. B. Vernberg (eds.), Pollution and Physiology of Marine Organisms. Academic Press, Inc., New York.

Sprague, J. B.   1969.   Measurement of pollutant toxicity to
    fish.   I. Bioassay method for acute toxicity.   Water Res.
    3: 793-821.
Tephly, T. R., G. Wagner, R. Sedman, and W. Piper.   1978.
    Effects of metals on heme biosynthesis and metabolism.
    Fed. Proc. 37: 35-39.
Umminger, B. L. and J. B. Mahoney.   1972.   Seasonal changes
    in the serum chemistry of the winter flounder, *Pseudopleuro-
    nectes americanus*.   Trans. Am. Fish. Soc. 101: 746-748.

THE USE OF BIVALVE MOLLUSCS IN
HEAVY METAL POLLUTION RESEARCH

Patricia A. Cunningham

Department of Biology
North Carolina Central University
Durham, North Carolina

INTRODUCTION

Heavy metals in trace amounts have been normal constituents
of the hydrosphere since the beginning of geologic time, added
principally through weathering of rocks and volcanic activity
(Klein and Goldberg, 1970). Environmental concentrations of
heavy metals have increased considerably, however, with the
development of highly industrialized societies (Johnels and
Westermark, 1969). Estuarine and coastal marine waters are
ultimately the repositories of heavy metals released from
anthropogenic activities, and metals in coastal marine
organisms reflect the increased concentrations in sea water
(Klein and Goldberg, 1970).

Bivalve molluscs possess several characteristics which
have made them attractive research models for the study of
heavy metal pollution: 1) bivalves inhabit estuarine and
coastal marine areas which are most susceptible to pollution;
2) their sessile nature prevents them from migrating away
from the source of pollution and individuals must therefore
adjust to increasing concentrations of metals or perish;
3) bivalves exhibit a relatively long life span which allows
their use in studies spanning several years; 4) the broad
geographic distribution of several intensively studied commercially valuable species permits their use as biological
indicators of widespread pollution; and 5) the ease of collection and high density of bivalves in coastal areas have
made them readily available. In addition, recent advances in
mariculture techniques of several commercially valuable species

have increased the availability of the embryos, larvae, and
juveniles of a known age, genetic parentage, and environmental
background for research purposes.

The orchestration of phases of study which has characterized
the scope and direction of research on other environmental
pollution problems (i.e., DDT and its derivatives, organophosphates, PCB's, oil, etc.) similarly occurred for heavy
metal pollution research. An in-depth study of heavy metal
pollution was initially stimulated by identification of
environmental damage caused by metals. Research interests
were then directed toward extensive field sampling programs
in various nations to identify coastal areas where contaminated
seafood products might cause public health problems (see
Phillips, 1977 for summary). Subsequent research has focused
on both laboratory and field studies of whole body residues
of various metals accumulated over a specified exposure
period and under various environmental conditions, studies
of pathways of accumulation through the food chain, and investigation of dose-response relationships in various target
species. Increasingly, research has focused on the pharmo-
kinetics of accumulation and elimination processes in various
bivalve tissues though the use of radioactive metal tracers,
and the identification of intracellular sites of heavy metal
activity in an attempt to determine the mechanisms of toxicity
for each metal. While early research concentrated on the
bivalve as a transmitter of metals to higher trophic levels
(i.e., man), present interest is concentrated on understanding
the ability of these organisms to accumulate high body
residues of metals without displaying any apparent deleterious
effects to the individual, and in identifying the metabolic
processes which may facilitate this resistance to metal
intoxication. Although adult bivalves may be capable of
tolerating high body residues of heavy metals, little attention
has focused on the effects these residues may have on gametogenesis, fertilization, and embryology.

The purpose of this review is threefold: 1) to summarize
research on the effects of heavy metals to marine and estuarine
bivalves; 2) to enumerate essential factors in the design of
experiments using bivalves; and 3) to identify promising areas
of study where research thus far has been neglected.

EFFECTS OF HEAVY METALS ON BIVALVES

The life cycle of many bivalves includes a period of
embryonic development initiated at fertilization from which a
straight-hinge larva is produced after approximately 48 hours.
This free-swimming larval stage generally lasts 2 to 4 weeks

depending on the bivalve species, and is concluded when the
larva settles out of the water column as a spat, thereby
assuming a sessile existence characteristic of the adult.
The period of larval settlement is a critical point in the
bivalve's life cycle, since retardation in larval growth can
prolong the pelagic stage, thereby increasing the chance of
larval loss through predation, disease, and dispersion, which
could possibly reduce recruitment into the population (Calabrese
and Nelson, 1974). Bivalves may be exposed to heavy metals
during any or all life stages; generally, however, each succeeding stage in development toward adulthood is increasingly
more resistant to higher environmental metal concentrations
(Cunningham, 1972; Thurberg et al. 1975).

Few studies on the effects of toxicants on invertebrate
embryonic stages have been performed. Recently, with development of improved techniques for artificial spawning of bivalves
and rearing of larvae in the laboratory (Loosanoff and Davis,
1963), experimentation has been directed toward studying
effects of heavy metals on embryos, larvae, and spat of
commercially important shellfish species.

Embryos

Brereton, Lord and Webb (1973) studied the effect of zinc
on embryonic development of *Crassostrea gigas* after intermittent
failures in rearing success occurred at a shellfish hatchery
in Conway, North Wales. Oyster embryos were exposed to both a
filtered sea water solution containing zinc (as $ZnSO_4$) and
to naturally occurring metal-rich mine waters containing zinc,
using percentage of embryos attaining normal development as a
criterion of toxicity. Embryos were slightly more sensitive
to natural mine waters than the prepared zinc solution, and
the authors suggested this may have been due to the additional
presence of 0.010 ppm lead and 0.015 ppm cadmium. These tests
confirmed that concentrations of zinc observed in sea water
used for oyster seed production (ranging from 0.010 - 0.470
ppm zinc) could have reduced embryonic survivorship in
hatchery stock.

Effects of several inorganic heavy metals on embryos of
Crassostrea virginica and *Mercenaria mercenaria* were examined
by Calabrese et al. (1973) and Calabrese and Nelson (1974),
respectively. Experimental designs for these studies were
identical, thereby permitting interspecific comparisons of the
48-hr LC_{50} responses (direct measurement of the LC_0 and LC_{100}
for all metals was also presented). Embryos of *C. virginica*
were found to be more sensitive to $AgNO_3$ and $Pb(NO_3)_2$, equally
sensitive to $HgCl_2$, and less sensitive to $ZnCl_2$ and $NiCl_2$
than *M. mercenaria* embryos. Embryos of both species, however,
exhibited sensitivity to five metals in the following order:
Hg>Ag>Zn>Ni>Pb. While these LC_{50} experiments conducted at a

constant temperature and salinity in a synthetic sea water
medium are not effective in providing criteria for absolute
toxicity of various heavy metals in natural marine environments,
where fluctuations in temperature, salinity, and the presence
of other toxicants may modify the embryo's resistance to a
particular metal, they do provide a means for comparing the
relative toxicities of various metals to embryonic stages of
two important species (Calabrese and Nelson, 1974).

Larvae

Several different criteria have been employed in assessing
the toxicity of metals to larvae of marine bivalves, including
determinations of LC_{50}, measurement of shell growth, and comparison of oxygen consumption among control and exposed
individuals. Wisely and Blick (1967) determined LC_{50} concentrations for larvae of *Mytilus edulis* and *Crassostrea*
commercialis using cessation of normal swimming movements as
a criterion of death. These authors reported survival of 50
percent of the exposed individuals to concentrations of 13
ppm mercury (*M. edulis*) and 180 ppm mercury (*C. commercialis*)
and 32 ppm copper (*M. edulis*), which far exceed concentrations
encountered in the natural environment. Due to the short
duration of their experiment (2 hours), however, resistance
of the larvae may have been due to their ability to withdraw
into their shells, thereby reducing penetration of metals into
their soft tissues. Using 5-day old *C. virginica* larvae,
Cunningham (1972) reported the effects of 5 concentrations of
mercury (as mercuric acetate) on oxygen consumption. After 24
hours exposure to the lowest concentrations tested (0.001,
0.01, and 0.1 ppm Hg), oxygen consumption was enhanced; at the
two highest concentrations (10 and 1 ppm Hg) consumption was
suppressed as compared to controls. In a similar test using
25-day old larvae, only exposure to 10 ppm mercury suppressed
oxygen consumption, while oxygen uptake in the 0.001, 0.01,
0.01 and 1.0 ppm groups was greater than or equal to the controls. Thurberg *et al.* (1975) assessed the effects of silver
(as silver nitrate) on oxygen consumption of 2 to 15-day old
larvae of the surf clam, *Spisula solidissima*. Oxygen consumption of all stages of larvae exposed to 0.05 ppm silver was
higher than controls and the difference was constant.
Effects of zinc on development of *C. gigas* larvae were
tested by Brereton, Lord and Webb (1973) in an estuary contaminated with metal-rich mine water. At 0.05 ppm zinc,
growth rates of 48-hr veligers were markedly slower than
controls, although development was normal; practically no umbo
development occurred at 0.15 ppm zinc, and veligers were
abnormal. The incidence of abnormal swimming behavior and
structural irregularities increased with increasing zinc

concentrations. These authors also tested concentrations of
zinc within the range found in the Conway Estuary and confirmed
that zinc contamination was contributing to the intermittent
failure of larval rearing success.

In an extensive study, Calabrese et al. (1977) reported
the effects of 8 to 12-day exposure of C. virginica and
M. mercenaria larvae to several metals. These experiments
measured growth of larvae upon exposure to the LC_{50} concentrations of various metal ions as compared to controls. For
C. virginica, LC_{50} concentrations of $HgCl_2$ (0.012 ppm Hg),
$AgNO_3$ (0.025 ppm Ag), $CuCl_2$ (0.033 ppm Cu), and $NiCl_2$
(1.2 ppm Ni) retarded shell growth to 49, 67, 68, and 45%
of control growth, respectively. For M. mercenaria, LC_{50}
concentrations of $HgCl_2$ (0.015 ppm Hg), $AgNO_3$ (0.032 ppm Ag),
$CuCl_2$ (0.016 ppm Cu), $NiCl_2$ (5.7 ppm Ni), and $ZnCl_2$
(0.195 ppm Zn) retarded shell growth to 69, 66, 52, 0, and 62%
of control growth, respectively. Nickel retarded shell growth
completely at 5.7 ppm, and even at the LC_5 concentration
(1.1 ppm Ni) growth was only 32% of control. For this species,
nickel appeared to be the least toxic metal tested, yet it had
the greatest inhibitory effect on growth. In addition, nickel-
exposed clam larvae were abnormal, with tissues extruding from
the shells, although normal swimming continued. The unusual
sublethal effects of nickel even at LC_5 concentrations clearly
demonstrate the weakness of basing environmental criteria on
LC_{50} concentrations.

Early studies by Prytherch (1934) on the effects of iron,
nickel, zinc, silver, tin, barium, lead, and copper on oyster
larvae settlement revealed that none of these metals, with the
exception of copper, stimulated an early onset in setting.
Furthermore, only concentrations from 0.01-0.60 ppm copper
stimulated setting, while no effect was observed at lower
concentrations; higher concentrations caused abnormal development of internal organs and death. Boyden, Watling, and
Thornton (1975) reported increased larval mortality of C.
gigas in increasing zinc concentrations and a coinciding
decline in the percentage of larvae that settled as spat after
a 5-day exposure period.

Spat and Juveniles

Although researchers have used LC_{50} determinations and
oxygen consumption as criteria for assessing the toxicity of
metals to spat and juveniles, primary interest has been
directed toward shell growth studies of these life stages.
Effects of inorganic heavy metal compounds on juvenile bay
scallops, *Argopecten irradians,* were assessed by Nelson et al.
(1976) using 96-hr LC_{50} as a criterion of toxicity. Results
indicated that juveniles (20-33 mm length) were most sensitive

to $AgNO_3$ (0.033 ppm Ag), followed by $HgCl_2$ (0.089 ppm Hg), $CdCl_2$ (1.48 ppm Cd), and $NaAsO_2$ (3.49 ppm As). Thurberg *et al.* (1975) reported that oxygen consumption increased significantly in juvenile *S. solidissima* exposed to 0.01 and 0.02 ppm silver as compared to controls.

Boyden, Watling and Thornton (1975) reported zinc concentrations of 0.25 and 0.50 ppm depressed shell growth in *C. gigas* spat to 78 and 51%, respectively, of growth exhibited by controls; however, after a 5-day cleansing period in ambient sea water, shell growth was 108% (0.25 ppm group) and 80% (0.50 ppm group) of the controls. Cunningham (1976) reported similar retardation of shell growth in juvenile *C. virginica* exposed to 0.01 and 0.10 ppm mercury for 47 days. After this lengthy exposure, shell growth in the 0.01 and 0.10 ppm groups was reduced to 67 and 23% of control growth. Reversibility of the inhibitory effect was demonstrated by the rapid recovery juveniles made after their return to ambient sea water. After 20 days of cleansing, the control and 0.01 ppm mercury groups exhibited comparable shell growth; however, it was not until after 34 days of cleansing had elapsed that the 0.1 ppm group recovered. Frazier (1976) compared shell growth between two populations of hatchery reared oysters that had been transported to two areas in Chesapeake Bay; a clean estuary and an estuary polluted by elevated metal concentrations. Although he reported no inhibition in shell growth over a 5 month period, a significant (13%) thinning in shells (mg shell/cm^2) occurred in oysters held in the polluted estuary.

Adults - Effects of Sublethal Heavy Metal Exposure

Adult bivalves are known for their ability to accumulate heavy metals in their tissues to several orders of magnitude above ambient sea water concentrations. Many researchers have reported elevated tissue residues without being able to identify significant increases in mortality or visible deleterious effects. Few studies, however, have documented physiological changes in adult bivalves exposed to sublethal concentrations.

Feeding Behavior. Dorn (1976) studied the effects of methyl mercury acetate concentrations on feeding behavior of the mussel, *M. edulis*. For this species, the mean feeding rate (diatoms consumed per mussel per day) declined rapidly as mercury concentrations increased. During a 24-hr exposure period, mussels consumed a mean of 6704, 660, 182, 186, 267, 123, 54, and 53 diatoms at mercury concentrations of 0, 0.4, 0.8, 1.2, 1.6, 2.0, 2.4, and 2.8 ppm, respectively. The author suggested that decreased feeding rates may have resulted from inhibitory effects of mercury on the nervous system which controls ciliary activity vital to the normal feeding process.

The mussels, however, may have closed their valves during the
test period to avoid exposure, thereby effectively disrupting
normal feeing processes.

Respiration and Heart Rate.   Effects of copper on survival,
heart rate, and respiration were determined for *M. edulis*
by Scott and Major (1972).   The toxicity threshold concentration for the copper complex (Cu++) was 0.2 ppm and 55% of
the mussels exposed to this concentration died within seven
days.   Although exposure to 0.2 ppm copper did not have any
effect on heart rate and respiration, significant cardio-
vascular and respiratory depression were observed at higher
concentrations.   Brown and Newell (1972) also reported that
copper induced a depression in oxygen consumption in *M. edulis,*
while exposure to comparable concentrations of zinc produced
no effect.   Another metal, silver, stimulated respiration of
surf clams, *S. solidissima,* gill tissue after exposure to
0.05 and 0.10 ppm silver (Thurberg *et al.* 1975).   In addition,
clams exposed to 0.05 ppm silver for 96 hours averaged 24.6
shell closures per hour, while controls averaged 12.0 closures
per hour.   Oxygen consumption in four other bivalve species
(C. virginica, M. mercenaria, M. edulis and *Mya arenaria)*
consistently increased as the silver concentration increased
from 0 to 1.0 ppm (Thurberg, Calabrese and Dawson, 1974).
From the experimental data available, it is clear that different heavy metals induce distinct responses in respiratory
processes.

FACTORS AFFECTING ACCUMULATION AND LOSS OF HEAVY METALS

Accumulation, distribution, and loss of metals from tissues
are influenced by intrinsic factors including age, size, weight,
reproductive condition, and heavy metal body burden attained,
and by extrinsic environmental factors, including the hydroclimate (particularly temperature and salinity), concentration,
duration of exposure, chemical form of the metal in sea water,
and the effect of one metal on the uptake of another metal.
In much of the research work performed to date, one or more of
these variables have not been considered, and yet all may have
a determinate role in the metabolism of each heavy metal in
bivalves.

INTRINSIC FACTORS

Age - Size - Weight.

Several authors have shown that the interrelated effects
of age, size, and weight of a bivalve have a definite influence
on the processes of accumulation and, presumably, depuration
of heavy metal residues. Generally, younger individuals in a
population accumulate more metal per gram than do older
individuals. Schulz-Baldes (1973) first described this
phenomenon for three weight classes of *M. edulis* exposed to
lead pollution in the Weser Estuary, Germany. Boyden (1974)
further reported that concentrations of lead, copper, zinc,
and iron in *M. edulis* decreased with increasing weight,
whereas concentrations of nickel and cadmium remained constant.
Bryan (1976) analyzed for cadmium concentrations in field
collected samples of the burrowing bivalve, *Scrobicularia
plana,* and found that tissue concentrations either remained
constant as body weight increased, as shown in samples collected
in the East Looe Estuary, or concentrations increased with
increasing body weight, as in the highly polluted Gannel and
Plym estuaries. Inconsistency in results for cadmium may have
been related to the metal itself or to a distinct response of
a particular bivalve species. A study by Mackay *et al.* (1975)
of cultured oysters, *C. commercialis,* from estuaries in New
South Wales reported that concentrations of copper, zinc, and
cadmium decreased with increasing wet weight and age (1.5, 2.5,
and 3.5 year old individuals). De Wolf (1975) found higher
mercury concentrations in larger specimens of *M. edulis*
collected at one location, but mussels from two other sites
exhibited no such trend. In a simulated natural environment
in which oysters, *C. virginica,* were exposed to 0.010 and
0.100 ppm mercury, Cunningham and Tripp (1975a) reported that
small oysters (less than 7.85 grams) contained significantly
higher mercury concentrations than large individuals (7.86
through 20.98 grams). Similarly, Phillips (1976a; 1976b)
reported that accumulation of zinc, cadmium, lead, and copper
in *M. edulis* was weight dependent. In the majority of studies
cited above, the concentration of heavy metals generally
decreased with increasing size, age, and weight of the bivalves
while the total body residue of heavy metals increased.

The fact that younger (smaller) bivalves accumulated
higher concentrations of metals in most cases was probably a
reflection of the more rapid metabolism which accompanies
growth in these individuals. Kennedy and Mihursky (1972)
reported that respiratory rates (one measure of metabolic
activity) decreased with increasing body size in three
Chesapeake Bay bivalves (*M. arenaria, Mulinia lateralis,* and

Macoma baltica). Dame (1972) reported that the instantaneous
growth rate for intertidal *C. virginica* decreased with
increasing body weight. Presumably, the more rapid growth
rate in younger bivalves permitted more rapid heavy metal
incorporation into the tissues. Although the effects of total
body weight on depuration rates have not been assessed, it is
likely that metals would be eliminated more rapidly from
smaller individuals after return to ambient sea water. The
faster growth rate of smaller bivalves as demonstrated by
Dame (1972), suggests a more rapid turnover of cellular
material and the subsequent increase in body weight would
dilute the metal concentration. Smaller bivalves also have a
larger surface to volume ratio than larger individuals;
therefore, proportionately more surface area would be available
for both the accumulation and removal of metals to occur.
Experiments have shown that age or size considerations must
be evaluated when choosing bivalves for pollution experiments.

Reproductive Condition

Many metal accumulation studies with bivalves have been
undertaken with little thought as to their reproductive condition at onset of the exposure period or the effect that
spawning during the exposure period may have on metal elimination. In experiments by Cunningham and Tripp (1973),
exposure of oysters, *C. virginica,* to mercury was initiated in
July during the period when oysters in Delaware Bay could
potentially begin spawning. During 1971, when this experiment
was conducted, the major spawning period for the test oysters
occurred between August 15 and August 30, near the end of the
accumulation period. Mercury concentrations in the 0.01 and
0.1 ppm exposure groups declined by 33 and 18%, respectively,
during this period despite continued exposure to mercury.
During the summer of 1972, Cunningham and Tripp (1975a)
repeated the earlier experiment and found that for the same
period the 0.10 ppm group exhibited a 6% decline in mercury
residues. Depuration of mercury was slightly more rapid
during the second experiment (1972), presumably because the
major spawning period occurred after depuration had begun.
Although Cunningham and Tripp (1975a) could provide no
definitive quantitative evidence that mercury loss occurred
in the gonad, spawning was probably the cause of the differences
in experimental results. Since the ripe oyster gonad may
comprise 31-41% of body weight (Galtsoff, 1964), it is
reasonable to assume that if mercury was accumulated in gonadal

tissue an appreciable loss might occur during spawning.   This
is of interest not only as a possible source of experimental
error, but because of implications of the metal-gamete interaction which might result in decreased reproductive potential.

A study on the transfer of five metals (Ag, Cd, Cu, Pb,
and Zn) from adult female oysters to their eggs was conducted
by Greig, Nelson, and Nelson (1975).   These authors examined
two oyster populations that contained different concentrations
of the five metals.   For cadmium and copper, the authors
reported that despite differences in concentrations in the two
adult groups, eggs from both populations contained comparable
concentrations.   This suggested that the amount of metal
transferred from the adults to the eggs was fairly constant
and was not dependent on the amount of metal available in the
adult.   To confirm the hypothesis, however, a larger sample
of adult oysters with a wider range of concentrations of these
metals would have to be examined.   The authors could draw no
conclusions from the silver, lead, and zinc analyses since
concentrations of these metals were similar for adults and
eggs in the two oyster populations.

Using median tolerance limit (TL_m) measurements as an
indicator of toxicity, Delhaye and Cornet (1975) reported an
increase in sensitivity of *M. edulis* to copper during the
spawning period.   The spawning period for *M. edulis* was
characterized by a period of high respiration (Delhaye and
Cornet, 1975; deVooys, 1976).   At 1 ppm copper, the TL_m value
was 9-10 days for January and February (months preceding
spawning), 6 days for March (beginning of spawning), and 2-3
days for April and May (height of spawning).   Delhaye and
Cornet (1975) concluded that the increase in copper toxicity
during spawning was more a result of increased metabolic rate
resulting in faster absorption of copper, rather than an
increase in sensitivity.   Spawning is a critical time in the
life-cycle of adult bivalves and consideration of this period
should be reflected in acceptable standards established to
protect any given species.

Sex

Early studies by Galtsoff (1964) of metals in *C. virginica*
revealed that female oysters contained higher concentrations
of manganese than males.   The fully developed ovary contained
manganese concentrations 15 times greater than those found in
the spermary, and manganese residues decreased rapidly after
spawning.   The author reported no such pattern of differential
uptake occurred for iron, copper or zinc.   Watling and Watling
(1976) reported differences in the concentrations of several
metals between sexes of the mussel, *Choromytilus meridionalis*.

Concentrations of zinc, copper, and manganese were higher in females than males; whereas concentrations of lead and bismuth were higher in males than females. Differences in metal uptake related to sexual differences would tend to be most pronounced during the reproductive period.

Total Body Residue

   Accumulation. Several authors have identified distinct phases in metal accumulation which appeared to be partially delineated by metal body residues attained by the bivalve. Shuster and Pringle (1969) defined three phases of accumulation in experiments with *C. virginica* exposed to zinc, copper, chromium, and cadmium. During Phase I, the rate of metal accumulation increased rapidly at a constant rate; however, during suceeding Phases II and III, the rate of uptake progressively declined. These authors suggested that either Phase I or Phases I and II may not be observed in some experiments, depending on the total metal body residue present at the initiation of the accumulation period. Schulz-Baldes (1974) exposed *M. edulis* to 8 different exposure concentrations of lead, and results revealed that rate of lead uptake was a linear function of the exposure concentrations of lead. More recent experiments by Mason, Cho, and Anderson (1976) suggested two distinct phases of metal uptake in oysters, *C. virginica*, exposed to mercuric chloride; a short-term logarithmic phase followed by a long-term linear phase. They hypothesized that the logarithmic phase prevailed until a heavy metal threshold concentration was attained. Below this threshold value, accumulation was reversible; above the threshold concentration, a second irreversible linear phase was initiated. These authors recommended that additional attention should be given to the initial logarithmic phase to better define conditions affecting the apparent shift to the irreversible accumulation phase. Identification of the threshold at which the change occurs should be made since it is of critical importance in considering effects of chronic sublethal metal exposure. At an exposure of 0.04 ppm mercury, Phase II was initiated after less than 40 hours exposure (Mason, Cho, and Anderson, 1976). The test was performed at several exposure concentrations and it was found that the shift to irreversible Phase II was a function of both the length of exposure and the exposure concentation.

   Results of several experiments presented conflicting descriptions of the process of accumulation. Observed differences may have been attributable to metabolic differences among bivalve species used or chemical differences among

the various heavy metals used. Differences in the sampling
frequency employed by these researchers during their uptake
studies may also explain some of the differences in the number
of phases identified.

   Depuration. The rate of metal loss can be defined in
terms of the biological half-life of a particular metal. It
is affected by the total body residue attained and can be
calculated over a definite time interval (depuration period).
The biological half-life ($B_{1/2}$) of a metal is defined as the
time required for half the accumulated tissue metal to be
eliminated as a result of biological processes, and can be
determined using the method of Renfro (1973). The equation
log y = a + bx was calculated for each experiment reviewed
in Table 1 using least-squares linear regression analysis,
   where y = % of metal retained in the tissues at time 0;
         a = y intercept;
         b = slope of the line;
         x = time (in days)
The slope of each line (b) was then substituted in the
following equation;
$$B_{1/2} = \frac{\log 2}{b},$$
where $B_{1/2}$ was equal to the biological half-life of the meta;
in days.
   Three characteristic patterns of elimination of metals
were discernable (Table 1):
   1. Pringle *et al.* (1968) found that lead loss in
C. virginica was characterized by an increase in the $B_{1/2}$
with an increase in the lead body residue. When tissue
residues were greatest, little lead was lost. This suggests
permanent deposition of the metal, and possible breakdown of
cellular physiological processes responsible for lead loss.
   2. The Schulz-Blades (1974) experiment, exposing *M. edulis*
to 8 different concentrations of lead, was characterized by a
stable $B_{1/2}$. As the internal lead residues increased, an
equilibrium was maintained by the proportionate increase in
rate of lead elimination. The highest lead concentrations in
M. edulis were found in the kidney, suggesting transport of
this metal via the blood to this organ, where excretion occurred.
   3. Cunningham and Tripp (1975a) monitored mercury loss in
C. virginica, which was characterized by a decrease in $B_{1/2}$
as internal concentrations of mercury increased. In both
temperature regimes, the 0.10 ppm mercury-exposed oysters
released a greater percentage of their tissue mercury (ug Hg/g)
than the 0.01 ppm mercury exposed oysters. In both cases, the
$B_{1/2}$ was about twice as long in the 0.01 ppm groups. This

pattern of loss could have occurred if mercury was associated
primarily with wandering amoebocytes and mucus which could be
rapidly eliminated from the body, and not associated with
major tissues.

EXTRINSIC FACTORS

Temperature

     Temperature affects many physiological processes in
molluscs. Galtsoff (1928) reported that the maximum pumping
rate for *C. virginica* occurred at 25°C; below this temperature
pumping rate decreased until a temperature of 7.6°C was reached,
at which time pumping ceased. Similar results were reported
in *Ostrea lurida* (Hopkins, 1933), *M. mercenaria* (Loosanoff,
1939), *M. edulis* (Cole and Hepper, 1954) and *M. baltica, M.
lateralis* and *M. arenaria* (Kennedy and Mihursky, 1972). A
study of *C. virginica* (Feng, 1965) found that heart-beat
frequency was also temperature dependent, being highest at
24°C and declining steadily to 10°C. Widdows (1973) found a
similar temperature dependent response in *M. edulis*. Water-
pumping rate and heart rate of bivalves are used as indicators
of general physiological fitness. Thus, with higher temperatures and increased metabolism, more rapid turnover of all
tissue constituents, including metals, would be anticipated.
This expectation has been demonstrated by Cunningham and
Tripp (1975a) using biological half-life as a parameter for
comparison. Residence time of accumulated tissue mercury in
oysters was shorter in individuals exposed to a summer
temperature regime ($25 \pm 2^\circ$C) than for a declining autumnal
temperature regime ($25 - 5^\circ$C).

Salinity

     Salinity is another environmental parameter which appears
to have a strong influence on the accumulation and toxicity
of metals to bivalves. Schulz-Baldes (1973) reported a
gradient in lead tissue concentrations for *M. edulis* which
increased with distance upstream in the Weser Estuary, Germany.
He suggested that dilution of polluted river water with
seawater might have been the cause of these results, but
salinity might also have had a direct affect on the rate of
uptake. A similar gradient in tissue residues of copper,
zinc, and cadmium in *C. commercialis* was also reported in a
field survey conducted by Mackay *et al.* (1975).

    Phillips (1976a) performed experiments with *M. edulis* to
evaluate the direct effects of salinity on uptake of various
heavy metals. He found no significant difference in uptake
of zinc from two salinity regimes (35o/oo and 15o/oo).
Copper and cadmium, however, were accumulated more rapidly at
15o/oo than at 35o/oo, while tissue residues of lead were
higher at 35o/oo. Effects of salinity on acute toxicity of
mercury, copper, and chromium were evaluated for *Rangia cuneata*
(Olson and Harrel, 1973); salinities of 1, 5.5 and 22o/oo
greatly affected the median tolerance limits of this clam to
metals. *Rangia* was most resistant to mercury and copper at
5.5o/oo, and to chromium at 22o/oo. Wide differences in the
TL_m reported in this study indicate that salinity should be
of prime consideration when establishing water quality criteria.

Chemical Form

    Few studies have been performed to determine the effect
of chemical form on accumulation of metals by bivalves.
Cunningham and Tripp (1975b) reported that the Biological
Concentration Factor (BCF) was almost six times greater for
oysters exposed to methyl mercury chloride than to inorganic
mercuric chloride. Although the exposure concentration of
methyl mercury was twice that for mercuric chloride, it was
apparent that environmental concentration was not solely
responsible for observed differences in body residues. These
results substantiated previous research of Kopfler (1974) who
reported that organic mercury concentrations were four times
those for inorganic mercury, even though exposure concentrations were identical (1µg/ℓ). George and Coombs (1977a)
reported that prior complexation of cadmium chloride with
EDTA, humic acid, alginic acid or pectin doubled the rate of
accumulation and the final tissue cadmium concentrations in
M. edulis, when compared to mussels exposed to cadmium chloride
only. George and Coombs (1977b) reported that the rate of
uptake, excretion, and tissue distribution of iron in
M. edulis were affected by the chemical form of iron present.
Prior complexation of iron with citrate, EDTA, and 1,
10-phenanthroline complexes produced an increase in the rate
of uptake and total body burden, as compared to ferric
hydroxide-exposed mussels; complexation of iron with ferri-
chrome and simple hydroxamate complexes produced a decrease
in the rate of uptake and total body residues of iron.

Heavy Metal Interactions

    Appreciable research has been performed to ascertain the
effects of a single heavy metal on a specific bivalve species,
often under laboratory conditions where synthetic seawater is

employed. Relatively little is known of the combined effects
of one metal on the uptake of another metal, or, as in nature,
a number of heavy metals on the accumulation and loss of a
specific metal under study. Analysis of interrelationships
in metal sequestering of zinc, copper, mercury, cadmium, and
manganese by *C. virginica* at six stations in Long Island Sound
indicated that significant interelemental correlation in the
triad Cu-Cd-Zn was universally present (Feng and Ruddy, 1974).
At one station, significant interelemental correlations in
accumulation between Cu-Hg and Zn-Hg were also discovered.
Mackay *et al.* (1975) found a positive correlation for the
triad Cu-Cd-Zn in the oyster, *C. commercialis* in New South
Wales, Australia. Tissue concentrations of lead were not
significantly correlated with any other metal tested, while
arsenic was negatively correlated with both Cu and Zn.   A
study by Frazier (1976) of interrelationships in metal uptake
in *C. virginica* revealed a significant correlation with the
triads Cu-Cd-Zn and Fe-Zn-Cd. All three studies above were
in agreement of the interrelationship of the Cu-Cd-Zn triad.
Frazier (1976) also reported coinciding losses in body residues of Zn (33%), Cu (50%), and Cd (33%) from mid-August
through mid-September. This appeared to be closely related
to gonadal development and spawning, either through direct
incorporation into the gametes or through functional requirements of the gonadal tissue during development.
     Further analysis of the interrelations among various metals
should be expanded to encompass other bivalve species and more
of the heavy metals which may be present in estuarine and
coastal marine environments. Whether the interrelationships
of the various metals that have been studied would be maintained if metal tissue residues attained toxic (near lethal)
limits is now known. Attention also needs to be given to the
effects other pollutants (i.e., DDT and its derivatives,
organophosphates, PCB, oil etc.) may have on metal accumulation,
toxicity, and metabolism by bivalves.

Concentration and Duration of Exposure

     Two factors, concentration and duration of exposure, have
been identified as having paramount roles in heavy metal accumulation processes (Pringle *et al.* 1968; Shuster and Pringle,
1969; Cunningham and Tripp, 1973; 1975a; Schulz-Baldes, 1974;
Thurberg, Calabrese and Dawson, 1974; Mason, Cho and Anderson,
1976; Nelson *et al.*, 1976; Zaroogian and Cheer, 1976).
Experiments performed thusfar have revealed several important
relationship;
     1) As exposure concentration increases, tissue residues of
metal increase until a saturation level is attained.

2) At higher exposure concentrations than those producing saturation, no further incorporation of metal occurs and increases in mortality (suggesting physiological impairment) have been reported (Cunningham and Tripp, 1973; Pringle *et al.*, 1968; Nelson *et al.*, 1976).

3) Generally, the longer the duration of exposure to a contaminated environment, the greater the amount of metal accumulated at least until saturation levels are reached.

4) Differences in results between experiments conducted by the same investigators under comparable environmental conditions vary greatly, as shown by results of duplicate experiments conducted for cadmium, copper, zinc, chromium, and lead uptake in *C. virginica* (Shuster and Pringle, 1969) and for mercury in *C. virginica* (Cunningham and Tripp, 1973; 1975a). These variations may have been related to slight differences in exposure conditions (i.e., temperature, salinity, number of organisms per experimental concentration, etc.) or genetic differences in the experimental stock.

Position in the Water Column

Several authors have reported variations in heavy-metal tissue residues in bivalves with reference to their depth in the water column. Nielsen (1974) reported that concentrations of cadmium, iron, lead, and zinc in the mussel, *Perna canaliculus,* varied with sampling depth at one station in New Zealand. The author suggested that variations could have resulted from differences in food availability, differences in the ratio of soluble to particulate metals or to the presence of hydrogen sulfide produced by sediment bacteria, which might have affected metal solubility. Concentrations of the same metals in mussels from another location characterized by greater water circulation exhibited no vertical gradient in metal residues. Mercury concentrations in *M. edulis* collected intertidally were higher than those collected subtidally DeWolf, 1975). Phillips (1976a) reported vertical gradients in concentrations of cadmium, lead, and zinc in *M. edulis* collected at 0.5, 1.5, 3.5, 6.5, and 9.5 meter depths. Concentrations of all three metals were generally higher in the upper three depths and lower in the bottom waters. The vertical gradients were most clearly defined in late winter, but were reduced or absent in late summer. The author suggested that the gradients were produced by exposure of mussels in surface waters to a metal-rich freshwater layer in winter; in summer, no vertical stratification was present.

Season

    Seasonal effects on the storage of metals in bivalves
have been partially identified and are related to the sexual
(reproductive) cycle, coupled with associated changes in body
weight. Pringle *et al.* (1968) found seasonal changes in the
concentrations of five metals in the oyster, *C. virginica.*
Bryan (1973) presented detailed seasonal profiles for changes
in zinc, lead, copper, cobalt, iron, manganese, and nickel in
tissues of the scallops *Pecten maximus* and *Chlamys operularis.*
In general, heavy-metal tissue residues in these scallops
were highest in autumn and winter; it was suggested that metal
concentrations were inversely related to phytoplankton abundance. Studies of variations in metal residues in *Mytilus*
galloprovincialis by Fowler and Oregioni (1976), however,
suggested that the seasonal maximum residue which appeared in
the spring was due to high water run-off which increased the
amount of available metals. Phillips (1976a) reached similar
conclusions with zinc, cadmium, lead, and copper residues in
M. edulis. These fluctuations were reciprocal to seasonal
changes in body weight; thus, the total metal content of each
individual changed little throughout the year. Weight changes
were, in turn, related to the sexual cycle, with minimum
weights in late winter or early spring coinciding with maximum
metal concentrations. Phillips (1976a) further noted that
seasonal fluctuations varied from site to site according to
proximity to polluted frewshwater inflows.

PATHWAYS FOR ACCUMULATION, STORAGE, AND LOSS OF HEAVY METALS

Tissue Distribution

    Mechanisms for accumulation, storage, and loss of metals
in bivalves comprise several alternative pathways, varying
with species, chemical form of the metal, and mode of assimilation (via food or direct uptake). A number of accumulation
experiments have indicated that metal concentrations were often
highest in the gill, digestive system, and kidney, with lower
concentrations in the mantle, gonad, and muscle tissue (Brooks
and Rumsby, 1967; Unlu, Heyraud and Keckes, 1970; Preston, 1971;
Schulz-Baldes, 1974; Fowler and Benayon, 1974; Cunningham and
Tripp, 1975b; George, Pirie, and Coombs, 1976; George and
Coombs, 1977b.) Depuration experiments have also established
the redistribution of metals occurred from superficial sites
(i.e., gill, digestive system, and mantle) to internal tissues,

notably gonad and adductor muscle (Schulz-Baldes, 1974;
Fowler and Benayon, 1974; Cunningham and Tripp, 1975b). The
extent of the redistribution of metals observed was partly a
function of the length of the depuration period.

As shown in Figure 1, there are four major sources of
metals in seawater available for uptake by bivalves;
1) dissolved inorganic metal ions, 2) organometallic ion
complexes (exemplified by complexation of cadmium ions with
humic acid, alginate and pectin as described by George and
Coombs, 1977a, 3) metal ions preconcentrated on phytoplankton
and detritus, and 4) metal ions complexed on inorganic sediment
particles (Moore, 1971). While accumulation of inorganic
metal ions from solution has been demonstrated by Preston
(1971), Cunningham and Tripp (1975b), and George and Coombs
(1977a), this is probably the least important source. Johnels

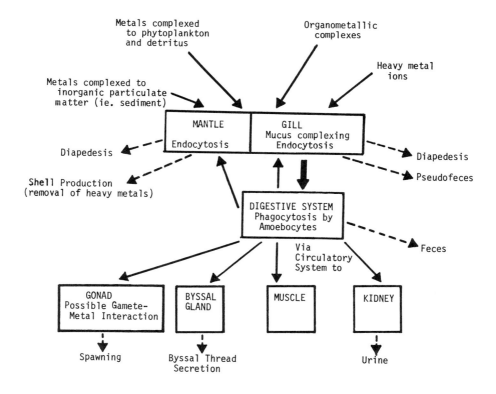

FIGURE 1. Schematic diagram of the possible pathways
of accumulation and mechanisms for loss of heavy metals in
bivalves.

Table 1. Biological half-life (B½) determinations for 3 heavy metal depuration experiments

Species and source	Metal and exposure period (days)	Seawater concentration during uptake (ppm)	Body residue (ppm wet weight)	Depuration period (days)	Water temperature (°C)	$B_{1/2}$ (days)
Crassostrea virginica (Pringle et al., 1968)	Pb 49	0.025	24	21	20	13.2
		0.050	32			33.3
		0.100	79			52.0
		0.200	203			143.0
Mytilus edulis (Schulz-Baldes, 1974)	Pb 35	0.005	33.7[b]	40	12	43.2
		0.010	65.7[b]			68.5
		0.050	395.4[b]			62.4
		0.100	558.8[b]			48.1
		0.200	1343.8[b]			66.5
		0.500	2479.8[b]			102.0
		1.000	4971.6[b]			70.3
		5.000	17625.2[b]			66.8
Crassostrea virginica (Cunningham and Tripp, 1975a)	Hg 45	0.010	12.1	80	25-5	35.4
		0.100	91.6			19.9
		0.010	5.4	25	25	16.8
		0.100	26.8			9.3

[a]From Cunningham and Tripp (1975a).

[b]Body residue determined as ppm dry weight.

et al. (1967) and Bryan (1971) suggested that, due to the great affinity of heavy metals for sulfhydryl groups and other active binding sites, coupled with the common occurrence of proteinaceous material in natural waters, the majority of metals would be bound to organic compounds, plankton, detritus, and bottom sediments, while only a small fraction would remain in solution. The exact percentage of total body burden derived from each source is uncertain.

Experimental results have shown that the gill and digestive system generally accumulate the highest concentrations of metals. The ability of these tissues to concentrate metals may be facilitated by the large surface area and the chemical nature of the mucus (a complex carbohydrate sulphate), which may act in ion exchange (Pringle *et al.,* 1968). Bivalve mucus has been suggested as being responsible for sequestering both particulate and soluble forms of heavy metals from solution (Brooks and Rumsby, 1967; Romeril, 1971; Pentreath, 1973). The primary function of the gills in many bivalves is in feeding and ions or food particles absorbed to the mucus quickly enter the mouth and are passed to the cells of the digestive gland for absorption (Fretter, 1953; Hobden, 1969). One means of elimination of certain metals may be facilitated by the production of pseudofeces. This occurs when mucus, heavily laden with food or sediment particles, reaches the mouth and is selectively ejected from the mantle cavity. Particulate metals may also enter directly into epithelial cells of the gill and mantle by endocytosis. Carmine and colloidal gold particles were found to be ingested by surface epithelial cells of *Macrocallista maculata* (Bevelander and Nakahara, 1966), while thorium was accumulated by *Pinctada radiata* and *Isognomon alatus* (Nakahara and Bevelander, 1967). Eventually, the tracers used in these studies became localized in lysosomes of the cells. In comparison to ingestion of particles via the feeding process, direct epithelial uptake is probably a minor process.

Several authors have suggested that the major pathway for absorption of heavy metals in bivalves is by ingestion of food particles and mucus laden with metals (Pringle *et al.,* 1968; Preston and Jefferies, 1969; Preston, 1971; Schulz-Baldes, 1974; Phillips, 1977), although ingestion of inorganic particulate matter is also thought to be significant (Raymount, 1972; Preston *et al.,* 1972; Boyden and Romeril, 1974). In experiments with *C. gigas* exposed to seawater containing different particulate content, Boyden and Romeril (1974) reported that the relative importance of metal uptake from solution and from inorganic particulate matter depended on the heavy metal used; however, uptake from both fractions occurred for all metals tested.

Accumulation of metals in gill, mantle, and digestive tissue after exposure to heavy-metal labelled food particles (Schulz-Baldes, 1974; Cunningham and Tripp, 1975b) is undoubtedly due to the nature of the bivalve feeding process. As water is drawn into the mantle cavity, rheological principles dictate that particles will settle out on gill and mantle tissue. These particles will be trapped in the mucus sheets covering these tissues and will then be transported to the digestive system. Similarly, metal ions in solution will pass over these same tissues, react with secreted mucus, and subsequently be passed to the digestive system. Once inside the digestive system, food and mucus complexed with metals may undergo assimilation. For one metal, iron, George, Pirie, and Coombs (1977) estimated that 30% of the ^{59}Fe reaching the digestive system was voided as feces. With respect to assimilation, Mathers (1972) traced the pathway of ^{14}C-labelled algae consumed by *Ostrea edulis*; heavy-metal laden algae particles presumably would follow similar digestion pathways. The digestive diverticulum appears to be the primary site of digestion of algae, whereas blood is the medium for transport of particles assimilated within the gut by wandering amoebocytes (Mathers, 1972; Galtsoff, 1964). Using autoradiography, Fretter (1953) found labelled amoebocytes in connective tissue and blood sinuses in *M. edulis* exposed to [90]Sr. Hobden (1969) reported high concentrations of ^{59}Fe in pedal blood sinuses in exposed individuals. Recently, Ruddell and Rains (1975) reported increased numbers of amoebocytes (basophils) correlated with increased zinc and copper residues in tissues of *C. virginica* and *C. gigas*, which further implicates amoebocytes as vehicles for heavy-metal transport.

Heavy metals accumulated from the environment are distributed throughout the bivalve body by transportation of metal-laden amoebocytes in the hemolymph. Several pathways by which bivalves may eliminate stored metals function continuously while other pathways facilitate metal loss during only a portion of the year.

1. Muscle. Heavy metal residues are generally lowest in muscle tissue after exposure (Unlu, Heyraud and Keckes, 1970; Preston, 1971; Cunningham and Tripp, 1975b; George, Pirie and Coombs (1976); George and Coombs, 1977a). Rapid elimination of stored metal residues from muscle has not been demonstrated (Cunningham and Tripp, 1975b).

2. Byssal Gland. Metals such as iron (Hobden, 1969; Pentreath, 1973; George, Pirie and Coombs, 1976) and strontium (Fretter, 1953) may be removed from bivalves, such as *Mytilus*, through secretion of the byssus. Presumably, this loss is due

to complexation of metals with proteinaceous byssal fibers.
Since byssal fiber production is a relatively continuous
process, this pathway could facilitate loss throughout the year.

3. *Kidney.* Tracer studies of the Mytilidae (i.e., *M.
edulis* and *M. californianus*) have revealed that metal concentrations in the kidney greatly exceed concentrations found in
other tissues. Young and Folsom (1967) found that 50% of
^{65}Zn in *M. californianus* was concentrated in the kidney,
which represented only 1.1% of the body weight. Concentration
factors for ^{115}Cd in *M. edulis* were 44: 11: 10: 4: 2 for the
kidney, viscera, gills, mantle, and muscle tissues, respectively (George and Coombs, 1977a). Concentration factors for
^{59}Fe in the kidney, viscera, mantle, gill and palp, and
muscle and foot were 65: 16: 9: 9: 5, respectively (George,
Pirie and Coombs, 1976). In scallops naturally exposed to low
metal concentrations, the highest metal residues appeared in
the kidney (Brooks and Rumsby, 1965; Bryan, 1973). Metal loss
could be facilitated on a continuous basis via excretion of
waste products.

4. *Gonad.* Metal residues occur in gonadal tissue of many
bivalve species; however, few researchers have provided
definitive evidence establishing a gamete-metal interaction
(Greig, Nelson, and Nelson, 1975). Heavy metals could be
associated with connective tissue or nutritive cells in the
gonad rather than with the gametes; however, regardless of
the association, the process of spawning could cause a substantial loss. The gonad of *C. virginica* may be 40% of the
body weight in some individuals and most of this tissue weight
would be eliminated during spawning (Galtsoff, 1964).

5. *Mantle.* In mantle tissue, metal concentrations may
increase directly from endocytosis or from transportation of
metals by amoebocytes. Two major mechanisms for loss of metals
from the mantle involve diapedesis and production of shell.
The continuous process of diapedesis involves the slow migration of amoebocytes from the blood through surface epithelial
tissue, where amoebocytes may be eliminated from the body
(Tripp, 1963). The mantle is also the primary organ involved
in shell production (Galtsoff, 1964). Experiments have shown
that metal concentrations in the shell increase as residues
in soft tissues increase (Unlu, Heyraud and Keckes, 1970;
Frazier, 1976; Kameda, Shimizu and Hiyama, 1968). Loss of
metals from soft tissues into the shell is a continuous process;
however, the greatest increase in shell growth generally occur
during periods of relatively high water temperature and abundant food supply, both before and after the spawning period
(Galtsoff, 1964).

6. *Gills*. During periods of exposure to metals, bivalve gills accumulate higher concentrations of metals than other tissues. Loss of metal residues in gills is extremely rapid once the bivalve has been moved to a "clean environment" (Cunningham and Tripp, 1975b; Schulz-Baldes, 1974). Residue elimination may be facilitated by diapedesis of metal-laden amoebocytes or by direct elimination of metal-laden mucus from the mantle cavity which effectively "scrubs" clean the gill surface (Cunningham and Tripp, 1975a).

7. *Digestive System*. Loss of metals from the digestive system may occur by elimination of unassimilated metals in the feces (George, Pirie, and Coombs, 1977) or by metal complexation to food particles which are phagocytized by wandering amoebocytes and transported from the digestive tract into the blood (Fretter, 1953; Hobden, 1969).

Possible pathways for the elimination of metals from major body tissues have been discussed relative to a generalized bivalve model (Figure 1). Certain of the pathways discussed for the model are more important in some species than in others, while other pathways are exclusive to a group of bivalves (i.e., byssal gland). Further research is needed to identify ofther possible mechanisms by which bivalves both accumulate and eliminate excessive body burdens of heavy metals.

Intracellular Distribution

Recent studies have been directed toward the intracellular activity of metals in bivalve systems. Differential centrifugation techniques, as well as electron microscopy and x-ray microprobe analysis, have all been invaluable tools in this area. Since each heavy metal could affect cellular metabolism in a distinct way, each is discussed separately.

Most research on intracellular effects of metals in bivalves has focused on Group III Transition Elements (zinc, mercury, and cadmium). Zinc occurs in high concentrations in molluscs under natural conditions and is an essential element involved in activation of numerous enzymes, such as carbonic anhydrase, alkaline phosphatase, carboxypeptidase A, malic dehydrogenase and α-D-mannosidase which have been reported in several bivalve species (Wolfe, 1970; Coombs, 1972). Two other transition metals, cadmium and mercury, are, however, the most toxic metals known.

Zinc. Subcellular fractionation of gill, mantle, and heart tissue from oysters, *Crassostrea angulata* and *O. edulis*, exposed to ^{65}Zn revealed greater than 50% accumulation in the nuclear and cell debris fraction (600 x g), while only 35% occurred in the digestive tissue (Romeril, 1971). In the

mantle, gill, and heart fractions, the labelled zinc was
concentrated in the insoluble tissue components associated
principally with tissue proteins. Wolfe (1970) reported that
^{65}Zn remained in the low speed fraction of *C. virginica* tissue
homogenates; however, of the zinc eluted with proteins of the
supernatant, 55% was associated with proteins with greater
than 300,000 molecular weight. Recent work on zinc distribution in the oyster, *O. edulis* further confirmed that zinc
accumulation was primarily in the subcellular fraction containing nuclei and cell debris, and that little zinc was
present in the supernatant portion in association with the
mitochondria or microsomal fractions (Coombs, 1972).

    Further studies of the nature of the zinc complex by
Coombs (1974) suggested two compartments: 1) a tissue residue,
cell-debris bound component similar to the zinc bound to the
cell debris fraction reported by Coombs (1972) and 2) a
soluble component similar to the dializable component of
Romeril (1971). The tissue-bound component was found to
contain two subspecies of complex; a firmly bound species
and a less firmly bound reversible species. The soluble
component was found to be weakly complexed to small (less
than 5,000 molecular weight) proteins, such as taurine,
lysine, ATP, and possibly homarine (Coombs, 1974). Although
some picture of zinc binding to bivalve proteins is emerging,
further research is needed.

Mercury

    Effects of mercury on shellfish have been intensively
studied as a result of the global nature of the mercury
pollution problem (Klein and Goldberg, 1970) and because
mercury contaminated shellfish have been implicated as the
causative agent in Minamata Disease (Kurland, Faro and Siedler,
1960). High concentrations of mercury accumulated in gill
tissue of oysters exposed directly to mercury contaminated
seawater or to contaminated algae (Cunningham and Tripp, 1975b)
indicate that this tissue may be physiologically stressed
during prolonged periods of exposure. In electron microscope
studies, Oura (1972) and Oura *et al.* (1972) found crystalline
inclusions in mitochondria of ciliated gill epithelial cells
excised from the clam, *Hormomya mutalilis*. These clams were
from Minamata Bay, Japan, an estuarine system to which
methyl mercury additions from industrial effluents were made
for almost a decade (Kurland, Faro and Siedler, 1960). Oura
et al. (1972) suggested, but did not prove that dense mitochondrial inclusions were a pathological response of this
organelle to methyl mercury. Oura (1972) further suggested
that the crystalline arrays of inclusions might have been related to alterations in the hydration state of the epithelial

cells or to disordered physiological metabolism in the mitochondria. Fowler, Wolfe, and Hettler (1975) in a similar
study exposed clams (*M. mercenaria*), to 0.1, 1.0 and 10.0 ppm
mercury (as mercuric chloride) for 6 days and then excised
mantle tissue. Although the ultrastructural appearance of the
mantle epithelial cells of clams exposed to 0.1 ppm mercury
was indistinguishable from that of the control, clams exposed
to higher concentrations exhibited increasing numbers of dense
cytosomes (structurally similar to mammalian lysosomes).
Energy dispersive x-ray microanalysis disclosed the presence
of high iron concentrations in relation to those of mercury
(Fe:Hg = 1:0.06) within the cytosomes; however, iron and mercury was one mechanism by which these cellular components pro-
tected other more sensitive organelles (i.e., mitochondria)
from metal toxicity, thus allowing the mollusc to rapidly accumulate metals without exhibiting increased mortality. Differences in results obtained by Oura (1972) and Oura *et al*.
(1972) and those of Fowler, Wolfe and Hettler (1975) may have
been due to differences between the two bivalve species used,
the tissue type examined or the toxic activity of the different
mercury compounds used.

Cadmium

    Cadmium, like mercury, has been shown to be highly toxic
to molluscs. George and Coombs (1977a) reported that cadmium
(as $CdCl_2$) was accumulated twice as rapidly from solution when
the cadmium was first complexed with an organic compound (i.e.,
EDTA, humic acid, pectin or alginate). These authors suggested that complexation of the metal ion might be a mechanism which effectively immobilized cadmium and prevented its
interaction with essential enzymes within the cell until
saturation concentrations were attained. Casterline and
Yip (1975) exposed oysters (*C. virginica*) to 0.1 ppm
cadmium for 6 days and found that 87.5% of the cadmium in
tissue homogenates was associated primarily with proteins in
the 105,000 x g supernatant fraction. Small amounts were found
in the nuclear, mitochondrial, and microsomal fractions.
Cadmium was associated with molecules of molecular weights
ranging from 9,200 - 13,800, while a non-dializable fraction
was associated with macromolecules greater than 50,000.
Noel-Lambot (1976) proposed the existence of a low molecular
weight cadmium-binding protein in mussel (*M. edulis*) cytosol
fraction, similar to thioneins found in mammalian systems, and
he suggested that this cadmium-binding protein might play a
protective role by limiting the amount of free cadmium in the
tissues. Engel and Fowler (1977) recently demonstrated the
existence of a low molecular weight metal-binding protein
selective for cadmium; however, amino acid analysis of this

protein isolated from *C. virginica* (conducted by Ridlington
and Fowler, 1977) revealed unusually high concentrations of
aspartic and glumatic acids and low concentrations of cysteine,
which is in marked contrast to the protein structure suggested
by Noel-Lambot (1976). Additional research in the analysis
of a cadmium-induced protein is necessary to determine whether
the observed differences reflect interspecific differences
within bivalves.

Copper

Copper accumulation in bivalves has been shown to cause
greening of the tissues (Galtsoff, 1964; Roosenburg, 1969;
Shuster and Pringle, 1969). Intracellular studies of the
activity of copper have been limited to work by Engel and
Fowler (1977) who exposed *C. virginica* to 0.1 ppm copper for
14 days. Electron microscope examination of gill epithelial
cells revealed the presence of rounded and swollen epithelial
cells, associated mitochondria containing rarified matrices,
and an increase in water content of the exposed tissue. These
authors found no specific copper-selective protein present in
the exposed oysters; instead, copper was associated with a
variety of high molecular weight proteins.

Iron

Several studies have reported iron accumulation in the
mussel, *M. edulis*. Hobden (1967; 1969) reported that the
highest iron concentrations were found in the byssal threads,
kidney, and digestive system, while Pentreath (1973) found
most iron in the digestive system, gills, and byssal gland.
George, Pirie and Coombs (1977) found iron most concentrated
in the viscera and kidney. Differences in the results of
these experiments may be related to differences in the chemical
form of iron used. Recently, George and Coombs (1977b)
reported that iron accumulation was greatest in the viscera,
gill, and palp tissues, although the chemical form in which
iron was available influenced the rate of accumulation, rate
of excretion, and the distribution in the tissues. Particulate
ferric hydroxide was phagocytized by gill and visceral cells
and stored in phagolysosomes. These iron containing vesicles
that migrated to the basal region of the cell, were voided
into capillaries, and were subsequently engulfed by amoebocytes
(George, Pirie and Coombs, 1976). Electron microscope studies
of subcellular localization revealed that iron occurred in
dense-bound aggregates in the gills or was distributed in
amorphous grey material (probably mucus) in membrane-limited
vescicles in the digestive system (George, Pirie and Coombs,
1977). No evidence of any diffusion of iron across the cell

membrane or identification of iron particles located free
within the cytoplasm was noted

It is clear from the preceding discussion that effects of
heavy metals within a cell may be distinctly different for
each metal and each tissue type under study. While research
has focused on the intracellular action of zinc, cadmium,
mercury, copper, and iron in bivalve tissues, studies of other
metals have been neglected entirely. The increased availability
and use of electron microscopy, coupled with electron microprobe analysis, will answer questions on structural alteration
in various tissues induced by heavy metals. Additional cell
fractionation studies are needed to identify the nature of
complexes these metals form with subcellular structures
(i.e., mitochondria, microsomes, and nuclei) and with components
of the cytosol (i.e., proteins, fatty acids, ATP, etc.).

USE OF BIVALVES AS INDICATORS OF POLLUTION

Bivalve Embryos as a Bioassay Tool

The use of embryos of the Pacific oyster, *C. gigas,* as a
bioindicator for assessing water quality was proposed by Woelke
(1972) after completion of 10 years of field bioassays. The
proposed criterion stated:

> Where marine water uses include fish and/or
> shellfish reproduction rearing, and/or harvesting,
> the percentage of abnormal 48-hr Pacific oyster
> embryos shall not exceed 5% in 95% of the samples
> and under no circumstances exceed 20% in a single
> sample. The criterion shall not apply if the
> salinity is less than 20o/oo. If in the bioassay
> control, percent abnormal exceeds 3%, the percent
> net risk defined as

$$\frac{\% \text{ treatment abnormal} - \% \text{ control abnormal}}{100 - \% \text{ control abnormal}} \cdot 100$$

> rather than the percent abnormal may be applied.

Selection of the Pacific oyster embryo as the bioassay
organism was based on test results which determined that:
1) the level of stress (concentration of toxicant) which
causes an increase in percentage of abnormal embryos after a
48-hr exposure would also have an adverse effect on other life
stages of the Pacific oyster and other bivalve larvae, as well

as an adverse effect on fishes, sea urchins, crustaceans,
algae, and other marine life

2) the level of stress that has no effect on the Pacific
oyster embryo in most cases would have no adverse effect on
other forms of marine life. The principal measurement employed
was the percentage of abnormal larvae found after 48-h of
exposure of oyster embryos to the toxicant under consideration.
As has been demonstrated for several heavy metals, the embryonic
stages of shellfish are generally more sensitive to environmental toxicants than larvae (Calabrese et al., 1977).

It is vital to the bioassay process that factors which
might affect the percentage of abnormal larvae produced be held
constant. Several factors which may influence normal development of the embryos include overcrowding of the experimental
containers and the hydroclimate (particularly termperature
and salinity) used in bioassay. Most of the experimental
variations produced by these factors can be minimized by strict
standardization of the bioassay test procedures (Woelke, 1968).
The feasibility of employing the Pacific oyster embryo as a
bioassay tool was substantiated by the extensive body of data
accumulated on the embryos' response to a variety of environmental pollutants (Woelke, 1972).

Adult Bivalves as Bioindicator Organisms

In recent years, several authors have stressed the need
for indicator organisms which could be employed to monitor
environmental contamination by heavy metals (Preston et al.,
1972; Haug, Melsom and Omang, 1974; Goldberg, 1975). The use
of bioindicator organisms has several advantages over measurement of metals in water or sediment. In both water and sediment samples, concentrations of metals may vary widely with
only small variations in depth, locality, or time of sampling
(Spencer and Brewer. 1969; Brewer, Spencer and Robertson, 1972;
Morris, 1974). Also, both are expensive and tedious to perform
since metal residues may be near the limits of detection even
using the most sophisticated techniques available.
Bioindicators obviate the need for multiple time series sampling
because the organism continuously concentrates metal, often
several orders of magnitude above the ambient concentration
(Brooks and Rumsby, 1965; Pringle et al., 1968). Thus, direct
analysis without preconcentration of the tissue sample is
possible.

In assessing the magnitude of metal pollution, a biological
indicator can be defined as an organism which may be used to
quantify relative concentrations of pollution by the measurement of metal concentrations in its tissues. Specific characteristics of such an indicator were presented by Butler et al.,
(1971):

1. The organism should accumulate metals without being killed by the environmental concentrations encountered.

2. The organism should be sedentary in order to be representative of the area of collection.

3. The organism should be abundant in the study area.

4. The organism should be relatively long-lived to allow the sampling of more than one age class or sampling over more than one season.

5. The organism should be of reasonable size to provide adequate tissue for analysis.

6. The organism should be easily collected and hardy enough to survive in the laboratory, thus allowing its use in controlled laboratory conditions. To these requirements, Haug, Melsom, and Omang (1974) added the following;

7. The organism should be able to tolerate brackish water conditions.

8. The organism should exhibit a high concentration factor for metals, allowing direct analysis without preconcentration of the sample.

9. A simple correlation must exist between the metal content of the organism and the ambient metal concentration in the surrounding environment. Phillips (1977) proposed that the major requirement to be fulfilled is:

10. All organisms in the survey must exhibit the same correlation between their metal contents and those in the surrounding waters at all locations studied and under all conditions.

Bivalves generally fulfill all of these requirements, and *M. edulis,* which has been employed in the metal pollution research throughout the world, has been proposed as a candidate bioindicator for global pollution studies (Phillips, 1976b).

Phillips (1976b) has outlined rigid requirements which must be satisfied in the establishment of an effective sampling program using bivalves as bioindicators. The list of requirements is based upon extensive study of metal accumulation in *M. edulis* in polluted marine and estuarine waters, and many of the requirements are proposed to minimize variations in metal residues among individual bivalves collected from natural populations. Five principal conditions should be satisfied in a sampling program:

1. All bivalves to be compared in field sampling studies should be collected in the same season over as short a period as possible.

2. The period of collection should be during the late winter since body residues are maximized during this season corresponding to body weight minimums. This procedure will allow the greatest analytical accuracy at low tissue concentrations and maximize differences between bivalves collected at different locations.

3. All bivalves used should be of similar shell length
and wet weight, and at least 10 individuals should be used in
residue analysis (minimum sample size).

4. All bivalves should be collected at the same depth
relative to the spring high water mark as differences in
vertical location at the same locality have been shown to
influence the accumulation of lead, zinc, and cadmium.

5. Bivalves should be collected from areas of similar
salinity for studies of cadmium and lead, and from areas of
similar water temperature in low-salinity regimes for studies
of cadmium.

Extensive field research by Phillips (1976a; 1976b) have
shown that *M. edulis* can be an effective bioindicator for
zinc, cadmium, lead, and iron; however, this species cannot
be used to monitor copper or manganese pollution, since the
kinetics of accumulation of these two metals under certain
environmental conditions are erratic. Until additional
research is performed on other bivalve species, it will not be
possible to evaluate their full potential as bioindicators.
The use of field collected bivalves as bioindicators of
metal pollution is, however, not without problems. In the past,
even when bivalves were exposed to metals under comparable
conditions, great variation in tissue residues often resulted
(Bryan, 1973; Cunningham and Tripp, 1973; 1975a). Several
factors which affect the dynamics of metal accumulation have
been identified. These include both intrinsic physiological
factors and external environmental factors. Despite these
individual variations, Phillips (1977) believes that bivalves
are perhaps the best indicators of pollution over a wide
variety of environmental conditions and has outlined requirements which would assist in reducing some of the variability
observed. Another means of reducing variability is presented
by Cunningham and Tripp (1975a). These authors pointed out that
when field collected bivalves are used to assess pollution,
neither the precise age of the individual or the length of time
it was exposed to a specific concentration of pollutant is known.
Recently developed mariculture techniques have facilitated
the production of bivalves artificially spawned and reared in
a clean environment. Hatchery reared stock can be placed in
various sample sites and, with periodic sampling, changes in
heavy-metal tissue residues or other physiological parameters
can be monitored. This method will allow adult bivalves of
similar physical, chemical, and biological background to be
used as bioassay organisms. Frazier (1976) recently used
hatchery reared *C. virginica* to follow the dynamics of manganese, iron, zinc, copper, and cadmium in soft tissues. These
oysters were not only hatchery reared, but were genetically

similar, thereby further reducing the variability expected
among individuals which has been demonstrated for field collected bivalve populations. The latter method for reducing
variability may provide the best means of increasing the
sensitivity of assays using bivalves as bioindicators of
heavy metal pollution.

CONCLUSION

A summary of heavy metal research utilizing bivalve
molluscs has been presented. In the past 5 years, use of new
techniques has increased our understanding of the kinetics
of metal uptake, tissue distribution, and release, but not all
metals have been equally well studied. To further improve
our understanding, genetically similar individuals with comparable histories need to be studied using the most sensitive
analytical techniques available (i.e., electron microscopy
coupled with microprobe analysis).

Several areas in which knowledge of the toxicological
effects of various metals could be expanded include:

I. Population Effects

1.   Expansion of research efforts to assess the effects
of chronic sublethal concentrations of metals on maintenance
of a population, including assaying genetic effects, fecundity
rates, and teratogeny.

II. Organism Effects

2.   The compilation of reliable and improved dose-response
data specifically directed toward the pathways for entry of
the metal to the species under consideration.

3.   Expansion of studies encompassing the effects of
intrinsic factors on accumulation of a particular metal by a
specific species.

4.   Identification of metal-induced mutagenic effects on
embryological and larval development, and effects of metal
intoxication on reproductive performance.

III. Organ and Tissue Effects

5.   Expansion of studies on the histopathological effects
of metal intoxication on bivalves which have not received
adequate attention.

6. Continuance of long-term studies on the effects of metal residues on critical target organs with regard to alterations in structure and function.

IV. Cellular and Subcellular Effects

7. A continuance in effort to evaluate metal residues and distribution in subcellular components (i.e., mitochondria, lysosomes, ribosomes, etc.) using electron microscopy, microprobe analysis, and subcellular fractionation methods.
8. Expansion of studies on the displacement of one metal by another, and the interaction among metals.
9. Identification of molecules with which metals react, including small molecules (i.e., amino acids) and macromolecules (i.e., enzymes or metal-binding proteins such as metallothionein and their effect on biochemical events within the cell.

Only when these kinds of data are available will it be possible to assess the impact of heavy metals on selected marine species and the implications for other biota, including man.

ACKNOWLEDGMENTS

I gratefully acknowledge the advice of Dr. M. R. Tripp during the preparation of this manuscript.

LITERATURE CITED

Bevelander, G. and Nakahara, H., 1966. Correlation of lysosomal activity and ingestion by the mantle epithelium. Biol. Bull., 131: 76-82.
Boyden, C. R. 1974. Trace element content and body size in molluscs. Nature, 251: 311-314.
Boyden, C. R. and Romeril, M. G. 1974. A trace metal problem in pond oyster culture. Mar. Pollut. Bull., 5: 74-78.
Boyden, C. R., Watling, H., and Thornton, I. 1975. Effect of zinc on the settlement of the oyster, *Crassostrea gigas*. Mar. Biol., 31: 227-234.
Brereton, A., Lord, H., and Webb, J. S. 1973. Effect of zinc on growth and development of larvae of the Pacific oyster, *Crassostrea gigas*. Mar. Biol., 19: 96-101.

Brewer, P. G., Spencer, D. W., and Robertson, D. E. 1972. Trace element profiles from the Geosec II test station in the Sargasso Sea. Earth Planetary Sci. Letters., 16: 111-116.

Brooks, R. R. and Rumsby, M. G. 1965. The biogeochemistry of trace element uptake by some New Zealand bivalves. Limnol. Oceanogr., 10: 521-527.

Brooks, R. R. and Rumsby, M. G. 1967. Studies on the uptake of cadmium by the oyster, *Ostrea sinuata*. Aust. J. Mar. Freshwat. Res., 18: 53-61.

Brown, B. and Newell, R. 1972. The effect of copper and zinc on the metabolism of the mussel, *Mytilus edulis*. Mar. Biol., 16: 108-118.

Bryan, G. W. 1971. The effects of heavy metals (other than mercury) on marine and estuarine organisms. Proc. Roy. Soc. (Series B)., 117: 389-410.

Bryan, G. W. 1973. The occurrence and seasonal variation of trace metals in the scallops *Pecten maximus* and *Chlamys opercularis*. J. Mar. Biol. Assoc. U. K., 53: 145-166.

Bryan, W. G. 1976. Some aspects of heavy metal tolerance in aquatic organisms. In Effects of Pollution on Aquatic Organisms, pp. 7-34, ed. by A. P. M. Lockwood. London: Cambridge University Press.

Butler, P. A., Andren, L., Bonde, G. J., Jernelov, A., and Reish, D. J. 1971. Monitoring Organisms. In F.A.O. Technical Conference on Marine Pollution and its Effects on Living Resources and Fishing, pp. 101-112, ed. by M. Ruvio. London: Fishing News Ltd.

Calabrese, A., Collier, R. S., Nelson, D. A. and MacInnes, J. A. 1973. The toxicity of heavy metals to embryos of the American oyster, *Crassostrea virginica*. Mar. Biol., 18: 162-166.

Calabrese, A., MacInnes, J. R., Nelson, D. A., and Miller, J. E. 1977. Survival and growth of bivalve larvae under heavy metal stress. Mar. Biol., 41: 179-184.

Calabrese, A. and Nelson, D. A. 1974. Inhibition of embryonic development of the hard clam, *Mercenaria mercenaria* by heavy metals. Bull. Environ. Contam. Toxicol., 11: 92-97.

Casterline, J. I. and Yip, G. 1975. The distribution and binding of cadmium in oyster, soybean, and rat liver and kidney. Arch. Environ. Contam. Toxicol., 3: 319-329.

Cole, H. and Hepper, S. 1954. The use of neutral red solution for the comparative study of filtration rates of lamelli-branchs. J. Cons. Perm. Int. Explor. Mer., 20:197-203.

Coombs, T. L. 1972. Distribution of zinc in oysters, *Ostrea edulis*, and its relation to enzyme activity and to other metals. Mar. Biol., 12: 170-178.

Coombs, T. L. 1974. The nature of zinc and copper complexes in the oyster, *Ostrea edulis*. Mar. Biol., 28: 1-10.

Cunningham, P. A. 1972. The effects of mercuric acetate on the adults, juveniles, and larvae of the American oyster, *Crassostrea virginica.*, M. S. Thesis, University of Delaware, Newark, Delaware. pp. 1-77.

Cunningham, P. A. 1976. Inhibition of shell growth in the presence of mercury and subsequent recovery of juvenile oysters. Proc. Natl. Shellfish. Assoc., 66: 1-5.

Cunningham, P. A. and Tripp, M. R. 1973. Accumulation and depuration of mercury in the American oyster, *Crassostrea virginica*. Mar. Biol., 20: 14-19.

Cunningham, P. A. and Tripp, M. R. 1975a. Factors affecting accumulation and removal of mercury from tissues of the American oyster, *Crassostrea virginica*, Mar. Biol., 31: 311-319.

Cunningham, P. A. and Tripp, M. R. 1975b. Accumulation, tissue distribution, and elimination of $^{203}HgCl_2$ and $CH_3^{203}HgCl$ in the tissues of the American oyster, *Crassostrea virginica*. Mar. Biol., 31: 321-334.

Dame, R. F. 1972. The ecological energies of growth, respiration, and assimilation in the intertidal American oyster, *Crassostrea virginica*. Mar. Biol., 17: 243-250.

Delhaye, W. and Cornet, D. 1975. Contribution to the study of the effects of copper on *Mytilus edulis* during the reproductive period. Comp. Biochem. Physiol., 50A: 511-518.

de Vooys, C. G. N. 1976. The influence of temperature and time of year on the oxygen uptake of the sea mussel, *Mytilus edulis*. Mar. Biol., 36: 25-30.

De Wolf, P. 1975. Mercury content of mussels from West European coasts. Mar. Pollut. Bull., 6: 61-63.

Dorn, P. 1976. The feeding behavior of *Mytilus edulis* in the presence of methyl mercury acetate. Bull. Environ. Contam. Toxicol., 15: 714-719.

Engel, D. W. and Fowler, B. A. 1977. Copper and cadmium induced changes in the metabolism and structure of molluscan gill tissue. In: Marine Pollution: Functional Responses, this volume, ed. by W. B. Vernberg, A. Calabrese, F. P. Thurberg, and F. J. Vernberg. New York: Academic Press.

Feng, S. Y. 1965. Heart rate and leucocyte circulation in *Crassostrea virginica*. Biol. Bull., 128: 198-210.

Feng, S. Y. and Ruddy, G. M. 1974. Zinc, copper, cadmium, manganese, and mercury in oysters along the Connecticut coast. In Investigations on Concentrations, Distributions

and Fates of Heavy Metal Wastes in Parts of Long Island
Sound. pp. 132-158, Groton, Connecticut: University of
Connecticut.

Fowler, B. A., Wolfe, D. A. and Hettler, W. F. 1975. Mercury
and iron uptake by cytosomes in mantle epithelial cells
of quahog clams, *Mercenaria mercenaria* exposed to mercury.
J. Fish. Res. Bd. Canada., 32: 1767-1775.

Fowler, S. W. and Benayoun, G. 1974. Experimental studies on
cadmium flux through marine biota. In Comparative Studies
of Food and Environmental Contamination. pp. 159-177,
Vienna: International Atomic Energy Agency.

Fowler, S. W. and Oregioni, B. 1976. Trace metals in mussels
from the North-West Mediterranean. Mar. Pollut. Bull.,
7: 26-29.

Frazier, J. M. 1976. The dynamics of metals in the American
oyster, *Crassostrea virginica*. II. Environmental effects.
Chesapeake Sci., 17: 188-197.

Fretter, V. 1953. Experiments with radioactive strontium-90
on certain molluscs and polychaetes. J. Mar. Biol. Assoc.
U. K., 32: 367-384.

Galtsoff, P. 1928. The effects of temperature on the
mechanical activity of the gills of the oyster. J. Gen.
Physiol., 11: 415-431.

Galtsoff, P. 1964. The American oyster, *Crassostrea virginica*.
Fish. Bull. Fish. Wildl. Serv. U. S., 64: 1-480.

George, S. G. and Coombs, T. L. 1977a. The effects of
chelating agents on the uptake and accumulation of cadmium
by *Mytilus edulis*. Mar. Biol., 39: 261-268.

George, S. G. and Coombs, T. L. 1977b. Effects of high
stability iron complexes on the kinetics of iron accumulation and excretion in *Mytilus edulis*. J. Exp. Mar. Biol.
Ecol., 28: 133-140.

George, S. G., Pirie, B. J. S., and Coombs, T. L. 1976.
The kinetics of accumulation and excretion of ferric hydro-
xide in *Mytilus edulis* and its distribution in tissues.
J. Exp. Mar. Biol. Ecol., 23: 71-84.

George, S. G., Pirie, B. J. S., and Coombs, T. L. 1977.
Absorption, accumulation, and excretion of iron complexes
by *Mytilus edulis*. In International Conference on Heavy
Metals in the Environment, pp. 887-900, ed. by T. C.
Hutchinson. Toronto: University of Toronto.

Goldberg, E. D. 1975. The mussel watch - a first step in
global marine monitoring. Mar. Pollut. Bull., 6: 111.

Greig, R. A., Nelson, B. A. and Nelson, D. A. 1975. Trace
metal content in the American oyster. Mar. Pollut. Bull.,
6: 72-73.

Haug, A., Melsom, S. and Omang, S. 1974. Estimation of heavy
    metal pollution in two Norwegian fjord areas by analysis
    of brown algae, *Ascophyllum nodosum*. Environ. Pollut.,
    7: 173-193.
Hobden, D. 1967. Iron metabolism in *Mytilus edulis*.
    I. Variation in total content and distribution, J. Mar.
    Biol. Assoc. U. K., 47: 597-606.
Hobden, D. 1969. Iron metabolism in *Mytilus edulis*.
    II. Uptake and distribution of radioactive iron. J. Mar.
    Biol. Assoc. U. K., 49: 661-668.
Hopkins, A. 1933. Experiments on the feeding behavior of the
    oyster, *Ostrea gigas*. J. Mar. Biol. Assoc. U. K., 13:
    560-599.
Johnels, A. G. and Westermark, T. 1969. Mercury contamination
    of the environment of Sweden. In Chemical Fallout,
    pp. 221-241, ed. by M. Miller and G. Berg. Springfield,
    Illinois: Charles C. Thomas Publishers.
Johnels, A. G., Westermark, T., Berg, W., and Perssom, P.,
    and Sjostrand, B. 1967. Pike (*Esox lucius*) and some
    other aquatic organisms in Sweden as indicators of mercury
    contamination in the environment. Oikos., 18: 323-332.
Kameda, K., Shimizu, M., and Hiyama, Y. 1968. On the uptake
    of ^{65}Zn and the concentration factor on zinc in marine
    organisms. 1. Uptake of ^{65}Zn in marine organisms. J. Rad.
    Res., 9: 50-62.
Kennedy, V. S. and Mihursky, J. A. 1972. Effects of tem-
    perature on the respiratory metabolism of three
    Chesapeake Bay bivalves. Chesapeake Sci. 13: 1-22.
Klein, D. and Goldberg, E. 1970. Mercury in the marine
    environment. Environ. Sci. Technol., 4: 765-768.
Kopfler, F. 1974. The accumulation of organic and inorganic
    mercury compounds by the Eastern oyster, *Crassostrea
    virginica*. Bull. Environ. Contam. Toxicol., 11: 275-280.
Kurland, L., Faro, S., and Siedler, H. 1960. Minamata disease:
    the outbreak of neurological disorder in Minamata, Japan
    and its relationship to the ingestion of seafood contami-
    nated by mercury. World Neurol., 1: 370-391.
Loosanoff, V. 1939. Effects of temperature upon shell move-
    ments of clams, *Venus mercenaria*, Biol. Bull., 76: 171-182.
Loosanoff, V. and Davis, H. C. 1963. Rearing of bivalve
    molluscs. Adv. Mar. Biol., 1:1-36.
Mackay, N. J., Williams, R. J., Kacprzac, J. L., Collins, A.J.,
    and Auty, E. H. 1975. Heavy metals in cultured oysters,
    Crassostrea commercialis from the estuaries of New South
    Wales. Aust. Mar. Freshwat. Res., 26: 31-46.
Mason, J. W., Cho, J. H. and Anderson, A. C. 1976. Uptake
    and loss of inorganic mercury in the Eastern oyster,
    Crassostrea virginica. J. Environ. Contam. Toxicol., 4:
    361-376.

Mathers, N. 1972. The tracing of a natural algal food
    labelled with a carbon-14 isotope through the digestive
    tract of *Ostrea edulis*. Proc. Malac. Soc. London., 40:
    115-124.
Moore, H. J. 1971. The structure of the latero-frontal cirri
    on the gills of certain lamellibranch molluscs and their
    role in suspension feeding. Mar. Biol., 11: 23-27.
Morris, A. W. 1974. Seasonal variation of dissolved metals
    in inshore waters of the Menai Straits. Mar. Pollut.
    Bull., 5: 54-59.
Nakahara, H. and Bevelander, G. 1967. Ingestion of particu-
    late matter by the outer surface cells of the mollusc
    mantle. J. Morph., 122: 139-146.
Nelson, D. A., Calabrese, A., Nelson, B. A., MacInnes, J.R.
    and Wenzloff, D. R. 1976. Biological effects of heavy
    metals on juvenile bay scallops *Argopecten irradians* in
    short term exposures. Bull. Environ. Contam. Toxicol.,
    16: 275-282.
Nielson, S. A. 1974. Vertical concentration gradients of
    heavy metals in cultured mussels. New Zealand J. Mar.
    Freshwat. Res., 8: 631-636.
Noel-Lambot, F. 1976. Distribution of cadmium, zinc, and
    copper in the mussel, *Mytilus edulis*. Existence of cadmium
    binding proteins similar to metallothioneins. Experientia,
    32: 324-326.
Olson, K. and Harrel, R. 1973. Effect of salinity on acute
    toxicity of mercury, copper, and chromium for *Rangia cuneata*
    Contr. Mar. Sci., 17: 9-13.
Oura, C. G. 1972. The crystalline and dense inclusions in
    mitochondria of shellfish gill epithelial cells. Okajimas
    Folia Anat. Jap., 48: 81-96.
Oura, C. G., Yasuzumi, Akahori, H., and Yonehara, K. 1972.
    The fine structure of intramitochondrial inclusions
    appearing in epithelial cells of the shellfish gill.
    Monitore Zool. Ital., 6: 147-153.
Pentreath, R. J. 1973. The accumulation from water of [65]Zn,
    [54]Mn, [58]Co, and ^{59}Fe by the mussel, *Mytilus edulis*.
    J. Mar. Biol. Assoc. U. K., 53: 127-143.
Phillips, D. J. H. 1976a. The common mussel, *Mytilus edulis,*
    as an indicator of pollution by zinc, cadmium, lead, and
    copper. I. Effects of environmental variables on uptake
    of metals. Mar. Biol., 38: 59-69.
Phillips, D. J. H. 1976b. The common mussel, *Mytilus edulis*
    as an indicator of pollution by zinc, cadmium, lead, and
    copper. II. Relationship of metals in the mussel to those
    discharged by industry. Mar. Biol., 38: 71-80.

Phillips, D. J. H. 1977. The use of biological indicator organisms to monitor trace metal pollution in marine and estuarine environments - a review. Environ. Pollut., 13: 281-318.

Preston, E. 1971. The importance of ingestion in chromium-51 accumulation by *Crassostrea virginica*. J. Exp. Mar. Biol. Ecol., 6: 47-54.

Preston, A. and Jefferies, D. F. 1969. Aquatic aspects in chronic and acute contamination situations. In Environmental Contamination by Radioactive Material. pp. 183-211. Vienna: International Atomic Energy Agency.

Preston, A., Jefferies, D. F., Dutton, J. W. R., Harvey, B. R., and Steele, A. K. 1972. British Isles coastal water: the concentration of selected heavy metals in sea water, suspended matter and biological indicators - a pilot survey. Environ. Pollut., 3: 69-82.

Pringle, B. H., Hissong, D. E., Katz, E. L., and Mulawka, S. 1968. Trace metal accumulation by estuarine molluscs. J. Sanit. Engineering Div. Am. Soc. Civ. Engineers., 94 (5970): 455-475.

Prytherch, H. F. 1934. The role of copper in setting, metamorphosis, and distribution of the America oyster, *Crassostrea virginica*. Ecol. Monographs., 4: 49-107.

Raymont, J. E. G. 1972. Some aspects of pollution in Southampton water. Proc. Roy. Soc. London. (Series B)., 180: 451-468.

Renfro, W. C. 1973. Transfer of [65]Zn from sediment by marine polychaete worms. Mar. Biol., 21: 305-316.

Ridlington, J. W. and Fowler, B. A. 1977. Isolation and partial characterization of an inducable cadmium-binding protein from American oyster. Pharmacol., 19: 180.

Romeril, M. G. 1971. The uptake and distribution of [65]Zn in oysters. Mar. Biol., 9: 347-354.

Roosenburg, W. H. 1969. Greening and copper accumulation in the American oyster, *Crassostrea virginica*. Proc. Natl. Shellfish. Assoc., 59: 91-103.

Ruddell, C. L. and Rains, D. W. 1975. The relationship between zinc and copper and the basophils of two crasso-streid oysters, *Crassostrea gigas* and *Crassostrea virginica*. Comp. Biochem. Physiol., 51A: 585-591.

Schulz-Baldes, M. 1973. Die miesmuschel *Mytilus edulis* als indikator fur die bleikonzentration im weserastuar und in der deutschen bucht. Mar. Biol., 21: 98-102.

Schulz-Baldes, M. 1974. Lead uptake from sea water and food, and lead loss in the common mussel, *Mytilus edulis*. Mar. Biol., 25: 177-193.

Scott, D. and Major, C. 1972. The effect of copper II on survival, respiration, and heart rate in the common blue mussel, *Mytilus edulis*. Biol. Bull., 143: 679-688.

Shuster, C. N. and Pringle, B. H. 1969. Trace metal
    accumulation by the American oyster, *Crassostrea virginica*.
    Proc. Natl. Shellfish. Assoc., 59: 93-103.
Spencer, D. W. and Brewer, P. G. 1969. The distribution of
    copper, zinc, and nickel in sea water of the Gulf of Maine
    and the Sargasso Sea. Geochim. Cosmochim. Acta., 33: 325-
    339.
Thurberg, F. P., Cable, W. D., Dawson, M. A., MacInnes, J. R.,
    and Wenzloff, D. R. 1975. Respiratory response of larval,
    juvenile and adult surf clams, *Spisula solidissima* to
    silver. In Respiration of Marine Organisms. pp. 71-82,
    ed. by J. J. Cech, D. W. Bridges and D. B. Horton. South
    Portland, Maine: Trigom Publishers.
Thurberg, F. P., Calabrese, A., and Dawson, M. A. 1974.
    Effects of silver on oxygen consumption of bivalves at
    various salinities. In Pollution and Physiology of Marine
    Organism. p. 67-78, ed. by F. J. Vernberg and W. B.
    Vernberg. New York: Academic Press.
Tripp, M. R. 1963. Cellular responses of mollusks. Ann. N. Y.
    Acad. Sci., 113: 467-474.
Unlu, M. Y., Heyraud, M., and Keckes, S. 1970. Mercury as a
    hydrospheric pollutant. I. Accumulation and excretion of
    $^{203}HgCl_2$ in *Tapes descussatus*. In F.A.O. Technical
    Conference on Marine Pollution and Its Effects on Living
    Resources and Fishing. pp. 1-6, Rome.
Watling, H. R. and Watling, R. J. 1976. Trace metals in
    Choromytilus meridionalis. Mar. Pollut. Bull., 7: 91-94.
Widdows, J. 1973. Effects of temperature and food on the
    heart beat, ventilation rate and oxygen uptake of *Mytilus
    edulis*. Mar. Biol., 20: 269-276.
Wisely, B. and Blick, R. A. P. 1967. Mortality of marine
    invertebrate larvae in mercury, copper, and zinc solutions.
    Aust. J. Mar. Freshwat. Res., 18: 63-72.
Woelke, C. E. 1968. Development and validation of a field
    bioassay method with the Pacific oyster, *Crassostrea gigas*
    embryo. Seattle, Washington: University of Washington.
Woelke, C. E. 1972. Development of a receiving water quality
    bioassay criterion based on the 48 hour Pacific oyster,
    Crassostrea gigas embryo. Washington Department of
    Fisheries, Seattle, 9: 1-93.
Wolfe, D. A. 1970. Zinc enzymes in *Crassostrea virginica*.
    J. Fish. Res. Bd. Canada., 27: 59-69.
Young, D. R. and Folsom, T. R. 1967. Loss of ^{65}Zn from the
    California sea mussel, *Mytilus californianus*. Biol. Bull.,
    133: 438-447.
Zaroogian, G. E. and Cheer, S. 1976. Accumulation of cadmium
    by the American oyster, *Crassostrea virginica*. Nature,
    261: 408-409.

PRELIMINARY OBSERVATIONS ON THE CYTOPATHOLOGICAL
EFFECTS OF COPPER SULFATE ON THE
CHEMORECEPTORS OF *CALLINECTES SAPIDUS*

Joel E. Bodammer

NOAA, National Marine Fisheries Service
Northeast Fisheries Center
Oxford Laboratory
Oxford, Maryland

INTRODUCTION

Behavioral and electrophysiological studies have emphasized both the acuity and importance of the chemical senses
in feeding and social interactions of crustaceans (McLeese,
1970, 1974; Hazlett, 1971; Mackie, 1973; Snow, 1973a; Fuzessary and Childress, 1975; Hindley, 1975; McLeese *et al.*,
1977). Recently, Pearson and Olla (1977) studying the feeding behavior of *Callinectes sapidus* have shown that the blue
crab will respond to solutions of clam extract in concentrations as low as 10^{-15} grams per liter.

Little is known, however, about the potentially toxic
effects of heavy metals on crustacean chemoreceptors. McLeese
(1975) has observed the following behavioral modifications in
lobsters (*Homarus americanus*) after exposure to copper at
various concentrations: a) solutions between 200-1000 mg/l
in Cu^{++} elicited avoidance behavior; b) continuous exposure
for 48 hours to solutions containing 40 to 100 µg/l of Cu^{++}
resulted in impaired or blocked feeding responses; and c)
normal feeding responses were regained by exposed animals
after their return to clean water for 48-96 hours. Results
similar to the above (i.e. blocked and regained physiological
responses) have also been obtained by Hara, Law, and Macdonald
(1976) in electrophysiological studies of olfaction in rainbow
trout (*Salmo gairdneri*) exposed to dilute copper solutions.

In the preceeding experiments, the investigators did not indicate whether copper had a pathological effect on the chemoreceptors.

Using McLeese's (1975) results as a guide, blue crabs in
this experiment were exposed for 48 hours to copper sulfate
solutions at several concentrations. The preliminary results
presented here show that copper may have a cytopathological
effect on crustacean chemoreceptors and that further studies
should include behavioral and/or physiological assessments of
function.

MATERIALS AND METHODS

Adult male, intermolt specimens of *C. sapidus* having an
average carapace width of 16 cm were collected from the Tred
Avon River (Maryland) and kept in continuous-flow tanks for
one week prior to the experiment. The animals were individually identified using the method of Newman and Ward (1973);
blood collected from the cardiac sinus appeared free of
bacteria or parasites.

The exposures were conducted for 48 hours at 20-22°C in
60-liter, aerated aquaria to which 30 liters of test solution
had been added. Hydrated copper sulfate reagent ($CuSO_4 \cdot
5H_2O$) was added to filtered, river water having a salinity of
12 °/oo to prepare toxicant solutions at concentrations of
1000, 100, 50, and 0 µg/l. One-liter water samples were
filtered, acidified with nitric acid to pH 2.0, and stored at
0-4°C prior to being analyzed by atomic absorption sepctro-
photometry.

Control tissues were obtained from all animals used in the
experiment by removing the left chemoreceptor antennule prior
to their being placed in one of the three toxicant-containing
aquaria or the clean-water tanks. Two animals were exposed at
each of the copper sulfate concentrations listed above, except
for the control group in which four animals (two per aquarium)
were used. Paired receptor organelles were taken from one
additional, unexposed animal for examination with the scanning
electron microscope (SEM).

Intact antennules were fixed for 24 hours in ice cold
(0-4°C), 4% glutaraldehyde adjusted to a pH of 7.4 with 0.1M
phosphate buffer. Following a brief rinse in buffer solution,
the specimens were brought to room temperatue and post-fixed
in 1% osmium tetroxide in 0.1M phosphate buffer at a pH of
7.4. The specimens studied by SEM were dehydrated and criti-
cally-point dried in accordance with the method of Anderson
(1951) before being plasma-coated with 15-20 nm of gold pal-
ladium alloy. The other specimens were dehydrated in a graded

series of ethanol and propylene oxide before infiltration and embedment in Spurr (1969) resin. Thick and thin sections of embedded material were made with glass and diamond knives. The specimens were examined with either a Zeiss EM 9S2 transmission electron microscope or AMR 1000 scanning electron microscope.[1]

RESULTS

Chemoreceptors--Normal and Control

The distal segment (Fig. 1) of the antennule in *C. sapidus* has an inner (IF) and outer (OF) flagellum of similar structure to those reported for a number of marine crustacea (Ghiradella, Case, and Cronshaw, 1968). One side of the outer flagellum bears a number of aesthetascs (A, Fig. 1) which are 0.7 to 1.0 mm in length and have diameters that vary along their proximal-distal axes (i.e. base =12-16 µm; tip = 2.5-4.0 µm).

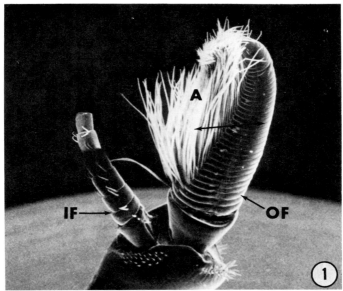

FIGURE 1. Scanning electron micrograph of the distal portion of a chemoreceptor antennule. The bar (limited by arrowheads) indicates the plane and height at which the aesthetascs (A) were sectioned. IF = inner flagellum; OF = outer flagellum. 30.6X.

[1]*Reference to trade names does not imply endorsement by the National Marine Fisheries Service, NOAA.*

An extension of the dark line shown on the outer flagellum (Fig. 1) illustrates both the angle and approximate level (height) at which the aesthetascs were sectioned. Consistent use of this procedure permitted the examination of numerous sensory hairs from animals in each exposure group and facilitated their comparison.

Images taken at the previously described level feature profiles of the cylindrical aesthetascs (Figs. 2,3) 9-11 µm in diameter. The sensory hair wall is composed of a coencentrically laminated cuticle (C, Fig. 2) that surrounds an electron lucent material containing small flocculent densities (FD, Fig. 4) and numerous vesicles (dendrites) of varying shape and size (D, Figs. 2,3). Within the dendrites, small circular neurotubules (NT, Figs. 2,4) 24-26 nm in diameter are found. The neurotubules (microtubules) and dendrites are comparable to those described for the chemoreceptors of other crustaceans (Ghiradella, Case, and Cronshaw, 1968; Snow, 1973b). Focal expansions in smaller cylindrical elements (FE, Fig. 4) are probably the source of at least some of the large dendrites observed in cross-sectional images (e.g. D, Figs. 2,3). Like the neurotubules, similar swellings in the dendrites of crustacean chemoreceptors have been noted previously (see above refs.). Ensheathment of the receptor cell dendrites by glial cell processes (Snow, 1973b) is absent at the plane of section used in this study.

The images described thus far were taken from chemoreceptor antennules removed prior to the animal's exposure in one of the three copper-containing solutions or control tanks. As is noted upon comparing the examples shown in Figures 2 and 3, the individual sensilla (aesthetascs) vary with regard to the number of dendrites present and in the quantitative distribution of neurotubules within them.

Micrographs of aesthetascs (Figs. 5,6) from specimens held for 48 hours in clean, unfiltered water resemble those sampled prior to any exposure (Figs. 2,3). This suggests that the animal's metabolic wastes had little effect on chemosensory cell fine structure in this experiment.

Chemoreceptors Exposed to 1000 µg/l Copper Sulfate

As illustrated in Figures 7 and 8, exposure to 1000 µg/l of copper sulfate solution results in changes affecting both the integrity and disposition of the dendrites within the aesthetascs. The cytopathological changes accompanying this exposure are as follows: (1) a condensation of the dendrites and flocculent extracellular material into a bandlike region (Fig. 7) or around the periphery of the sensillum (Fig. 8); (2) a consistent loss of the large dendritic profiles; and (3) the occurrence of electron dense filamentous structures

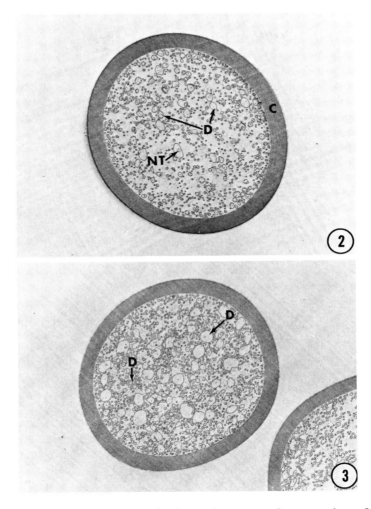

FIGURES 2 and 3. Transmission electron micrographs of sensilla from antennules that were not exposed to any test solutions. Note the variability in the diameters of the dendrites (D) and their random distribution. C = cuticle; NT = neurotubules. Fig. 2, 5760X; Fig. 3, 5545X.

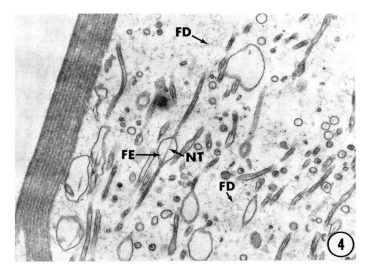

FIGURE 4. High magnification image of an oblique section
of the dendrites illustrating how their diameters may be
greatly increased by focal expansions (FE) in their limiting
membranes. FD = flocculent densities; NT = neurotubules.
18,366X.

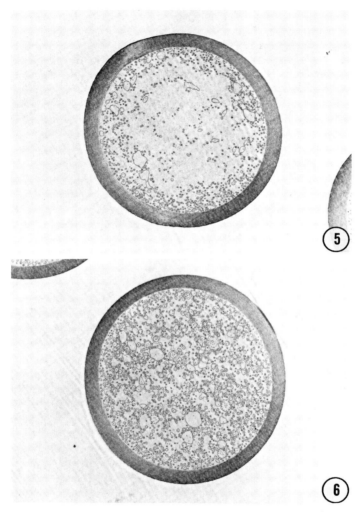

FIGURES 5 and 6. Electron micrographs of sensilla from animals exposed for 48 hours to clean water. The condition and distribution of their dendrites appears "normal" when compared to unexposed aesthetascs (e.g. Figs. 2,3). Fig. 5, 5545X; Fig. 6, 5603X.

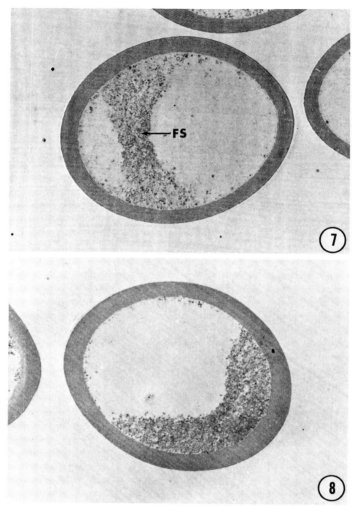

FIGURES 7 and 8. As is illustrated by these examples, exposure to 1000 μg/l of copper sulfate disrupts the structure of the dendrites and affects their distribution within the sensilla. FS = filamentous structure. Fig. 7, 5549X; Fig. 8, 5595X.

(FS, Fig. 7) not present in the normal or control preparations. At higher limits of resolution, the filamentous material referred to above appears to be membranous (M, Fig. 9) and may represent fragments of formerly intact dendrites. Even in those instances where tubular structure remains, the limiting membrane appears thickened and lacking in clarity.

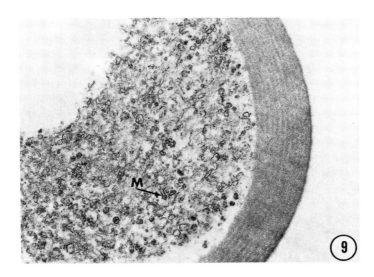

FIGURE 9. High resolution image of a portion of an
aesthetasc from the 1000 µg/1 test. The peripherally condensed material contains a number of membrane fragments (M)
which may represent portions of disrupted dendrites. 18,050X.

Chemoreceptors Exposed to 100 µg/1 Copper Sulfate

Changes in the structure of chemoreceptors after exposure
to 100 µg/1 copper sulfate solution (Figs. 10, 11) resemble
those found at 1000 µg/1 but appear less pronounced. A condensation of the extracellular, flocculent material occurs
(FD, Figs. 10, 11) either as a central band or peripherally;
however, many of the dendrites (vesicles) and their inherent
neurotubules can be resolved. Perhaps as a result of their
compression into the limited space occupied by the condensed
material, the large dendritic profiles (D, Figs. 10, 11) often
appear less circular than those in the control material (D,
Figs. 2, 3).

It should be noted that some of the aesthetascs examined
in this exposure group sustained little damage and could not
be distinguished from the 50 or 0 µg/1 specimens.

Chemoreceptors Exposed to 50 µg/1 Copper Sulfate

Exposure to 50 µg/1 of copper sulfate solution has little,
if any, noticeable effect on the morphology of blue crab
chemoreceptors. The majority of the sensilla resemble the
control specimens while only a few (Figs. 12, 13) show some
indication of either condensation or margination of

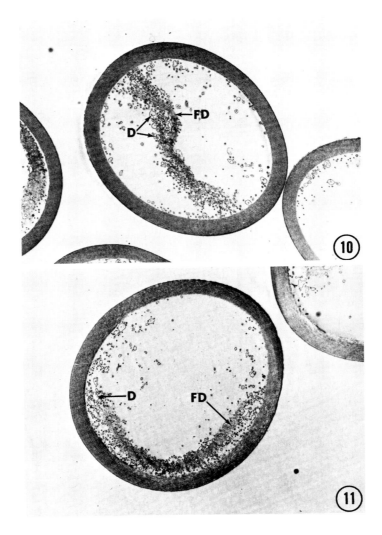

FIGURES 10 and 11.   Electron micrographs of aesthetascs
after exposure to 100 µg/l copper sulfate solution.   Both the
dendrites (D) and flocculent dense (FD) material are condensed
centrally or along the margin of the cuticle.   Fig. 10, 6000X;
Fig. 11, 5680X.

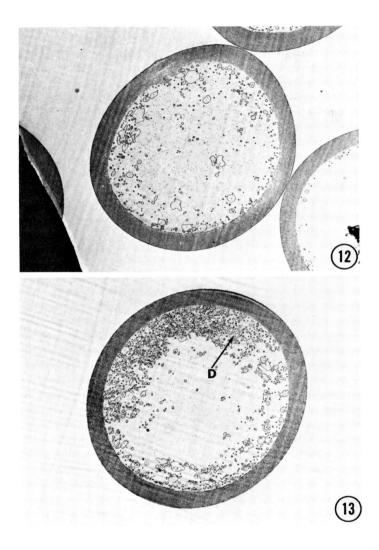

FIGURES 12 and 13. As indicated by these micrographs, exposure to 50 µg/l copper sulfate solution does not markedly alter the structure of receptor cell dendrites; however, some of them (e.g. D, Fig. 13) appear less circular than those in the controls. The slight margination of aesthetasc contents in these examples could not be differentiated from many of the control specimens (e.g. Figs. 2, 5). Fig. 12, 5800X; Fig. 13, 5227X.

aesthetasc contents. Both large and small diameter dendrites
and their associated neurotubules are present; however, some
of the sensory cell processes (D, Fig. 13) appear less circular.

DISCUSSION

     Presently, information on the cytopathological effects of
heavy metals on the sensory systems in aquatic animals is
limited. Gardner and LaRoche (1973) have shown that brief
exposure to copper chloride solutions produces severe
degenerative and hyperplastic change in cells of teleost
lateral line and olfactory systems. Similar results have also
been reported by Gardner (1975) on the effects of other heavy
metals, crude oil, and the pesticide methoxychlor on teleost
sensory epithelia. The present investigation indicates that
copper may have a necrotizing effect on the dendrites of blue
crab chemoreceptors. Regrettably, the limited scope of this
study precludes answering important questions regarding
cellular modifications that may be present in other parts of
the sensory cells (i.e. soma or axon).
     At 1000 µg/l, copper sulfate solutions completely disrupt
the integrity and disposition of the dendrites within all
the sensilla that were examined. In this instance, the damage
sustained by the receptor cells is extensive and suggests that
they are probably nonfunctional. It is not known at the
present time whether the observed lesions resulted directly
from an interaction between the toxicant and receptor cell
processes, or whether they occurred as a result of other
alterations in homeostasis. Thurberg, Dawson, and Collier
(1973) have shown that copper can abolish osmoregulatory
capability in decapod crustaceans. Perhaps the compression
and distortion of the dendrites observed here is the result
of an edematous condition occurring in a restricted space
of small dimensions (B. A. Fowler, pers. comm.). It would be
of interest to compare these types of changes with those
obtained by severing the antennulary nerves and examining the
sensory cells during retrograde degeneration.
     At lesser concentrations (100 and 50 µg/l) the effects of
copper sulfate on the receptor cell dendrites are less intense
and variable between aesthetascs in the 100 µg/l group, while
the vast majority of sensilla from the 50 µg/l test appear
"normal". Any statement regarding the function of receptors
exposed at these levels would be inappropriate at this time.
These results reflect to a certain extent the observations of
McLeese (1975) and others (Hara, Law, and Macdonald, 1976)

that the loss of olfactory or chemoreceptive capacity
resulting from the presence of heavy metals is a gradual
event that does not simultaneously affect all of the sensory
cells to the same degree.

The difficulties in analyzing the fine structural information on specimens from the lower toxicant concentrations
(i.e. 50 and 100 µg/l) clearly points out the need to augment
studies of this kind with appropriate behavioral or
electrophysiological measures. The use of such ancillary
methods may demonstrate that conventionally applied electron
microscope techniques cannot resolve the subtle modifications
or perturbations that probably occur in the nerve membrane
when the reactive groups of receptor molecules bind to certain
species of heavy metals. That such a phenomenon is responsible for the observed loss in chemoreceptive function has been
proposed by Hara, Law, and Macdonald (1976) in their study
on the depressive effects of mercury and copper on olfactory
sensitivity in the trout, *Salmo gairdneri*. This interpretation receives strong support from the better studied effects
of heavy metals on enzyme activity and biochemical interactions in other macromolecular systems (Eichhorn, 1975).

Considerable difficulty was encountered during this study
in attempting to expose the animals to a known concentration
of toxicant for a given period of time. Analyses of the
clean water used to prepare the toxicants and for the 48-hour
control test indicated that it contained 0.56 ppb of copper
and was within normal limits (L. V. Sick, pers. comm.). Copper concentrations in the 50 µg/l experiment were 66.0 and
93.6 ppb at 24 and 48 hours respectively. In the 100 and
1000 µg/l tests, copper concentrations at 24 and 48 hours
were only one-fifth and one-fiftieth of their anticipated
value. Failure to retain for analysis both the paper used to
filter the toxicant solutions sampled during the course of
the experiment and the tissues from exposed animals were
significant oversights. Presently, there are no adequate
explanations for these unusual results.

During the experiment, the test solutions in the 100 and
1000 µg/l exposures became cloudy and a yellow-green precipi-
tate was noted. Preliminary studies by P. T. Johnson
(pers. comm.) have indicated that blue crabs secrete, via
glands in the hindgut and oesophagus, a mucous-like (PAS+)
substance in response to cadmium, and perhaps this occurred
here as well. That such secretions may have been involved
in reducing the copper concentrations of the toxicant solutions is suggested from research on fish mucins (Chow,
Patterson, and Settle, 1974; Varanasi, Robisch, and Malins,
1975) which are known to bind tightly to heavy metals. Future

work may demonstrate that the blue crab and other crustaceans utilize this secretory mechanism to mitigate the effects of hostile environments.

ACKNOWLEDGMENTS

I want to thank Dr. Lowell V. Sick for analyzing the water samples and for his critique of the exposure methods. Dr. Bruce A. Fowler's critical review of this paper and comments regarding methodology were appreciated. The technical assistance of Ms. Jane T. Wade in preparing the specimens for observation is acknowledged.

LITERATURE CITED

Anderson, T. F. 1951. Techniques for the preservation of three-dimensional structure in preparing specimens for the electron microscope. Trans. N.Y. Acad. Sci., 13: 130-133.

Chow, T.J., C.C. Patterson and D. Settle. 1974. Occurrence of lead in tuna. Nature, 251: 159-161.

Eichhorn, G. L. 1975. Active sites of biological marcomolecules and their interaction with heavy metals. In: Ecological Toxicology Research, pp. 123-142. ed. by A.D. MacIntyre and C.F. Mills. New York and London: Plenum Press.

Fowler, Bruce A. 1978. Personal communication.

Fuzessary, Z.M., and J.J. Childress. 1975. Comparative chemosensitivity to amino acids and their role in the feeding activity of bathypelagic and littoral crustaceans. Biol. Bull., 149: 522-538.

Gardner, G.R. 1975. Chemically induced lesions in estuarine or marine teleosts. In: The Pathology of Fishes, pp. 657-694, ed. by W.E. Ribelin and G. Migaki. Madison, Wisconsin: University of Wisconsin Press.

Gardner, G. R. and G. LaRoche, 1973. Copper induced lesions in estuarine teleosts. J. Fish. Res. Board Can., 30: 363-368.

Ghiradella, H. T., J. F. Case and J. Cronshaw, 1968. Structure of aesthetascs in selected marine and terrestrial decapods: Chemoreceptor morphology and environment. Am. Zool., 8: 603-621.

Hara, T. J., Y. M. C. Law, and S. Macdonald, 1976. Effects of mercury and copper on the olfactory response in rainbow trout, *Salmo gairdneri*. J. Fish. Res. Board Can., 33: 1568-1573.

Hazlett, B. A.   1971.   Antennule chemosensitivity in marine
     decapod crustacea.   J. Anim. Morphol.   Physiol., 18:
     1-10.
Hindley, J. P. R.   1975.   The detection, location, and
     recognition of food by juvenile banana prawns,
     Panaeus merguiensis de Man.   Mar. Behav. Physiol.,
     3:   193-210.
Mackie, A. M.   1973.   The chemical basis of food detection
     in the lobster *Homarus gammarus*.   Mar. Biol., 21:   103-
     108.
McLeese, D. W.   1970.   Detection of dissolved substances by
     the American lobster (*Homarus americanus*) and olfactory
     attraction between lobsters.   J. Fish. Res. Board Can.,
     27:   1371-1378.
McLeese, D. W.   1974.   Olfactory response and fenitrothion
     toxicity in American lobsters *(Homarus americanus)*.
     J. Fish. Res. Board Can., 31:   1127-1131.
McLeese, D. W.   1975.   Chemosensory response of American
     lobsters *(Homarus americanus)* in the presence of copper
     and phosphamidon.   J. Fish. Res. Board Can., 32:   2055-
     2060.
McLeese, D. W., R. L. Spraggins, A. K. Bose, and
     B. N. Pramanik, 1977.   Chemical and behavioral studies
     of the sex attractant of the lobster *(Homarus americanus)*.
     Mar. Behav. Physiol., 4:   219-232.
Newman, M. W.   and G. E. Ward, (Jr.), 1973.   A technique
     for the immobilization of the chelae of blue crabs and
     identification of individual animals.   Chesapeake Sci.,
     14:   68-69.
Pearson, W. and B. L. Olla, 1977.   Chemoreception in the blue
     crab, *Callinectes sapidus*.   Biol. Bull., 153:   346-354.
Sick, L. V.   1977.   Personal communication .
Snow, P. J.   1973a.   The antennular activities of the hermit
     crab, *Pagurus alaskensis* (Benedict).   J. Exp. Biol,
     58:   745-765.
Snow, P. J. 1973b.   Ultrastructure of the aesthetasc hairs
     of the littoral decapod, *Paragrapsus gaimardii*.
     Z. Zellforsch., 138:   489-502.
Spurr, A. R.   1969.   A low viscosity epoxy resin embedding
     medium for electron microscopy.   J. Ultrastruct. Res.,
     26:   31-43.
Thurberg, F. P., M. A. Dawson, and R. S. Collier, 1973.   Ef-
     fects of copper and cadmium on osmoregulation and oxygen
     consumption in two species of estuarine crabs.
     Mar. Biol., 23:   171-175.
Varanasi, U., P. A. Robisch and D. C. Malins, 1975.
     Structural alterations in fish epidermal mucus produced
     by water-borne lead and mercury.   Nature, 258:   431-432.

COPPER AND CADMIUM INDUCED CHANGES IN THE
METABOLISM AND STRUCTURE OF MOLLUSCAN GILL TISSUE

David W. Engel

National Marine Fisheries Service, NOAA
Southeast Fisheries Center
Beaufort Laboratory
Beaufort, North Carolina

Bruce A. Fowler

Laboratory of Environmental Toxicology
National Institute of Environmental Health
Sciences, Research Triangle Park, North Carolina

INTRODUCTION

Coastal and estuarine waters of the Atlantic and Gulf
coasts have been heavily impacted by continuing urbanization
and industrialization. Because of these activities there has
been an increase in the quantities of trace metals being introduced into the marine and estuarine environments from anthropogenic sources (Helz *et al.* 1975). Some of these metals such
as copper and cadmium can be highly toxic to aquatic organisms.
Copper can be released into the environment from a number
of sources: sewage effluents, industrial wastes, antifouling
paints, and steam electric generating plants. The toxicity of
copper to aquatic species is well documented for molluscs
(Shuster and Pringle, 1969; Nelson *et al.* 1976; Davenport 1977;
and Eisler 1977); for fish (Baker 1969; Eisler and Gardner 1973;
and Reish *et al.* 1975); and for crustacea (Kerkut and Munday
1962; Thurberg *et al.* 1973; and Andrew *et al.* 1977). In these
investigations, copper was shown to affect survival, tissue
morphology and physiological processes of organisms.
Cadmium additions to coastal waters come primarily from
industrial sources such as zinc smelting and electroplating.

239

Cadmium has been linked to kidney damage (Adams *et al.* 1969) and Itai-Itai disease in man (Douglas-Wilson 1972). In the aquatic environment, molluscs have been shown to concentrate cadmium, but they are undamaged at extremely high levels. Although *Mytilus edulis* accumulated large amounts of cadmium (60 ppm from the environment), they showed no observable effect (Nickless *et al.* 1972). In laboratory experiments, however, oysters suffered heavy mortalities when the body burden of cadmium exceeded 100 ppm (Shuster and Pringle 1969). Also, in our laboratory we have measured oxygen consumption rates on excised clam gill tissue following acute cadmium exposure (0.1 to 1.0 ppm Cd) and have demonstrated little effect on the rate of oxygen consumption by the tissue (Engel, unpublished data).

The American oyster, *Crassostrea virginica,* was chosen for this investigation because of its importance to man as food and its ability to concentrate a variety of metals (Eisler, *et al.* 1972; Huggett, *et al.* 1974; Kopfler and Mayer 1973; Shuster and Pringle 1969; and Roosenberg 1969). It also fulfills many of the requirements for a biological indicator organisms of trace metal pollution (Phillips 1977). Previous investigations at our laboratory have shown that although oxygen consumption rates of excised gill tissue of some molluscs were elevated by a short-term copper exposure, the oyster tissue was relatively unaffected. Acute exposure to cadmium also had little measurable effect upon gill-tissue oxygen consumption at the concentrations tested. The present study examines the physiological and cellular effects of accumulated copper and cadmium resulting from protracted rather than acute exposures. The purpose of these experiments was to examine changes in (1) metal accumulation (2) oxygen consumption and (3) fine structure of gill tissue and (4) to investigate synthesis of metal binding proteins in the whole oyster as a function of accumulated copper and cadmium.

MATERIALS AND METHODS

All of the oysters used in these experiments were collected in a marsh near the National Marine Fisheries Service's, Beaufort Laboratory, Beaufort, North Carolina from January through September 1977. The animals were maintained in the laboratory in flowing seawater for 24 hours to clear the gut, and then placed in 16 liter exposure-tanks to acclimate for 72 hours at a salinity of 34o/oo and temperature of $\sim20^0$C. After acclimation the oysters were either exposed to copper or cadmium (added as $CuSO_4 \cdot 5H_2O$ and $CdCl_2$) in a flowing water system with a flow rate of 4 liters of water per hour (Engel *et al.* 1976), and the controls were treated in a similar manner, but not exposed to metal. The concentrations of metals in the water were checked regularly using atomic absorption spectrophotometry and varied no more than 20% from the desired levels, which were 50 and 100 ppb copper and 100 and 600 ppb cadmium. All tanks were vigorously aerated and the pH of the water ranged from 8.0 to 8.2.

Oysters used in respiration experiments were removed from the tanks after 1, 3, 5, 7, and 14 days, and then were allowed to flush in flowing seawater for 24 hours. All respiration measurements were made on excised oyster gill tissue from the same area of the gill. One oyster was used in each measurement with ten oysters per sampling time and each experiment was replicated twice except for the 600 ppb cadmium group. The respiration rates were measured using standard manometric techniques and a Gilson Respirometer*. Gill tissue was dried at $90^{\circ}C$ for 24 hours and weighed, and all respiration rates were expressed as microliters of oxygen consumed per milligram dry weight per hour.

Metal concentrations in gill tissue were determined on the same gill pieces used for the respiration measurements. Two pieces of dried tissue were pooled for each measurement. The tissue was wet ashed with concentrated HNO_3, taken to dryness, diluted to desired volume with 0.25N HCl, and then measured for copper and cadmium concentrations by atomic absorption spectrophotometry. A National Bureau of Standards bovine liver standard was run through the same procedure as our experimental samples; the values were within 95% of the expected value.

To determine whether metal-induced changes in tissue oxygen consumption were correlated with alterations in cellular structure, segments of mantle and gill epithelium were dissected from the same oysters used in the oxygen consumption experiments for examination with the electron microscope. The tissue was fixed, processed, and embedded for electron microscopy as previously described (Fowler *et al.* 1975). Thin sections of tissue (600-900 Å thick) were cut with a Porter-Blum MT-1 ultramicrotome and examined in a Phillips EM 300 operated at 80 kev.

The molecular binding of copper and cadmium in exposed oysters was assessed by gel chromatography and protein purification procedures. For these studies, whole oysters were homogenized and the homogenates centrifuged at 27,100 X g for 15 minutes in a Sorval RC-5 centrifuge. Supernatant material was decanted and heat treated in a shaker bath at $60^{0}C$ for 10 minutes. The heat treated supernatant was centrifuged at 27,100 X g and the resulting supernatant was subjected

Reference to trade name does not imply endorsement of the product by the National Marine Fisheries Service, NOAA.

to G-75 Sephadex chromatography, followed by DEAE chromatography and disc gel electrophoresis as described by Winge *et
al.* (1975) and Ridlington and Fowler (1977).

RESULTS

Metal Accumulation and Tissue Oxygen Consumption

The rate of copper accumulation by the oyster gill tissue
was a function of the copper concentration (Figure 1). Linear
regression analysis of these data demo-strated there was a
significant difference (P < 0.05) between the rates of copper
accumulation at 50 and 100 ppb, and these rates of uptake of
copper by the oyster gills were different by a factor of 2.
Also, some mortality of oysters occurred in the 100 ppb group,
but none at either 50 ppb or in the controls. This mortality
may have been caused by accumulated copper. The surviving
oysters exposed at 100 ppb copper for 14 days were severely
affected by the copper which may have contributed to the
observed cariation in measured copper concentrations.

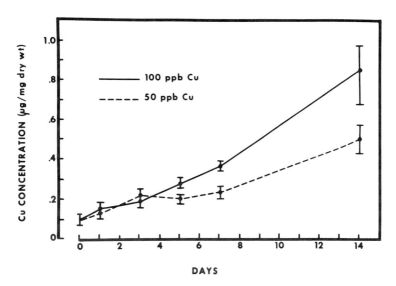

*FIGURE 1. The accumulation of copper by the gills of
oysters exposed to 50 and 100 ppb copper for 1, 3, 5, 7 and 14
days. Each point represents a mean of 10 determinations ± 1
standard error. Control copper concentrations in gill tissue
remained constant throughout the experiments and a mean value
is given at day 0.*

Oxygen consumption rates of excised oyster gills were dependent upon both time of exposure and copper concentration (Figure 2). Linear regression analysis demonstrated that the rate of increase in oxygen consumption with time was significantly greater (P < 0.05) for oysters exposed to 100 ppb than for either the controls of 50 ppb group, but there was no significant difference (P > 0.05) between the 50 ppb group and control oysters. The most obvious difference between treatments occurred after 14 days exposure, at which time the tissue concentration of copper had reached 0.8 µg/mg dry weight.

The uptake of cadmium by oysters was both time and concentration dependent (Figure 3). The concentration of cadmium in oyster gill tissue from animals exposed to 600 ppb appeared to approach saturation between the 7th and 14th days, and the rate of accumulation was significantly greater than at 100 ppb. The accumulation of cadmium by oyster gill at 100 ppb increased linearly through 14 days. The cadmium concentration for control oyster gill was reported as zero, because it was < 0.05 which was below our detection limit where we had an obvious loss of precision. Only five oysters in the group exposed to 600 ppb survived until the 14th day while the other treatment group and the control had no mortalities.

Cadmium concentration of the water and the time of exposure affected rates of oxygen consumption by oyster gill tissue (Figure 4). Direct comparison of the oxygen consumption rates between the two cadmium doses was complicated by large differences in the oxygen consumption rates of the two sets of controls. Linear regression analysis demonstrated that the oxygen consumption rates of gill from oysters exposed to 600 ppb were significantly affected by cadmium (P < .01) and time of exposure (P < .01), but that there was no significant interaction between time of exposure and concentration. Oysters exposed to 100 ppb had gill-tissue oxygen consumption rates that did not differ significantly from the controls with respect to concentration or time of exposure, except at 5 days when unexplained low control values were obtained.

Electron Microscopy

Ultrastructural examination of gill tissue from oysters exposed to 100 ppb cadmium for 14 days showed no significant changes in ciliated epithelial cells, but demonstrable changes did occur in oysters exposed to 100 ppb copper for 14 days. The ultrastructure of gill epithelial cells from control oysters at 14 days is shown in Figure 5. These cells are columnar with numerous microvilli and cilia at the apical surface. The cytoplasm is rarified and elongated mitochondria are most prominent in the apical region. Not only was there to detectable ultrastructural alteration of these cells

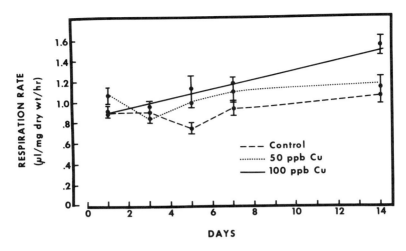

FIGURE 2. Respiration rates of excised oyster gill tissue from animals exposed to 0, 50 and 100 ppb copper for 1, 3, 5, 7 and 14 days. Each point represents a mean respiration rate of 10 to 20 oysters ± standard error.

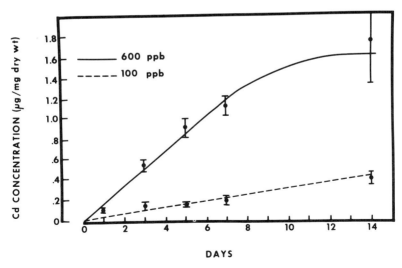

FIGURE 3. The accumulation of cadmium by the gills of oysters exposed to 100 and 600 ppb cadmium for 1, 3, 5, 7 and 14 days. Each point represents a mean of 10 determinations ± 1 standard error. Control cadmium concentrations in gill tissue was below the level of detection for our population of oysters.

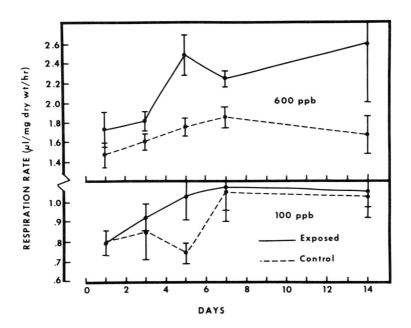

FIGURE 4. *Respiration rates of excised gill tissue from animals exposed to 100 and 600 ppb cadmium for 1, 3, 5, 7 and 14 days. Each point at 100 ppb represents a mean of 20 animals and at 600 ppb 10 animals ± 1 standard error.*

following cadmium exposure, but there was also no demonstrable change in the hydration state of the tissue. In contrast, gill epithelial cells from copper-treated animals at 14 days were rounded and swollen (Figure 6). Mitochondria with rarified matrices were also present in epithelial cells from these

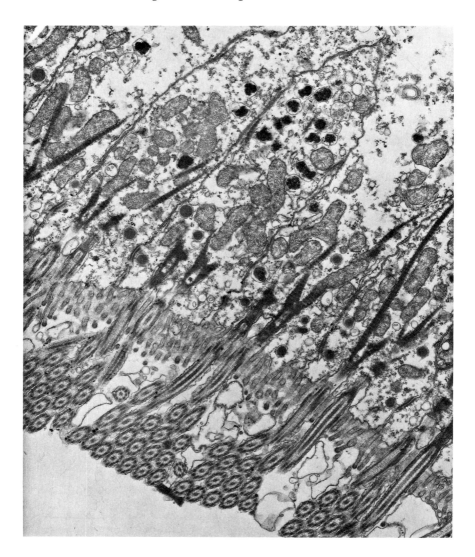

FIGURE 5. An electron micrograph of columnar gill epithelial cells from a control oyster showing numerous mitochondria in the apical region of the cells. The magnification was 25,047X.

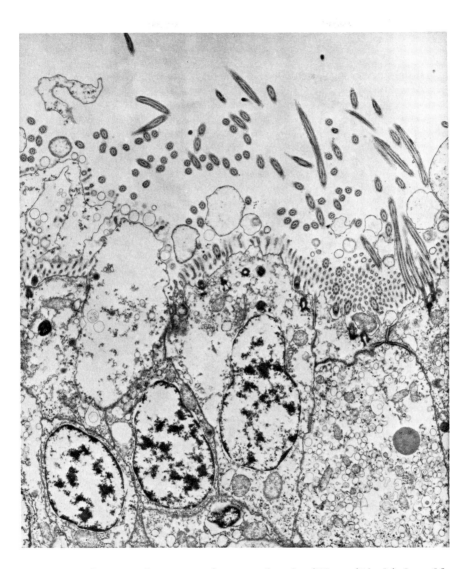

FIGURE 6. An electron micrograph of gill epithelial cells from an oyster exposed to 100 ppb copper for 14 days. The micrograph shows apical swelling, cytoplasmic vesiculation and mitochondria with rarified matrices 11,463X.

animals (Figure 7). Additional evidence for copper-induced
edema of gill tissues was observed when water contents of
exposed and control tissues were calculated from dry weight-
wet ratios (Figure 8). While there was considerable variation
within treatment groups, a general increase in water content

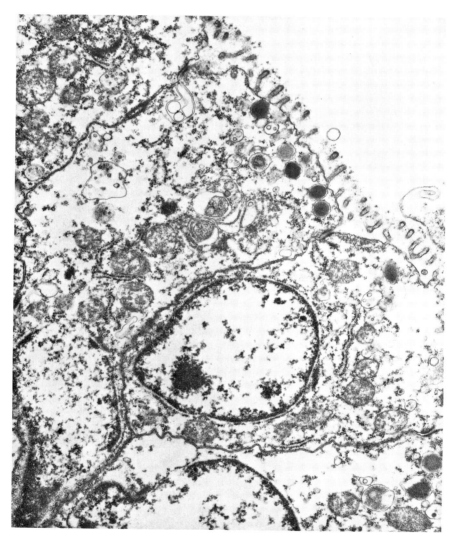

*FIGURE 7. A higher magnification electron micrograph of
gill epithelial cells from oysters exposed to 100 ppb copper
for 14 days. The micrograph shows the breakdown of the mitochondrial matrices in greater detail 25,047X.*

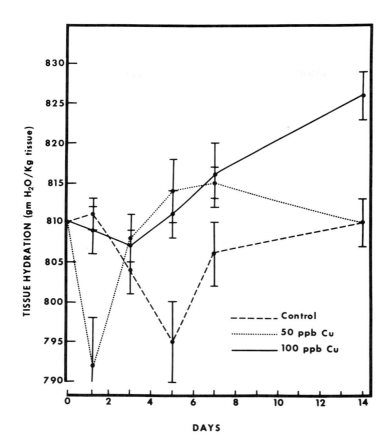

FIGURE 8. The percent hydration of oyster gill tissue from animals exposed to 0, 60 and 100 ppb copper as a function of time exposure. Each mean represents a mean of 10 oyster gills ± 1 standard error.

with time of exposure was noted for gills from oysters treated with 100 ppb copper, which indicated edema. Such changes in water content are consistent with ultrastructural findings on the same animals.

Protein Isolation and Characterization

In experiments made on whole oysters exposed to 100 ppb cadmium a cytoplasmic protein was demonstrated. This was a low molecular weight metal-binding protein selective for cadmium, and synthesis of the protein was found to be time dependent (Figure 9). It should be stressed that the protein is synthesized *de novo* since none was present before exposure

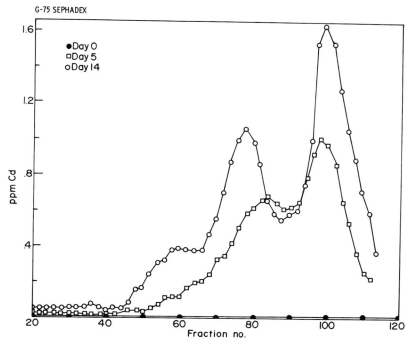

FIGURE 9. A series of G-75 Sephadex column chromatographic elution profiles for protein from oysters exposed for
0, 5 and 14 days to 100 ppb cadmium. The two protein peaks
increased in height with the duration of cadmium exposure and
fall in the low molecular weight region.

to cadmium. Ridlington and Fowler (1977) performed an amino
acid analysis on this isolated protein and found unusually
high concentrations of aspartic and glutamic acids and low
concentrations of cysteine. This suggests that the chelation
properties of the protein are probably related to the carbonyl
group of these dicarboxylic acids rather than the sulfhydryl
group of cysteine as seen in mammalian metallothionines.

    In the oysters exposed to copper no specific copper-
selective protein could be identified. Instead, copper was
associated with a variety of different high molecular weight
proteins.

DISCUSSION

    The effects of metal pollutants on the tissue morphology
and physiology of marine and estuarine organisms, e.g. the
oyster, amy be independent upon the tissue concentration of the

metal and the rate of its accumulation. In the present study
the test concentrations of both copper and cadmium showed no
significant immediate effects on tissue oxygen consumption or
mortalities, but at 11-14 days, when sufficient copper and
cadmium had accumulated, such effects were evident. The level
of accumulated copper which we found to be lethal was around
1.0 µg/mg dry weight. Roosenburg (1969) measured copper concentrations in oysters collected near the Chalk Point, Maryland
steam electric station. He found copper concentrations up to
1.2 µg/mg dry weight of oyster. Such concentrations in his
investigation were "sublethal", but similar concentrations in
our studies caused sever pathological changes, which could be
accounted for if the rate of accumulation was a controlling
factor. The accumulation of cadmium in oysters exposed to
600 ppb approached saturation between 7 and 14 days, and caused
mortalities by the 14th day. Since it has been demonstrated
that oysters produce a cadmium binding protein in response to
an exposure to 100 ppb cadmium (Ridlington and Fowler 1977),
the dose of 600 ppb may have simply saturated the system allowing the excess cadmium to bind to critical sites and cause
pathological effects.

Studies with copper indicate that this metal was able to
enter tissues, such as the gills, and alter cell morphology as
well as oxygen consumption rates. The cellular swelling and
mitochondrial damage observed by electron microscopy suggest
that the observed increases in gill-tissue respiration may be
related to increased cellular or mitochondrial membrane permeability. Further studies are needed to clarify the mechanism
of this phenomenon and its impact on control of respiration
including possible effects on cellular ion transport.

The ability to detoxify substances may be of substantial
importance to the survival of an organism in an environment
contaminated with heavy metals. The production of a low
molecular weight, cadmium-binding protein probably functions
as a detoxifying feedback mechanism in which cadmium is bound
up in a "metabolically inert" complex, thereby preventing
cadmium from binding with other more sensitive sites. Although
a binding protein was induced in oysters, other investigators
have demonstrated an apparent upper limit to the tissue concentration of cadmium, 96.5-125.8 ppm wet weight (Shuster and
Pringle 1969). In their experiments cadmium was the only metal
of the four tested (Cd, Cu, Cr, and Zn) that caused significant
mortalities, and the time required for 100% mortality (16 and
13 weeks) was dose dependent (0.1 and 0.2 ppm). Thus, this
bound cadmium must have some finite limit with regard to intracellular storage. Interestingly, in our experiments copper
did not appear to elicit a similar response from the oysters,
i.e. induction of a low molecular weight metal-binding protein.
However, in experiments done by Shuster and Pringle (1969) with

oysters, 20-week exposures to 0.05 ppm copper resulted in oysters
with copper concentrations of 943.8 to 1125.0 ppm wet weight,
which did not cause excessive mortality. Also, they reported
that these oysters exuded or "bled" a blue-green gelatinous
meterial. This material may have been protein bound copper,
which the oysters were eliminating. Thus, once these metals
enter the oyster, they may be complexed and stored intracell-
lularly or expelled from the animal. The complexation of
metals may also alter the bioavailability of these metals to
higher trophic levels.

In recent experiments conducted on the effects of trace
metals upon aquatic species, it was demonstrated that the free
metal ion concentration in the medium may be responsible for
the observed metal toxicity. Sunda and Guillard (1976) demonstrated that the accumulation and toxicity of copper in algae
was dependent upon the free cupric ion activity rather than
on total dissolved copper concentrations. Similar results
have been demonstrated for the acute toxicity of cadmium to
grass shrimp, *Palaemonetes pugio,* (Sunda *et al.*in press), and
the toxicity of copper to a freshwater cladoceran, *Daphnia
magna* (Andrew *et al.* 1977). Such investigations were designed
to determine the mechanism of trace metal toxicity and were
conducted using both organic and inorganic ligands to control
metal complexation in the experimental media. These data
suggest that the active chemical form in trace metal bioavailability and toxicity is the free metal ion whose concentration may be significantly less than the total metal concentration, depending upon the level of complexation.

In our experiments with the accumulation and effects of
copper and cadmium on oyster gill tissue, the total element
concentration and pH of the exposure media were controlled,
but we were aware that dissolved complexing agents could alter
the availability of the two metals to the oysters. Our prime
concern was the chelation capacity for copper of our seawater
supply, and this was measured by Sunda (personal communication)
using a bacterial assay system similar in design to that described by Davey and coworkers (1973). His measurement indicated that our lowest concentration, 50 ppb, was at least 5
times higher than the chelation capacity of our seawater.
Thus, the bioavailability of copper to oysters would not be
affected significantly by the dissolved complexing agents in
the water. Cadmium ion concentration was not of great concern,
because cadmium should be complexed orimarily by chloride ion
(Sunda *et al.* in press), and this level of chloride complexation
would of course be constant at a constant salinity (34o/oo).
Therefore, the concentration of free cadmium ion in solution

should have remained constant due to the stable salinity. So, by controlling the total element, pH, and salinity we effectively controlled both free cupric and cadmium ion concentrations in the exposure seawater.

Since it has been demonstrated that accumulated copper and cadmium can affect the physiological balance and intracellular morphology of oysters, further experiments are needed to study the role of metal speciation on accumulation and toxicity, how rate of accumulation of copper and cadmium affect physiological response, and how both cellular morphological changes and biochemical reaction rates may be linked.

Finally, data from the above studies are of interest with respect to understanding how marine food organisms such as the oyster are capable of acting as vectors for human exposure to toxic metals like cadmium without displaying overt signs of toxicity themselves. It is apparent that knowledge of intracellular uptake, binding, and distribution of metals is a key factor in understanding this phenomenon. Studies concerning the bioavailability of toxic metals to humans following ingestion of contaminated marine organisms, relative to other potential exposure routes, also are needed badly to assess possible deleterious human health effects.

SUMMARY

Experiments were conducted to determine the physiological and cellular effects of accumulated copper and cadmium on the gill tissue of Eastern oyster, *Crassostrea virginica,* as a measure of damage to the whole animal. Measurements were made of (1) metal accumulation, (2) oxygen consumption, (3) fine structure of gill tissue, and (4) induction of metal-binding proteins specific for cadmium or copper. The oysters were exposed to either 50 or 100 ppb copper, or 100 or 600 ppb cadmium continuously for up to 14 days. The gill tissue from exposed oysters accumulated both copper and cadmium as a function of time of exposure. Copper caused significant increases in the rate of oxygen consumption only in the 100 ppb exposure group and 600 ppb cadmium also caused similar changes, but the data were more variable. Comparisons of the fine structures of gill tissue from oysters exposed to either 100 ppb copper or 100 ppb cadmium for 14 days showed that only copper caused intracellular alterations. Exposure to 100 ppb cadmium, however, caused the induction of a cadmium-binding, low-molecular weight protein. This study demonstrated that accumulated copper and cadmium caused significant physiological and cellular changes in the gill tissue of the oyster.

ACKNOWLEDGMENTS

This research was supported in part by the U. S. Environmental Protection Agency under an interagency agreement relating to the Federal Interagency Energy/Environmental Research and Development Program, and in part through an interagency agreement between the Energy Research and Development Administration and the National Marine Fisheries Service. The authors thank Dr. Webster Van Winkle of Oak Ridge National Laboratory and also Mr. William Bowen of the Beaufort Laboratory for their assistance in conducting this investigation. Southeast Fisheries Contribution Number 78-04B.

LITERATURE CITED

Adams, R. G., J. F. Harrison, and P. Scott. 1969. The development of cadmium-induced proteinuria, impaired renal function and ostemalacia in alkaline battery workers. Q. J. Med. 38: 425-443.

Andrew, R. W., K. E. Biesinger, and G. E. Glass. 1977. Effects of inorganic complexing on the toxicity of copper to *Daphnia magna*. Water Res. 11: 309-315.

Baker, J. T. P. 1969. Histological and electron microscopial observations on copper poisoning in the winter flounder (*Pseudopleuronectes americanus*). J. Fish. Res. Board. Can. 26: 2785-2793.

Davenport, J. 1977. A study of the effects of copper applied continuously and discontinuously to speciments of *Mytilus edulis* (L.) exposed to steady and fluctuating salinity levels. J. Mar. Biol. Assoc. U. K. 57: 63-74.

Davey, E. W., M. J. Morgan, and S. J. Erickson. 1973. A biological measurement of copper complexation capacity of seawater. Limnol. Oceanogr. 18: 993-997.

Douglas-Wilson, I. 1972. Cadmium pollution and Itai-Itai disease. Lancet 1972: 382-383.

Eisler, R. 1977. Acute toxicities of selected heavy metals to the softshell clam, *Mya arenaria*, **Bull.** Environ. Contam. Toxicol. 17: 137-145.

Eisler, R., and G. R. Gardner. 1973. Acute toxicology to an estuarine teleost of mixtures of cadmium, copper and zinc salts. J. Fish. Biol. 5: 131-142.

Eisler, R., Zaroogian, G. E. and Hennekey, R. J. 1972. Cadmium uptake by marine organisms. J. Fish Res. Board Can. 29: 1367-1369.

Engel, D. W., S. M. Warlen, R. M. Thuotte and C. W. Lewis.
    1976. A comparison of a flow-through bioassay system with
    a static system, p. 412-422. In Annual Report of National
    Marine Fisheries Service, Beaufort Laboratory, Beaufort,
    N. C. to Energy Research and Development Administration,
    1976.

Fowler, B. A., D. A. Wolfe, and W. F. Hettler. 1975. Mercury
    and iron uptake by cytosomes in mantle epithelial cells of
    quahog clams (*Mercenaria mercenaria*) exposed to mercury.
    J. Fish. Res. Board Can. 32: 1767-1775.

Helz, G. R., R. J. Huggett, and J. M. Hill. 1975. Behavior
    of Mn, Fe, Cu, Zn, Cd and Pb discharged from a wastewater
    treatment plant into an estuarine environment. Water Res.
    9: 631-636.

Huggett, R. J., O. P. Bricker, G. R. Helz, and S. E. Sommer.
    1974. A report on the concentration, distribution, and
    impact of certain trace metals from sewage treatment plants
    on Chesapeake Bay. Chesapeake Research Consortium Publ.
    No, 31. 17 p.

Kerkut, G. A., and K. A. Munday. 1962. The effect of copper
    on the tissue respiration of the crab *Carcinus maenas*.
    Cah. Biol. Mar. 3: 27-35.

Kopfler, F. C., and J. Mayer. 1973. Concentration of five
    trace metals in the waters and oysters (*Crassostrea
    virginica*) of Mobile Bay, Alabama. Proc. Natl. Shellfish.
    Assoc. 63: 27-34.

Nelson, D. A., A. Calabrese, B. A. Nelson, J. R. MacInnes, and
    D. R. Wenzloff. 1976. Biological effects of heavy metals
    on juvenile bay scallops, *Argopecten irradians*, in short-
    term exposures. Bull. Environ. Contam. Toxicol. 16: 275-
    282.

Nickless, G., R. Stenner, and N. Terrille. 1972. Distribution
    of cadmium, lead and zinc in the Bristol Channel. Mar.
    Pollut. Bull. 3: 188-190.

Phillips, D. J. H. 1977. The use of biological indicator
    organisms to monitor trace metal pollution in marine and
    estuarine environments - A review. Environ. Pollut. 13:
    281-317.

Reish, D. J., T. J. Kauwling, and A. J. Mearns. 1975. Marine
    and estuarine pollution. J. Water Pollut. Control Fed.,
    47: 1617-1635.

Ridlington, J. W., and B. A. Fowler. 1977. Isolation and
    partial characterization of an inducible cadmium-binding
    protein from American oysters. Pharmacologist 19: 180.

Roosenburg, W. H. 1969. Greening and copper accumulation
    in the American oyster, *Crassostrea virginica*. Chesapeake
    Sci. 10: 241-252.

Shuster, C. N., Jr., and B. H. Pringle. 1969. Trace metal
    accumulation by the American oyster, *Crassostrea virginica*.
    Proc. Natl. Shellfish. Assoc. 59: 91-103.

Sunda, W. G., and R. R. L. Guillard. 1976. The relationship
    between cupric ion activity and the toxicity of copper to
    phytoplankton. J. Mar. Res. 34: 511-529.

Sunda, W. G., D. W. Engel, and R. M. Thuotte. In Press.
    Effect of chemical speciation on the toxicity of cadmium
    to grass shrimp, *Palaemonetes pugio*: Importance of free
    ion concentrations. Environ. Sci. Technol.

Thurberg, F. P., M. A. Dawson, and R.S. Collier. 1973. Effects
    of copper and cadmium on osmoregulation and osygen con-
    sumption in two species of estuarine crabs. Mar. Biol. 23:
    171-175.

Winge, D. R., R. Premkumar, and K. V. Rajagopalan. 1975.
    Metal-induced formation of metallothionines in rat liver.
    Arch. Biochem. Biophys. 190: 242-

PART III

Pesticides and PCBs

PESTICIDES: THEIR IMPACT ON
THE ESTUARINE ENVIRONMENT

DelWayne R. Nimmo

U.S. Environmental Protection Agency
Environmental Research Laboratory
Sabine Island
Gulf Breeze, Florida

Synthetic organic pesticides have been present in our
coastal waters for nearly four decades. The successful control of arthropods as vectors of disease by the pesticide,
DDT, during and after World War II insured subsequent use of
related chemicals and introduction into the environment.
Only after the book, "Silent Spring" (Carson, 1962), was
published did the general public become aware of the seemingly
pervasive nature of some of these pesticides and their undesirable effects on non-target species or ecosystems. One
year later, Butler and Springer (1963) described the potential
hazards of pesticides to coastal areas in a paper entitled
"Pesticides--A New Factor in Coastal Environments." The
authors noted that earlier research (1945-1951) was concerned
with species in inland waters, but more recent attention had
been directed towards the effects of pesticides in coastal
waters. Although the environmental awareness movement was in
its infancy, usage of organic pesticides continued to increase
throughout the 1960's. Toxicologists occasionally used marine
species to test the effects of selected chemicals, but few
addressed the potential ecological effects.
    Not until the late 1960's was attention given to the pesti-
cide problem generally and the impact on our coastal waters
specifically. "Pesticides and the Living Landscape" (Rudd,
1964) directed attention to the potential ecological problems
of pesticides. The book, "Since Silent Spring" (Graham, 1970),
also addressed many related subjects, including a history of
the Nation's response to the pesticide problem and the inadequate control of chemical pesticides other than DDT. However,
in my opinion, the publication that inspired critical

examination of the effects of pesticides on the Nation's
estuaries and marine waters was "DDT Residues in an East Coast
Estuary: A Case of Biological Concentration of a Persistent
Insecticide" (Woodwell et al., 1967). As a result, this
Nation was thrust into the 1970's "decade of environmental
awareness," during which the use or misuse of pesticides
became a prime concern.

FACTORS INFLUENCING THE USE OF PESTICIDES

     Even with the environmental awareness of the 1970's, it
is still difficult to determine the prime cause for the decline
in use of DDT.   Perhaps by looking at past annual-use trends,
we can discover some of the reasons. Despite warnings of
hazards during the early 1960's, production of chlorinated
hydrocarbon pesticide types continued to increase by 6% per
annum (von Rumker et al., 1975). In 1969-70, some decreases
in production were noted (probably entirely due to declining
sales of DDT), but overall sales rebounded in the next two
years, with at least 7% increase per year. Use of DDT actually
began to decline as early as 1965 (von Rumker et al., 1975),
according to domestic sales receipts. Two factors could have
operated in concert: public awareness of hazards of DDT or
actions of the pesticide regulatory agencies.
     Although restrictions by the Environmental Protection
Agency (EPA) concerning the use of certain pesticides have
been instituted, relatively few pesticide registrations have
been cancelled (e.g., Kepone); most have been restricted to
small-volume usage or to specialized locations or restricted
uses, such as subsurface termite control. EPA's classification of chemicals in restricted or general-use categories, or
its RPAR (rebuttal presumption against re-registration policy),
gives notice to the manufacturers and the public that cur-
tailment might follow. Chemicals must be registered before
use is permitted and less than 100 new chemicals have received
registration labels since 1972 (Betz, personal communication,
Office of Pesticide Programs, EPA, Washington, D.C.). RPAR is
a non-adversary process in which the evidence of risk or hazard
versus the benefit of a chemical is accumulated over a period
of months. The RPAR process is not supported by the chemical
manufacturers (Anonymous; Farm Chemicals, 1977a) because the
process presumes that the chemical is "harmful" and, therefore
makes it necessary for the company to prove there is no
potential for harmful effects, or that any risk can be minimized to an acceptable level. The risk/benefit assessment of
RPAR may eventually decide in favor of the use of the chemical,
but some manufacturers believe that RPAR can cause sales to

decrease because charges of risk or hazard (whether eventually
founded or unfounded), often cast a cloud of doubt in the minds
of the public about the propriety of using the chemical.

HOW THE PESTICIDE PROBLEM IS PERCEIVED

In national surveys of environmental problems, pesticides
are ranked very high by some, but not necessarily all, elements
of our society. For example, environmental problems in various
freshwater streams in Texas were ranked by categories of
pollutants (Hann, 1975) and pesticides were rated only ninth,
preceded by such problems as pathogenic bacteria, organics,
inorganic ions, nutrients, sediments, temperature changes,
heavy metals, and radionuclides. On the other hand, at the
conference where the above data were presented, Li (1975)
showed that agricultural operations were responsible for 7.8%
of the fish kills in the Nation's water from 1963-1972. Of
the total fish killed, 37% occurred in estuarine water; of the
fish killed by agricultural operations, 64.9% resulted from
pesticides.

WHAT HAPPENS WHEN THE USE OF A PESTICIDE IS HALTED?

Abatement of pesticide use in or near an estuary does not
guarantee that the effects of the pesticide already in the
system will subside quickly. Historically, it has been difficult to document cases in which causal effects can be attributed to a specific pollutant, despite frequent circumstantial
evidence. To validate possible effects, we have tended in the
past to study chemicals that degrade slowly and, therefore,
leave residues or metabolites.

In a recent publication, Klass and Belisle (1977) showed
that after six years, DDT and metabolites in nine estuarine
species were reduced from between 84 and 99%. DDE, a metabolite of DDT, was still present in 1973 but at reduced levels
when compared to those in 1967. In contrast, Butler (in press)
noted that total concentrations of DDT and metabolites from
North Carolina fish did not decrease appreciably, but during
the five-year study, there was a shift of DDT from 31 to 0%
while DDE increased from 44 to 82%. The longevity of persistent chemicals in an estuary has been demonstrated in oysters
monitored monthly for a polychlorinated biphenyl (PCB) in
Escambia Bay, Florida (Wilson and Forester, 1978). PCB's
in oysters showed definite decreases over time during the
first four years (1969 to 1972), but the rate of decrease

diminished during the four succeeding years (1973 to mid-
1976). Thus, concentration of certain persistent chemicals
may diminish after curtailed usage, but rates of decline are
evidently functions of rates of degradation or metabolism and
rates of export out of estuaries. Unfortunately, in the case
of pesticides, such as Kepone or DDT, the degradation rates
can be extremely slow.

INCREASED HERBICIDE USAGE

From 1966 to 1971, the percentage increase in use of herbicides has surpassed that of other chemicals (von Rumker et al.,
1975) and many changes in agriculture and other markets fore-
cast additional increases. Changing methods of agriculture,
movement of people to areas offering more leisure activities
out-of-doors, increased gardening, and newly constructed high-
ways all contribute to increased reliance on herbicides.
Increase in herbicide usage is also reflected in expenditures
for expansion of airport runways, electric utilities, land-
scaping for large industrial sites, golf courses, etc.; in
1976, herbicides sales totaled an estimated $105 million on
the retail market (Anonymous; Farm Chemicals, 1977b).
Increases in herbicide use could come from changes in
agricultural practices. A new practice, "no-till or
restricted-till" farming, consists of one crop planted with
tilling, followed by a second planting without breaking and
tilling the soil (Triplett and Van Doren, 1977). Instead of
plowing under the previous cover crop, a special planter opens
the soil by cutting through the mulch and plants the seed.
Herbicides are applied before or during planting to control
competing foliage. In comparison with conventional tillage, a
saving results in labor, energy, moisture, and soil due to
less erosion. Obviously, this practice increases reliance on
chemical--herbicides to control the weeds and insecticides to
control the increased populations of insects residing in the
cover crop.
Present estimates indicate that about 90% of all herbicides
used by the American farmer are applied to field crops: corn,
50%; soybeans, 17%; cotton, 11% (von Rumker et al., 1975).
Under typical conditions, less than an estimated 5% of the
amount of the herbicides applied enter aquatic areas from runoff water (Unger and Bailey, 1977). These authors, however,
presented data on two herbicides, dichlobenil and atrazine,
that appear in surface runoff at concentrations as high as 7
and 11% of that applied, respectively.
In general, herbicides are considered less toxic to estuarine species than other pesticides and data supporting this

view were presented by Butler (1965). For example, he showed
the organophorphorus-type insecticides to be about 1000 times
more toxic than the herbicides to shrimp. We should be
cautious, however, about generalizations because some herbicides are very toxic to aquatic organisms. For example, the
herbicide trifluralin is acutely and chronically toxic to a
freshwater daphnid and the fathead minnow at concentrations of
a few parts per billion (μg/ℓ) (Macek et al., 1976). Recently,
Parrish et al. (in press) showed deleterious effects of trifluralin on reproduction of an estuarine fish (*Cyprinodon
variegatus*) exposed to 1.3 μg/ℓ in a chronic test. The authors
concluded that to protect this estuarine species throughout
its reproductive phase, the maximum acceptable toxicant concentration (MATC) should be less than 1.3 μg/ℓ.

USE OF PESTICIDES IN INDUSTRY

   Some chemicals, not generally recognized as pesticides,
but produced in tremendous quantities for decades, could
inflict more damage on the estuarine environment than many
pesticides currently under study. For example, von Rumker
et al. (1975) reported that in 1972, the annual production of
creosote as a wood preservative was 970 million pounds;
pentachlorophenol (PCP), 38 million pounds; and inorganic preservatives, 14.5 million pounds. Approximately 8% were used
on marine pilings. PCP uses include treatment of railroad
crossties and utility poles that may come in direct or
indirect contact with water. Pentachlorophenol, as the
sodium salt, is also used as a wide spectrum fungicide-
bactericide to retard discoloration of freshly sawed lumber,
and in the formulation of drilling lubricants for oil exploration (von Rumker et al., 1975). One-third of the PCP sodium
salt produced is used to inhibit algae and fungi in air-
conditioning cooling towers. Small amounts of PCP are also
used as a herbicide.

NEW PESTICIDES

   The purpose of this section is to mention some of the new
chemicals coming into use; a few are labeled for restricted
use. Earlier I mentioned that the controlled use of pesticides
can sometimes pose a minimal risk to the environment. The
current problem to toxicologists, however, is the advent of
new pesticides to replace those used historically. Most

of these are types of which we have little data as to their
toxicity, bioconcentration, fate, or metabolism.

Among those likely to be used more widely in the future
are the pyrethroids that are related to the naturally occurring
insecticides, pyrethrins. The pyrethroids are more stable
photochemically than pyrethrins because of their molecular
structure, thereby retaining the features that give the molecule its insecticidal activity and low mammalian toxicity.
One pyrethroid that shows promise of greater use is Permethrin®
(Elliott, 1976). Its toxicity ratio of insect/rate was 1400,
compared to that of another highly toxic insecticide, parathion, that had a ratio of only 11. Currently, at least three
pyrethroid products are under pre-registration review by EPA.

The second group that appears to be a viable alternative
to many insecticides currently in use is the insect growth
regulators. This group contains two types that are distin-
guishable by their physiological action. Altosid®, also known
as Methoprene, is an example of the juvenile-hormone type;
Dimilin® is an example of a chemical that interferes with the
molting process.

Altosid, a chemical analogue to a juvenile hormone of
insects (Siddall, 1976), prevents the emergence of the adult
from the pupal stage. A positive feature of the insecticide
is that it needs to be present in the environment only during
a critical life history stage of the target species. Absence
of this chemical in the environment for weeks or months
decreases chances of delayed toxicity to, or bioconcentration
in, other species (e.g., fishes). Altosid is registered for
flood-water mosquito control in non-crop areas and for certain
uses on rice and pastures (Betz, personal communication, Office
of Pesticide Programs, EPA, Washington, D.C.); registration
labels are pending for other uses.

Another insect growth regulator (IGR), Dimilin, was approved by EPA only for control of the gypsy moth. It represents
a group of chemicals that interfere with the formation of
insect cuticle during molt, but whether this is the single mode
of action is questioned (Wright, 1976). Wright observed that
when reciprocal crosses of either males or females of the
stable fly exposed to Dimilin mated with the opposite untreated
sex, reduced fecundity and fertility resulted. Cunningham
(1976) found similar results in mating pairs of brine shrimp,
Artemia salina, after exposure to differing concentrations of
Dimilin. Another possible concern with Dimilin and other
insect growth regulators is their persistence. Metcalf et al.
(1975) showed that they are moderately persistent in a laboratory model ecosystem, especially in killing invertebrates.
Chironomid midges were effectively controlled in a California
residential lake for 8-10 weeks after a single application of
0.1 pound per acre (Mulla et al., 1976).

We have recently completed some life-cycle toxicity tests in which the mysid, *Mysidopsis bahia,* was exposed to Dimilin, using the methods of Nimmo *et al.* (1977). We determined survival of mysids and effects of the chemical on growth, reproduction and behavior in flowing water. The 96-hr LC50 was 2.06 µg/ℓ (95% confidence limits, 1.63-2.69 µg/ℓ). We also found that an estimated 75 ng/ℓ (parts-per-trillion) significantly reduced the number of young produced by each female mysid.

CONCLUSIONS

It is difficult to predict which pesticides will likely endanger coastal waters in the future. Judging from history, current or past contamination of estuaries by pesticides cannot be ascribed solely to agricultural uses or to the control of disease vectors. In many instances, pesticide pollution came inadvertently or mistakenly from point sources. Examples of accidental introduction are: (1) endrin in the Mississippi River in the early 1960's; (2) DDT released from the Los Angeles outfall into Pacific waters; and most recently (3) Kepone entering the James River in Virginia. We need to place more emphasis on predicting or monitoring obvious potential sources of accidental contamination. The recent toxic substances legislation will make this possible.

Some might infer that the non-point sources of pesticides should not be of major concern to ecologists. The argumentative scenario could be based on the following: (1) considering the successful control of hundreds of pests with hundreds of chemicals or formulations during the last four decades, estuarine environments have sustained imperceptible damage; or (2) the assimilative capacity of biological systems is greater than those often suggested. I counter these two notions with documented effects of pesticides on various biota after the chemicals had been applied in prescribed amounts (Table 1). These cases as summarized involve either (1) short-term physiological effects, such as depressed brain acetylcholinesterase activity in fishes after malathion was applied to estuarine waters in Louisiana, or (2) longer term effects, such as the development of pesticide-tolerant fish populations in Mississippi. Another example of a long-term effect is the well-known thinning of brown pelican eggshells that is directly associated with DDE (Blus *et al.*, 1972).

I also point out that laboratory studies, in some instances, may underestimate the potential for environmental damage in the field. In one instance, our laboratory studies on contamination in Escambia Bay, Florida, with three organisms

TABLE I.   Summary of effects observed on field populations
of animals after exposure to pesticides

Observed effects and organisms	Pesticide	Reference
Depressed acetylcholinesterase activity in brain tissue of fish	Malathion	Coppage and Duke (1971)
Failure of eggs to develop in seatrout	DDT	Butler et al. (1970)
Documenting the presence of pesticide-tolerant fish	Strobane and Chlordane	Ferguson (1968)
Failure to re-establish populations of bass	Several	Bingham (1970)
Eggshell thinning of brown pelican eggs	DDE	Blus et al. (1972)

underestimated the bioconcentration potential of the polychlorinated biphenyl (Aroclor®1254) by one or two orders of
magnitude below levels found in the tissues of the same three
organisms captured from the Bay (Nimmo, 1976).   The problem
was also addressed by Livingston (1976), who noted that laboratory toxicity tests (particularly with the organochlorine
pesticides) have not considered many contributing and synergistic factors present in the field; e.g., reduced oxygen
levels resulting in increased DDT toxicity to various freshwater fishes.

In the 1970's, investigators in the field of aquatic
toxicology have discovered sophiscated, sensitive methods of
determining effects of pesticides in various biota.   Such
advances include determinations of Kepone-induced pathology
in fishes (Couch et al., 1977), collagen/Vitamin C indicators
of chronic organochlorine stress in fishes (Mehrle et al.,
1977), and routine use of life-cycle toxicity tests on fish
(Parrish et al., 1978) and a crustacean (Nimmo et al., 1977).
Determining the effects of a pesticide on the feeding activity
of a polychaete is now an accepted test (Rubenstein, 1977).
Procedures are also available to assess the movement of a
pesticide through a laboratory food chain (Bahner et al.,
1977)   and methods to determine impacts of toxicants on

communities of benthic organisms (Tagatz *et al.*, 1977) are
now "standard" procedures. It is possible that a model of
pesticide flux in a river and associated estuary will soon be
announced. The future direction of research to meet the
environmental challenges of new chemicals must be resolved
by research groups from both academia and industry, as well
as government. Whatever our role, forecasting the impact or
safety of new chemicals, developing new methods to assess
effects, or planning new research efforts--the increased
use of such potentially harmful chemicals, as herbicides,
wood preservatives, and second or third generation insecticides,
should make us more alert as ecologists concerned with the
environment.

ACKNOWLEDGMENTS

I thank Dr. Nelson R. Cooley for technical assistance on
the manuscript. Gulf Breeze Contribution Number 366.

LITERATURE CITED

Anonymous, 1977a. A Trojan horse name RPAR. Farm Chemicals.
    140(8): 13-20.
Anonymous, 1977b. Don't overlook the professional market.
    Farm Chemicals. 140(3): 56.
Bahner, L.H., A.J. Wilson, Jr., J.M. Sheppard, J.M. Patrick,
    Jr., L.R. Goodman, and G.E. Walsh. 1977. Kepone®
    bioconcentration, accumulation, loss, and transfer
    through estuarine food chains. Chesapeake Science 18:
    299-308.
Bingham, C.R. 1970. Comparison of insecticide residues
    from two Mississippi oxbow lakes. Proc. 23rd Annu. Conf.
    Southeast. Assoc. Game and Fish Comm. Mobile, Ala.
    p. 275-280.
Blus, L.J., C.D. Gish, A.A. Belisle, and R.M. Prouty. 1972.
    Logarithmic relationship of DDE residues to eggshell
    thinning. Nature 235: 376-377.
Butler, P.A. 1965. Effects of herbicides on estuarine fauna.
    (Presented at the 18th Annual Meeting of the Southern
    Weed Conference, Jan. 19-21, Dallas, Texas) pp. 3.
Butler, P.A. 1977. Bioaccumulation of DDT and PCB in marine
    fishes. Second Symposium on Aquatic Toxicology ASTM,
    Cleveland, OH, Oct. 31-Nov. 1, 1977 (In Press).

Butler, P.A., R. Childress, and A.J. Wilson, Jr. 1970. The
    association of DDT residues with losses in marine produc-
    tivity. *In* FAO Technical Conference on Marine Pollution
    and Its Effects on Living Resources and Fishing. Rome,
    Italy, FIR: MP/70/E-76. pp 1-13.

Butler, P.A., and P.F. Springer. 1963. Pesticides--a new
    factor in coastal environments. Trans. 28th N. M. Wildl.
    Nat. Resour. Conf. pp 378-390.

Carson, R.L. 1962. Silent Spring. Houghton Mifflin.
    Boston. pp. 368.

Coppage, D.L., and T.W. Duke. 1971. Effects of pesticides
    in estuaries along the Gulf and Southeast Atlantic Coasts.
    In: Proc. 2nd Gulf Coast Conf. Mosquito Suppression
    and Wildl. ,anag. New Orleans, LA. pp. 24-31.

Couch, J.A., J.T. Windstead, and L.R. Goodman. 1977. Kepone-
    induced scoliosis and its histological consequences in
    fish. Science 197:585-587.

Cunningham, P.A. 1976. Effects of dimilin (TH6040) on
    reproduction in the brine shrimp, *Artemia salina*.
    Environ. Entomol. 5: 701-706.

Elliott, M. 1976. Properties and applications of pyrethroids.
    Environ. Health Prospect. 14: 3-13.

Ferguson, D.E. 1968. Characteristics and significance of
    resistance to insecticides in fishes. Reservoir Fishery
    Resources Symposium. Athens, GA, pp. 531-536.

Graham, F., Jr. 1970. Since Silent Spring. Houghton
    Mifflin Co. Boston. pp. 332.

Hann, R.W. 1975. Industrial waste pollution and Gulf Coast
    estuaries. *In*: Estuarine Pollution Control and Assessment
    1: 319-329. Office of Water Planning and Standards, EPA,
    Washington, DC.

Klass, E.E. and A.A. Belisle. 1977. Organochlorine pesticide
    and polychlorinated bephenyl residues in selected fauna
    from a New Jersey salt marsh--1967 vs. 1973. Pest. Monit.
    J. 10: 149-158.

Li, M. 1975. Pollution in nation's estuaries originating
    from the agricultural use of pesticides. *In*: Estuarine
    Pollution Control and Assessment 2: 451-466. Office of
    Water Planning and Standards, EPA, Washington, DC.

Livingston, R.J. 1976. Dynamics of organochlorine pesticides
    in estuarine systems: effects on estuarine biota.
    Estuarine Processes 1: 507-522.

Macek, K.J., M.A. Lindberg, S. Sauter, K.S. Buxton, and P.A.
    Costa. 1976. Toxicity of four pesticides to water fleas
    and fathead minnows. EPA-600/3-76-099 pp. 57. U.S.
    Environmental Protection Agency., Washington, D.C.

Mehrle, D.M., F.L. Mayer, and W.W. Johnson. 1977. Diet
    quality in fish toxicology: effects on acute and chronic

toxicity. *In*: Aquatic Toxicology and Hazard Evaluation, ASTM **STP** 634. F.L. Mayer and J.L. Hamelink, eds. American Society for Testing and Materials.

Metcalf, R.L., P. Lu, and S. Bowlus. 1975. Degradation and environmental fate of 1-(2,6-difluorbenzoyl)-3-(4-chlorophenyl) area. J. Agric. Food Chem. 23: 359-364.

Mulla, M.S., W.L. Kramer, and D.R. Barnard. 1976. Insect growth regulators for control of chironomid midges in residential-recreational lakes. Jour. Econ. Entomol. 69: 285-291.

Nimmo, D.R. 1976. Statement before the Committee on Merchant Marine and Fisheries, House of Representatives, on the matter of Polychlorinated Biphenyls, Jan. 28, 29, 30, 1976. Serial No. 94-24: 138-166.

Nimmo, D.R., L.H. Bahner, R.A. Rigby, J.M. Sheppard, and A.J. Wilson, Jr. 1977. *Mysidopsis bahia:* an estuarine species suitable for life-cycle toxicity test to determine the effects of a pollutant. Aquatic Toxicology Hazard Evaluation, STM STP 634. F.L. Mayer and J.L. Hamelink, eds., American Society for Testing Materials.

Parrish, P.R., E.E. Dyar, J.M. Enos, and W.G. Wilson. 1978. Chronic toxicity of chlordane, trifluralin, and penta-chlorophenol to sheepshead minnows (*Cyprinodon variegatus*). EPA-60D/3-78-010 pp. 53. U.S. Environmental Protection Agency, Washington, D.C.

Rubinstein, N.I. 1978. Effect of sodium pentachlorophenate on the feeding acitivity of the lugworm, *Arenicola cristata* Stimpson. *In:* Pentachlorophenol: Chemistry, Pharmacology and Environmental Toxicology. K.R. Rao, ed., Plenum Press. New York and London. pp. 175-179.

Rudd, R.L. 1964. Pesticides and the Living Landscape. Univ. Wisconsin Press, Madison WI. pp. 320.

Siddall, J.B. 1976. Insect growth regulators and insect control: a critical appraisal. Environ. Health Prospect. 14: 119-126.

Tagatz, M.E., J.M. Ivey, J.C. Moore, and M. Tobia. 1977. Effects of pentachlorophenol on the development of estuarine communities. J. Toxicol. Environ. Health 3: 501-506.

Triplett, G.B., Jr., and D.M. Van Doren, Jr., 1977. Agriculture without tillage. Scientific American 236: 28-33.

Unger, S.G., and G.W. Bailey. 1977. Environmental implications of trends in agriculture and silviculture. Vol. 1: Trend identification and evaluation. Ecological Research Series EPA-600/3-77-121.

von Rumker, R., E.W. Lawless, and A.F. Meiners. 1975. Production, distribution, use and environmental impact potential of selected pesticides. EPA 540/1-74-001. U.S. Environmental Protection Agency, Washington, D.C.

Wilson, A.J. and J. Forester. 1978. Persistence of Aroclor 1254 in a contaminated estuary. Bull. Environ. Contam. Toxicol. 19: 637-640.

Woodwell, G.M., C.F. Wurster, Jr., and P.A. Isaacson. 1967. DDT residues in an east coast estuary: a case of biological concentration of a persistent insecticide. Science: 156: 822-823.

Wright, J.E. 1976. Environmental and toxicological aspects of insect growth regulators. Environ. Health Prospect. 14: 127-132.

FATTY ACID COMPOSITION OF PHOSPHOLIPIDS IN THERMALLY
ACCLIMATING SCULPINS (*LEPTOCOTTUS ARMATUS*) TREATED
WITH POLYCHLORINATED BIPHENYLS (AROCLOR 1254)[1]

Richard S. Caldwell
Elaine M. Caldarone
Barbara A. Rosene

Department of Fisheries and Wildlife
Oregon State University, Marine Science Center
Newport, Oregon

INTRODUCTION

The physiological adaptation of poikilothermic organisms
to changing environmental temperatures is a complex and multi-
faceted process which occurs primarily at the biochemical level.
These adaptations enable organisms to achieve a degree of
temperature independence beyond that expected from thermo-
dymanic considerations (Hochachka and Somero 1973). Numerous
investigations have provided convincing evidence that control
of the lipid composition of cellular membranes, especially the
degree of fatty acid unsaturation, is an important part of the
adaptation process. It has been shown repeatedly that organisms
from cold environments, and those that are cold acclimated in
the laboratory, have more unsaturated lipids than their warm
adapted counterparts (Caldwell and Vernberg 1970; Farkas and
Herodek 1964; Johnston and Roots 1964; Kemp and Smith 1970;
Knipprath and Mead 1965; Selivonchick *et al.* 1977). It is
now generally accepted that such lipid modifications perform
an essential role in preserving specific biophysical properties
of biological membranes during periods of temperature change
and that the control of these properties is essential for
maintaining optimal membrane function (Hazel 1973; Hochachka
and Somero 1973; Prosser 1973).

Fulco (1972) has shown that the bacterium, *Bacillus
megaterium*, is capable of Δ 5 desaturation of saturated fatty

[1]*Technical Paper No. 4751, Oregon Agricultural Experiment
Station.*

271

acids when cultured at 20 C, but not at 35 C, and that the
response is dependent upon the presence of a specific Δ 5
desaturase induced only at low culture temperatures. In
higher organisms the control of lipid unsaturation is thought
to reside in part in the activity of microsomal fatty acid
desaturases. Ninno *et al.* (1974) found that liver microsomes
from fish, *Pimelodus maculatus,* acclimated to 16 C for three
weeks exhibited rates of desaturation of palmitic, oleic,
linoleic, linolenic and eicosa-8,11,14-trienoic acids at
35 C 2.0 to 4.2 times greater than control fish acclimated
to 29 C, a result consistent with the hypothesis that induction of desaturase activity occurred at the lower acclimation
temperature.

The fatty acid desaturating system in animals consists of
a microsomal electron transfer chain composed of three sequential components: 1) either of two flavorprotein containing
pyridine nucleotide cytochrome b_5 reductases, one specific
for NADPH and the other for NADH; 2) cytochrome b_5; and
3) a cyanide sensitive factor (CSF) postulated to be the
actual site of oxygen activation and fatty acid desaturation
(Oshino and Sato 1972; Schenkman *et al.* 1976). The CSF is
thought to be the rate limiting step in desaturation (Oshino
and Sato 1972).

The environmentally important mixed function oxidase (MFO)
system is a multicomponent electron transfer chain which contains cytochrome P450, requires reduced pyridine nucleotide and
O_2 as cofactors, and is located primarily in liver microsomes.
The induction of this system in fish upon exposure to a variety
of xenobiotic chemicals is well documented (Burns 1976; Chambers
and Yarbrough 1976; Gruger *et al.* 1977; Hill *et al.* 1976).
The functional and chemical similarities between the fatty
acid desaturating system and the MFO system, and the fact that
both are found in microsomes, suggested to us that adaptive
responses of the MFO system to pollutant exposure might result
in abnormal functioning of the fatty acid desaturases. Either
of two general responses could be envisioned: 1) the fatty
acid desaturating system could be co-induced with the MFO
system resulting in increased desaturase activity, or 2) fatty
acid desaturase activity could decline concomitant with MFO
induction, either because of competition in the synthesis of
desaturase and MFO components or competition between the MFO
and desaturase system for common cofactors.

Although the precise nature of the interaction between these
two microsomal electron transfer systems remains poorly known,
it is clear that interactions do exist (Schenkman *et al.* 1976).
Ninno *et al.* (1974) found that administration of 150 ppm
dieldrin in the diet of rats resulted within three days in a
substantial and persistent increase of Δ 9 desaturating activity as well as induction of the MFO system in the liver

microsomes, a result suggesting that the MFO and fatty acid
desaturating systems may be co-induced. Oshino (1972) showed
that daily injection of starved rats for four days with the
MFO inducer phenobarbital resulted in a four-fold increase in
cytochrome P450, a two-fold increase in cytochrome b_5, but
an inhibition of the normal response to refeeding (i.e. induction of CSF and stearyl CoA desaturase activity). Although
Oshino's data indicates that co-induction of the MFO system
and at least the cytochrome b_5 portion of the fatty acid
desaturase system may occur, blockage of the refeeding response
of the CSF and stearyl CoA desaturase activity suggests that
competitive interactions between the two systems may also exist.

Aroclor 1254, a mixture of polychlorinated biphenyls (PCB's)
with 54% chlorine content, is known to be a potent MFO inducer
in animals (Hill *et al*. 1976; Litterst and vanLoon 1972). In
this paper we report on studies of the fatty acid composition
of phospholipids of thermally acclimating sculpins, *Leptocottus
armatus*, treated with this Aroclor. These studies represent
an initial step in our investigations of the effects of MFO
inducing chemicals on that aspect of thermal adaptations involving the lipid biochemistry of marine poikilotherms.

MATERIALS AND METHODS

Juvenile staghorn sculpins, *Leptocottus armatus*, were
collected by beach seine from shallow water areas of Yaquina
Bay, Oregon. Before utilization in experiments, sculpins were
held for a minimum of one week in laboratory aquaria supplied
with flowing seawater at a temperature of 13.5 C and fed 3
times per week with a standard pelleted fish diet (Oregon Moist
Pellet, Bioproducts, Inc., Warrenton, Oregon) containing polyunsaturated fatty acids characteristic of marine fish oils.

Treatment of fish with Aroclor 1254 (Lot No. KA 625 obtained
as a gift from Monsanto, Corp.) was by intraperitoneal injection
(1 µl/g fish) of an ethanol solution of the PCB to give a dosage
of 30 mg/kg. Controls were injected with an equal volume of
ethanol. Lipid analyses were conducted on fish from two separate experiments. In the first, 15 fish were placed in each
of three 60-liter glass aquaria (initial group mean weights
ranged from 9.3 to 11.6 g). Two aquaria contained ethanol
injected fish (controls) and the third held PCB injected fish.
Each of the aquaria was supplied with 400 ml/min of filtered
laboratory seawater at 13.5 C. On the third day after injection
the temperature of two of the aquaria (one containing control
fish and the other the PCB treated fish) was gradually lowered
over a 12 hour period to 7 C. At the same time, the temperature
of the third aquarium was raised gradually to 20 C. After 12

days of exposure to these temperatures, fish were removed
from the aquaria, their digestive tracts removed, and the
carcasses frozen for later analysis.   In the second experiment,
14 ethanol injected fish (initial mean weight 6.1 g) were
placed in one aquaria, 13 PCB injected fish (initial mean weight
5.6 g) in another, and the temperatures were immediately raised
over a period of 10 to 12 hours to a final temperature of 19.2
C. Following 12 days exposure to this temperature the fish
were removed and stored for analysis as in the first experiment.
Frozen storage did not exceed 2 weeks in either experiment.
During both experiments the oxygen content of the aquaria water
was maintained at air saturation by vigorous aeration, the salinity ranged from 32 to 34o/oo, and the photoperiod was 15
hours of light and 9 hours of darkness.   Fish were fed Oregon
Moist Pellet three times per week during the thermal acclimation period in the first experiment and twice per week during
the second experiment.

Lipids were extracted from sculpins with chloroform :
methanol (2:1, v:v) according to the method of Folch *et al.*
(1957) employing 0.034% $MgCl_2$ as the wash medium.   The washed
extracts were evaporated to dryness under vacuum at 30-40 C
in a rotary evaporator.   The residues were dissolved in a
small volume of acetone, taken to dryness to remove residual
water by co-distillation with acetone, and redissolved in 5
ml of chloroform.

Approximately 40 mg of the total lipid extract from each
fish was separated into neutral lipid and phospholipid fractions by chromatography on 1 X 7 cm columns of silicic acid
(Bio Rad, Bio-Sil A, 100-200 mesh) previously conditioned
with chloroform.   The neutral lipids were eluted from the
columns with 30 ml of chloroform followed by 20 ml of chloroform : methanol (19:1, v:v).   Phospholipids were recovered by
further elution with 20 ml of chloroform : methanol (2:1, v:v)
followed by 75 ml of methanol.   Solvents were evaporated from
the phospholipid fraction under vacuum at 35 C and the residue
was taken up in 2 ml of chloroform.   Recovery of phosphorus
averaged 99% of that applied to the columns.   Of the recovered
phosphorous, 99% was in the phospholipid fraction, which contained only negligible contamination by neutral lipid based
on further analysis by silica gel thin layer chromotography.
Phospholipid phosphorus was measured by a modification of the
method of Bartlett (1959).

Phospholipid fatty acids were methylated with BF_3-methanol
reagent according to the procedures of Morrison and Smith (1964).
The fatty acids were chromatographed on a Hewlett Packard Model
5711A gas chromatograph equipped with a hydrogen flame detector.
Columns were 1.83 m X 3 mm stainless steel packed with 10%
SP-2330 on 100/120 mesh Chromosorb W AW.   Nitrogen was employed
as the carrier gas at a flow of 20 ml/min.   The oven temperature

was 210 C, and the injection port and detector temperatures
were 250 and 300, respectively. Peak areas were estimated as
the product of peak height and peak width at one half peak
height. Chromatography of quantitative methyl ester standards
(Supelco, Inc., RM-3; Nu-Chek-Prep, Inc. #17-A) confirmed that
the detector response was proportional to mass for the range
of fatty acid chain lengths employed (Horning et al. 1964)
permitting expression of the analytical results for fatty acids
as weight percent. Quantitative analysis of the standards
agreed with the stated composition data with a relative error
less than 4% for major components (> 10% of total mixture)
and less than 12% for minor components (< 10% of total mixture).
Peaks were identified by comparison with standards (Supelco,
Inc. RM-3, PUFA #1 and PUFA #2) and checked using log plots
of retention time against chain length and by chromatographic
separation of esters by degree of unsaturation on $AgNO_3$ impreg-
nated thin layer plates (Privett et al. 1963), All solvents
used for lipid analyses were glass redistilled prior to use
except hexane which was Burdick and Jackson spectroquality grade.
All lipid analytical work was carried out under a N_2 atmosphere
and lipid materials were kept regrigerated during short periods
of storage not exceeding a few weeks.
   Statistical comparisons were made using Student's two-
tailed t test.

RESULTS

   A comparison of the fatty acid composition of the phospholipids of sculpins held for 12 days at 7 C with those held at
20 C showed a pattern of change typical of thermally acclimating
poikilotherms (Table 1). The major changes were an increase
of palmitic acid (16:0) and a decrease of eicosapentaenoic
acid (20:5 ω3) in the lipids of warm adapted fish relative to
the fish at 7 C. Changes in the levels of these two acids,
which alone accounted for nearly a third of the total phospholipid fatty acids, decreased the level of unsaturation in the
warm compared to the cold acclimated fish. The levels of
arachidonic acid (20:4 ω6) and docosapentaenoic acid (22:5 ω3)
also decreased in the warm adapted fish, but the remaining
major polyunsaturated acid, docosahexaenoic acid (22:6 ω3) did
not change. Other acids showing changes in level were myristic
acid (14:0), which increased, and linoleic acid (18:2 ω6)
which decreased at 20 C. The changes in these acids which together accounted for only about 3% of the total fatty acids,
contributed further to decreasing unsaturation at higher temperatures.

TABLE 1. Fatty acid composition of total phospholipids
from cold and warm acclimated sculpins

| Fatty acid | Weight percent | | Student's |
	7 C Acclimated	20 C Acclimated	t
14:0	0.9 ± 0.1	1.3 ± 0.2	6.69[a]
16:0	15.6 ± 1.5	17.5 ± 1.6	3.31[a]
16:1 ω7	3.1 ± 0.3	3.1 ± 0.3	0.15
18:0	7.3 ± 0.4	7.5 ± 0.4	1.13
18:1 ω9	16.3 ± 0.9	16.9 ± 0.7	1.76
18:2 ω6	2.2 ± 0.6	1.6 ± 0.4	-2.91[a]
18:3 ω3	0.4 ± 0.1	0.4 ± 0.1	0.49
20:1 ω9	3.1 ± 0.6	3.4 ± 0.2	1.73
20:4 ω6	6.8 ± 0.7	5.6 ± 0.5	-4.93[a]
20:5 ω3	16.7 ± 1.6	14.3 ± 0.8	-4.90[a]
24:0	0.5 ± 0.1	0.5 ± 0.2	0.76
24:1 ω9	1.8 ± 0.3	1.8 ± 0.3	-0.23
22:5 ω3	4.9 ± 0.7	3.6 ± 0.5	-5.43[a]
22:6 ω3	15.6 ± 1.5	16.6 ± 2.2	1.37
Others	4.8	5.9	

Data given are weight percent as fatty acid methyl esters
± one standard deviation. N=15 for 7 C acclimated fish and
N=13 for 20 C acclimated fish.

[a]Means significantly different at the 0.01 level (26 degrees
of freedom)

Treatment of sculpins with 30 mg/kg of Aroclor 1254 prior
to transfer for 12 days to low (7 C) or high (19.2 C) acclimation temperatures modified the patterns of fatty acids compared
to controls also acclimated to these temperatures (Tables 2
and 3, respectively). There were some notable similarities
in the effects associated with Aroclor 1254 treatment at the
two temperatures, although there were in several instances
differences in the actual acids that changed and in the level
of significance of some of the changes. The most pronounced
similarity was a large decrease of 22:6 ω3 in the PCB treated
fish at both temperatures. At 7 C, the phospholipids of the
Aroclor 1254 fish contained 12.0% by weight of this major
polyunsaturated acid compared to 15.6% in the controls, and
at 19.2 C the PCB treated fish averaged 11.8% of 22:6 ω3 by
weight while the controls averaged 14.6%. Since this acid
contains the highest number of double bonds per mole of any of
the acids identified, such a change would substantially reduce
the unsaturation level of the whole phospholipid fraction if

TABLE 2. Effect of Aroclor 1254 on the fatty acid composition of total phospholipids from cold (7 C) acclimating sculpins

Fatty acid	Weight percent		Student's t
	Controls	Aroclor 1254 treated	
14:0	0.9 ± 0.1	0.6 ± 0.1	-5.57[b]
16:0	15.6 ± 1.5	16.9 ± 1.6	2.36[a]
16:1 ω7	3.1 ± 0.3	2.9 ± 0.4	-1.31
18:0	7.3 ± 0.4	8.4 ± 0.8	4.59[b]
18:1 ω9	16.3 ± 0.9	17.6 ± 1.7	2.63[a]
18:2 ω6	2.2 ± 0.6	2.9 ± 0.8	2.76[a]
18:3 ω3	0.4 ± 0.1	0.3 ± 0.1	-3.32[b]
20:1 ω9	3.1 ± 0.6	2.6 ± 0.5	-2.35[a]
20:4 ω6	6.8 ± 0.7	6.9 ± 0.8	0.61
20:5 ω3	16.7 ± 1.6	15.8 ± 1.8	-1.53
24:0	0.5 ± 0.1	0.5 ± 0.1	-0.55
24:1 ω9	1.8 ± 0.3	2.0 ± 0.3	2.05[a]
22:5 ω3	4.9 ± 0.7	5.0 ± 0.8	0.30
22:6 ω3	15.6 ± 1.5	12.0 ± 2.8	-4.42[b]
Others	4.8	5.6	

Data given are weight percent as fatty acid methyl esters ± one standard deviation. N=15 for control fish and N=14 for Aroclor 1254 treated fish.

[a]Means significantly different at the 0.05 level (27 degrees of freedom)
[b]Means significantly different at the 0.01 level (27 degrees of freedom)

TABLE 3.   Effect of Aroclor 1254 on the fatty acid composition of total phospholipids from warm (19.2 C) acclimating sculpins

| Fatty acid | Weight percent | | Student's t |
	Controls	Aroclor 1254 treated	
14:0	1.1 ± 0.2	0.9 ± 0.2	-2.16[a]
16:0	18.0 ± 1.5	18.6 ± 1.7	0.72
16:1 ω7	3.1 ± 0.4	3.5 ± 0.5	2.39[a]
18:0	7.8 ± 0.3	8.1 ± 0.3	1.93
18:1 ω9	17.6 ± 0.9	20.0 ± 1.0	5.88[b]
18:2 ω6	5.2 ± 1.2	5.0 ± 1.7	-0.29
18:3 ω3	0.4 ± 0.0	0.5 ± 0.1	1.52
20:1 ω9	2.9 ± 0.3	2.7 ± 0.4	-1.42
20:4 ω6	5.0 ± 0.7	5.6 ± 1.1	1.69
20:5 ω3	12.5 ± 1.0	12.9 ± 0.8	0.94
24:0	0.5 ± 0.1	0.5 ± 0.1	0.92
24:1 ω9	1.5 ± 0.3	1.7 ± 0.4	1.24
22:5 ω3	3.2 ± 0.6	3.0 ± 0.6	-1.07
22:6 ω3	14.6 ± 1.6	11.8 ± 2.1	-3.74[b]
Others	6.6	5.2	

Data given are weight percent as fatty acid methyl esters ± one standard deviation. N=13 for control fish and N=11 for Aroclor 1254 treated fish.

[a]Means significantly different at the 0.05 level (22 degrees of freedom)
[b]Means significantly different at the 0.01 level (22 degrees of freedom)

the acid was replaced by saturated or monounsaturated fatty acids. At the lower acclimation temperature (7 C) increases were seen in the levels of palmitic, stearic, oleic (18:1 ω9) and linoleic acids in the PCB treated fish compared with controls. Except for linoleic acid, all of these also increased in level in PCB treated fish at the high acclimation temperature; however, only the increase in level of oleic acid was statistically significant at this temperature.

Figure 1 summarizes graphically the direction and extent of change of each fatty acid as a function of warm acclimation and as a function of Aroclor 1254 treatment at low and high acclimation temperatures. It is clear that Aroclor 1254 treatments resulted in responses generally similar to that found with warm acclimation, i.e. decreases in levels of long chain polyunsaturated fatty acids and compensatory increases in short chain saturated and monounsaturated fatty acids. Calculation of the average double bond contents of fatty acids from treated and control fish revealed that the changes resulted in decreased

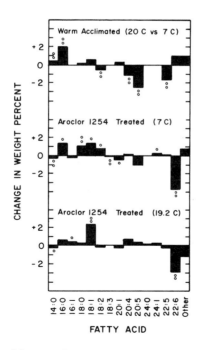

FIGURE 1. Effect of warm acclimation, Aroclor 1254 treatment at low temperature, and Aroclor 1254 treatment at high temperature on the change in weight percent of each fatty acid (as methyl esters) relative to the total fatty acid content. Circles indicate that differences are significant at the 0.05 (o) or 0.01 (ᵒₒ) levels.

double bond contents averaging between 0.13 and 0.21 double
bonds per molecule for the three treatments (Table 4), and
that the decreased double bond content following PCB treatments was approximately equivalent to that caused by an
increase of 13 C in the exposure temperature.

Although the effects of Aroclor 1254 treatment and warm
acclimation on the pattern of change of phospholipid fatty
acids were generally similar, it is also evident that there
were distinctive differences (Figure 1). The most obvious of
these was that Aroclor treated fish at both exposure temperatures
were characterized by reduced levels of 22:6 ω3, but no change
in the levels of 20:4 ω6, 20:5 ω3 and 22:5 ω3, while acclimated
fish experienced no significant change in the level of 22:6 ω3,
but reduced levels of 20:4 ω6, 20:5 ω3 and 22:5 ω3. In addition, warm acclimated fish were characterized by increased
levels of myristic and palmitic acids and little change in the
18 carbon acids. In contrast, PCB treated fish at the lower
temperature had increases in both stearic and oleic acids as
well as palmitic acid, but lower levels of myristic acid. The
PCB treated fish at 19.2 C had a similar response pattern to
those at 7 C but not all of the acid changes were statistically
significant.

Tables 5 and 6 summarize observations on two other parameters examined in the experimental fish; the phospholipid
phosphorus content and the initial and final weights of fish
at each treatment. The total phospholipid phosphorus content
of fish from the first experiment was essentially identical
in all three treatments, averaging about 300 μg P_i/g fish
(Table 5). In the second experiment, however, the PCB treated
fish contained significantly less phospholipid phosphorus, 258
μg P_i/g fish, than the controls, 320 μg/g.

Fish in the second experiment also were initially smaller
than those in the first test and tended to lose weight during
the experimental period (Table 6). The rate of loss was higher
in the PCB treated fish in experiment II, and the weight gain
was less in the PCB treated fish in experiment I than in their
respective controls. All fish were observed to feed in the
five treatments, but the rate of food consumption may have been
lower in those treated with Aroclor 1254. In an earlier preliminary experiment (not reported in this paper) fish that were
injected with 100 mg/kg of PCB, about 3 times that employed in
the present experiments, were decidedly less active than ethanol
injected controls and appeared to consume less food.

TABLE 4. Effect of temperature and Aroclor 1254 treatment on the double bond content of fatty acids

Treatment comparison	Mean number of double bonds per molecule	Difference
Effect of warm acclimation		
7 C acclimated	2.53	
20 C acclimated	2.36	−0.17
Aroclor 1254 effect at low temperature		
7 C control	2.53	
7 C Aroclor 1254	2.32	−0.21
Aroclor 1254 effect at high temperature		
19.2 C control	2.20	
19.2 C Aroclor 1254	2.07	−0.13

Values computed from weight percent data (Tables 1-3) after conversion to mole percent. The minor components designated "others" were assumed to have an average molecular weight of 300 and 2 double bonds per molecule.

TABLE 5. *Phospholipid phosphorus content of whole sculpins*

Treatment	Phospholipid phosphorus (μg P_i/g fish)	Student's t	d.f.
EXPERIMENT I			
20 C control	301 ± 16 (14)		
		0.04	25
7 C control	300 ± 45 (13)		
		-0.12	25
7 C Aroclor 1254	298 ± 33 (14)		
EXPERIMENT II			
19.2 C control	320 ± 46 (12)		
		-4.16[a]	22
12.2 C Aroclor 1254	258 ± 24 (12)		

Phosphorus data are means ± one standard deviation. The numbers of observations are shown in parentheses. Concentration data are based on fish wet weights.

Student's test comparisons are made against the 7 C control in Experiment I.

[a]Significantly different at the 0.01 level

TABLE 6. *Weight changes during temperature acclimation of control and Aroclor 1254 treated sculpins*

Treatment	Initial weight (g)	Final weight (g)	Percent weight change
EXPERIMENT I			
20 C control	11.6 ± 4.3 (15)	14.7 ± 5.5 (15)	+27
7 C control	9.3 ± 3.3 (16)	9.7 ± 4.3 (16)	+4
7 C Aroclor 1254	10.2 ± 4.2 (15)	10.3 ± 4.6 (14)	+1
EXPERIMENT II			
19.2 C control	6.1 ± 2.3 (14)	5.8 ± 2.0 (13)	−5
19.2 C Aroclor 1254	5.6 ± 0.8 (13)	4.8 ± 1.0 (12)	−14

Weight data are means ± one standard deviation. The numbers of observations are shown in parentheses.

DISCUSSION

In the present study we have found that treatment of sculpins with a sublethal dose of Aroclor 1254 is associated with
a modification of the fatty acid pattern of the fish's phospholipids. The nature of the changes, compared to controls, is
similar regardless of whether the fish have been acclimated to
low or high temperatures immediately following the PCB administration and in both cases the resultant fatty acid changes
are in the direction of more saturated lipids. Furthermore,
the extent of change of saturation following PCB treatment is
comparable to that seen between untreated fish acclimated to
7 and 20 C (Table 4) suggesting that PCB treatment at this
level may adapt the fish, with respect to the biophysical
properties of the cell membranes, to a temperature approximately
13 C above the acclimation temperatures.

The extent to which these changes may affect the physiology
of sculpins is uncertain. Smith and Kemp (1971) have demonstrated that the time courses for changes in net amino acid
transport rates and membrane fatty acid composition of goldfish intestinal membranes during temperature adaptation are
similar and have suggested that the lipid changes may be responsible for the altered transport rates. Experimental modification of fatty acid unsaturation in yeast mitochondria has
been shown to alter the oxidative phosphorylation efficiency
of these organelles (Proudlock et al. 1969). A number of
other studies have demonstrated that the degree of unsaturation
of lipids in biological membranes has an influence on the catalytic properties of membrane bound enzymes as indicated by
shifts in the temperature of breaks in Arrhenius plots. Invar-
iably the breaks occur at lower temperatures when the degree
of unsaturation of the lipids is increased. For example,
Thomson et al. (1977) reported that an Arrhenius plot discon-
tinuity in eel gill microsomal Na^+,K^+-ATPase occurred at 20 C
in seawater adapted eels acclimated to 12 C, but occurred at
about 12-13 C in seawater adapted eels acclimated to 5 C.

The saturating effect resulting from Aroclor 1254 treatment
was found to be qualitatively different from that occurring as
a result of warm acclimation. Thus, in the Aroclor treated
fish, 22:6 ω3 was found to decrease instead of 20:4 ω6, 20:5 ω3
and 22:5 ω3 as was found in the warm acclimated fish. In
addition, the acids which showed a compensatory increase in
concentration tended also to differ; mainly 18 carbon acids
rather than 14 and 16 carbon acids. It is apparent, therefore,
that Aroclor 1254 and temperature differ in their specific
effects on fatty acid metabolism. Unfortunately little information is available on the enzymatic mechanisms normally
involved in the control of unsaturation levels during cold or

warm acclimation in fish or other poikilotherms, but an
observation of Ninno et al. (1974) may be pertinent. These
workers compared the rates of desaturation of five fatty acids
representing Δ 9, Δ 6 and Δ 5 desaturations in fish (Pimelodus
maculatus) acclimated to 16 and 29 C for three weeks. The
rates of Δ 6 desaturation of oleic, linoleic and α-linolenic
acids and of Δ 9 desaturation of stearic acid in warm acclimated
fish were almost exactly half those found in the cold acclimated
fish, but the Δ 5 desaturation of eicosa-8,11,14-trienoic
acid (20:3 ω6) in the warm fish was only 25% of that found in
the cold fish. These results indicate that the Δ 5 desaturations become less prominent relative to Δ 6 and Δ 9 desaturations in warm acclimating Pimelodus. It is possible, therefore,
that lower levels of 20:4 ω6 and 20:5 ω3 would occur in the
lipids of warm adapted fish as these two acids are the products
of Δ 5 desaturation reactions. A decrease in the level of
22:5 ω3, the chain elongation product of 20:5 ω3, may also be
expected if the level of its precurser declines. These expec-
tations are consistent with our observations of the fatty acid
changes occurring during warm acclimation of sculpins and
with patterns of change noted previously in the mitochondrial
lipids of acclimating goldfish and yellow bullheads (Caldwell
and Vernberg 1970.)

One interpretation of our observations that Aroclor 1254
did not significantly affect the levels of 20:4 ω6, 20:5 ω3
or 22:5 ω3, but significantly lowered the level of 22:6 ω3
is that the PCB failed to affect the activity of the Δ 5
desaturase enzyme and instead may have selectively inhibited
either the activity or synthesis of the Δ 4 desaturase which
is responsible for the conversion of 22:5 ω3 to 22:6 ω3. As
a result synthesis of this hexaenoic acid may have been insufficient to keep peace with its normal rate of breakdown. The
lack of observed compensatory rises in the levels of 20:5 ω3
and 22:5 ω3 could be explained by product inhibition of the
Δ 5 desaturase by these acids since their levels remained
normal.

The compensatory increases of 18 carbon fatty acids, especially oleic acid, in the Aroclor treated fish may be indicative of an attempt to compensate for the decline of lipid
unsaturation due to the loss of docosahexaenoic acid. Animals
are capable of elaboration of the oleic acid family of unsatu-
rates from endogenous precursers (acetate) and it is well
known that this ω9 family of acids increases in essential fatty
acid deficiency usually resulting in the buildup of eicosa-5,
8,11-trienoic acid (20:3 ω9) (Brenner and Nervi; Castuma
et al. 1972). That 20:3 ω9 did not seem to occur in the PCB
treated fish may be due to preference of the Δ 5 desaturase for
the more unsaturated substrates of the ω6 and ω3 families of
acids (Brenner 1974) which were still present in normal amounts.

Our working hypothesis in these studies was that fatty
acid patterns may be affected by treatment of fish with an
MFO inducing chemical (Aroclor 1254) as a result of interactions between the MFO and fatty acid desaturase systems.
We do not, in this paper, present any direct evidence that
treatment of sculpins with Aroclor 1254 resulted in induction
of the MFO system or affected fatty acid desaturation. Nevertheless, we have found that Aroclor 1254 treated fish exhibited
a modified pattern of phospholipid fatty acids. The nature
of the changes suggests that this PCB may specifically affect
the Δ 4 desaturase causing a decrease in its activity. This
proposed effect would differ in type from that observed by
Ninno *et al.* (1974) who reported an increase in Δ 9 desaturating activity in the livers of rats fed dieldrin; a treatment
which also resulted in induction of the MFO system in the liver
microsomes.

Our suggested mechanism for the Aroclor 1254 effect on
the fatty acid pattern of phospholipids in sculpins may, however, be similar to mechanisms of DDT and/or phenobarbital
effects in rats. Darsie *et al.* (1976) reported that liver
microsomes from rats fed o,p'-DDT exhibited decreased rates
of desaturation of palmitate and eicosa-8,11,14-trienoic
acid, Δ 9 and Δ 5 desaturations respectively, and exhibited
charactersitic essential fatty acid deficiency symptoms even
though dietary intake of linoleate was adequate. The desaturation of palmitate by normal microsomes was also significantly
inhibited *in vitro* by addition of low concentrations of o,p'-DDT
to the incubation medium. Darsie *et al.* (1976) did not determine the effect of fed o,p'-DDT on the MFO system, but their
results would seem to indicate a direct and general inhibitory
effect of DDT on some component of the desaturase system.
Oshino (1972) found that phenobarbital treatment of fasted
rats cause induction of the cytochrome b_5 component of the
desaturase system along with the MFO system, but the normal
response to refeeding, an increase in the level of CSF and
stearyl CoA desaturase activity, was blocked. A possible
interpretation of Oshino's results is that induction of the
CSF involved in Δ 9 desaturations is inhibited because of
competition in its synthesis with components of the MFO system.

Either of the above two mechanisms--direct inhibition of
desaturase or competitive inhibition of desaturase synthesis--
could account for the changes in the fatty acid patterns that
we observed in Aroclor 1254 treated sculpins except that the
effect would appear to be specific for the Δ 4 desaturase
since Aroclor treatment resulted in a decrease only in 22:6 ω3
among the 20 and 22 carbon polyunsaturated fatty acids. Since
the specificity of desaturase reactions may reside at the
level of CSF, and since the CSF is probably the rate limiting

step in desaturation (Oshino and Sato 1972), it follows that
the effects of Aroclor 1254 may be limited to the CSF responsible for Δ 4 desaturation. Our current studies are attempting
to more precisely elucidate these relationships.

SUMMARY

Treatment of juvenile sculpins *(Leptocottus armatus)* with
a sublethal dose of the polychlorinated biphenyls (PCB's)
mixture, Aroclor 1254, resulted in a modification of the
pattern of fatty acids in whole body phospholipids during 12
days of acclimation to either 7 or 19.2 C compared to ethanol
injected controls. At both acclimation temperatures, the
phospholipids of the PCB treated fish, compared to controls,
contained more saturated fatty acids and the magnitude of the
change in saturation level at both temperatures was comparable
to the differences seen between cold (7 C) and warm (20 C)
acclimated fish not treated with PCB's. The specific effects
associated with PCB treatment were a decline in the level of
docosahexaenoic acid (22:6 ω3) and compensatory increases in
short chain saturated and monounsaturated acids including
palmitic acid (16:0), stearic acid (18:0) and oleic acid
(18:1 ω9). Based on the observed effects due to PCB treatment
in this study and the observed and known adaptive responses
of fatty acid composition associated with temperature acclimation of poikilotherms, it is suggested that the Aroclor
treated fish may be functionally adapted to a temperature
approximately 13 C higher than their acclimation temperature.
The observed effects on the fatty acid composition of the
phospholipids of PCB treated fish is interpreted in the context
of effects of the pollutant on the microsomal fatty acid
desaturating system of vertebrates.

ACKNOWLEDGMENT

This research was supported in part by the Oregon State
University Agriculture Experiment Station and by Grant No.
R805625010 from the U. S. Environmental Protection Agency.

LITERATURE CITED

Bartlett, G. R. 1959. Phosphorus assay in column chromatography. J. Biol. Chem. 234: 466-468.
Brenner, R. R. 1974. The oxidative desaturation of unsaturated
fatty acids in animals. Mol. Cell. Biochem. 3: 41-52.
Brenner, R. R. and A. M. Nervi. 1965. Kinetics of linoleic
and arachidonic acid incorporation and eicosatrienoic
depletion in the lipids of fat-deficient rats fed methyl
linoleate and arachidonate. J. Lipid Res. 6: 363-368.
Burns, K. A. 1976. Microsomal mixed function oxidases in
an estuarine fish, *Fundulus heteroclitus,* and their induction as the result of environmental contamination. Comp.
Biochem. Physiol. 53B: 443-446.
Caldwell, R. S. and F. J. Vernberg. 1970. The influence of
acclimation temperature on the lipid composition of fish
gill mitochondria. Comp. Biochem. Physiol. 34: 179-191.
Castuma, J. C., A. Catala and R. R. Brenner. 1972. Oxidative
desaturation of eicosa-8,11-dienoic acid to eicosa-5,8,
11-trienoic acid: comparison of different diets on oxidative desaturation at the 5,6 and 6,7 positions. J. Lipid
Res. 13: 783-789.
Chambers, J. E. and J. D. Yarbrough. 1976. Xenobiotic biotransformation systems in fishes. Comp. Biochem. Physiol.
55C: 77-84.
Darsie, J., S. K. Gosha and R. T. Holman. 1976. Induction of
abnormal fatty acid metabolism and essential fatty acid
deficiency in rats by dietary DDT. Arch. Biochem. Biophys.
175: 262-269.
Farkas, T. and S. Herodek. 1964. The effect of environmental
temperature on the fatty acid composition of crustacean
plankton. J. Lipid Res. 5: 369-373.
Folch, J., M. Lees and G. H. Sloane-Stanley. 1957. A simple
method for the isolation and purification of total lipids
from animal tissues. J. Biol. Chem. 226: 497-509.
Fulco, A. J. 1972. The biosynthesis of unsaturated fatty
acids by Bacilli. IV. Temperature-mediated control mechanisms. J. Biol. Chem. 247: 3511-3519.
Gruger, E. H., Jr., M. M. Wekell, P. T. Numoto and D. R.
Craddock. 1977. Induction of hepatic aryl hydrocarbon
hydroxylase in salmon exposed to petroleum dissolved in
seawater and to petroleum and polychlorinated biphenyls,
separate and together, in food. Bull. Environ. Contam.
Toxicol. 17: 512-520.
Hazel, J. R. 1973. The regulation of cellular function by
temperature-induced alterations in membrane composition.
In: Effects of Temperature on Ectothermic Organisms, pp.
55-67, ed. by W. Wieser, New York: Springer-Verlag.

Hill, D. W., E. Hejtmancik and B. J. Camp. 1976. Induction of hepatic microsomal enzymes by Aroclor 1254 in *Ictalurus punctatus* (Channel catfish). Bull. Environ. Contam. Toxicol. 16: 495-502.

Hochachka, P. W. and G. N. Somero. 1973. Strategies of Biochemical Adaptation. Philadelphia: W. B. Saunders Co.

Horning, E. C., E. H. Ahrens, Jr., S. R. Lipsky, F. H. Mattson, J. F. Mead, D. A. Turner and W. H. Goldwater. 1964. Quantitative analysis of fatty acids by gas-liquid chromatography. J. Lipid Res. 5:20-27.

Johnston, P. V. and B. I. Roots. 1964. Brain lipid fatty acids and temperature acclimation. Comp. Biochem. Physiol. 11: 303-309.

Kemp, P. and M. W. Smith. 1970. Effect of temperature acclimatization on the fatty acid composition of goldfish intestinal lipids. Biochem. J. 117: 9-15.

Knipprath, W. G. and J. F. Mead. 1965. Influence of temperature on the fatty acid pattern of muscle and organ lipids of the rainbow trout (*Salmo gairdneri*). Fish. Indust. Res. 3:23-27.

Litterst, C. L. and E. J. van Loon. 1972. Enzyme induction by polychlorinated biphenyls relative to known inducing agents. Proc. Soc. Exptl. Biol. Med. 141: 765-768.

Morrison, W. R. and L. M. Smith. 1964. Preparation of fatty acid methyl esters and dimethylacetals from lipids with boron fluoride-methanol. J. Lipid Res. 5:600-608.

Ninno, R. E., M. A. P. De Torrengo, J. C. Castuma and R. R. Brenner. 1974. Specificity of five- and six-fatty acid desaturases in rat and fish. Biochim. Biophys. Acta. 360: 124-133.

Oshino, N. 1972. The dynamic behavior during dietary induction of the terminal enzyme (cyanide sensitive factor) of the stearyl CoA desaturation system of rat liver microsomes. Arch. Biochem. Biophys. 149: 378-387.

Oshino, N. and R. Sato. 1972. The dietary control of the microsomal stearyl CoA desaturation enzyme system in rat liver. Arch. Biochem. Biophys. 149: 369-377.

Privett, O. S., M. L. Blank and O. Romanus. 1963. Isolation analysis of tissue fatty acids by ultramicro-ozonolysis in conjunction with thin-layer chromatography and gas-liquid chromatography. J. Lipid Res. 4: 260-265.

Prosser, C. L. 1973. Comparative Animal Physiology, 3rd edition. Philadelphia: W. B. Saunders Co.

Proudlock, J. W., J. M. Haslam and A. W. Linnane. 1969. Specific effect of unsaturated fatty acid depletion on mitochondrial oxidative phosphorylation in *Saccharomyces cerevisiae*. Biochem. Biophys. Res. Commun. 37: 847-852.

Schenkman, J. B., I. Jansson and K. M. Robie-Suh. 1976. The many roles of cytochrome b_5 in hepatic microsomes. Life Sci. 19: 611-624.

Selivonchick, D. P., P. V. Johnston and B. I. Roots. 1977. Acyl and alkenyl group composition of brain subcellular fractions of goldfish (*Carassius auratus* L.) acclimated to different environmental temperatures. Neurochem. Res. 2: 379-393.

Smith, M. W. and P. Kemp. 1971. Parallel temperature-induced changes in membrane fatty acids and in the transport of amino acids by the intestine of goldfish (*Carassius auratus* L.). Comp. Biochem. Physiol. 39B: 357-365.

Thomson, A. J., J. R. Sargent and J. M. Owen. 1977. Influence of acclimatization temperature and salinity on ($Na^+ + K^+$) -dependent adenosine triphosphatase and fatty acid composition in the gills of the eel (*Anguilla anguilla*). Comp. Biochem. Physiol. 56B: 223-228.

COMPARATIVE METABOLISM OF PARATHION
BY INTERTIDAL INVERTEBRATES

R. L. Garnas[1]
D. G. Crosby

Environmental Toxicology Department
University of California at Davis
Davis, California

INTRODUCTION

Our coastal waters are an important resource not only for
recreational, industrial, and fishery purposes but also as a
means of disposing of wastes. With increasing pressures on
land disposal methods, it must be anticipated that this use of
the marine environment will be further exploited in the future.
It is a legitimate use of the sea to add to it man's various
waste products; much of what we discharge is already accepted,
decomposed, and recycled (Portmann, 1974). But the use of the
oceans as a resource can be endangered if we affect too seriously its natural balance.

Because of its great size, the sea has been looked upon as
a bottomless sink, capable of absorbing all of the wastes that
man is able to produce. If every pollutant could be dispersed
evenly throughout the entire volume of water soon after discharge, even the more toxic chemicals would give little cause
for alarm. The danger arises because mixing is incomplete,
and because pollution is most intense in estuaries and coastal
waters near to centers of population (Waldichuk, 1974).

Assessment of the effect of this pollution on marine organisms is complicated by the numbers of chemicals and species
involved. After an extended examination, it becomes apparent
that no consistent body of knowledge presently exists from

[1]*Present Address: U. S. Environmental Protection Agency,
Environmental Research Laboratory, Gulf Breeze, Florida 32561*

which to make accurate, detailed predictions as to the fate of these materials in aquatic organisms (Crosby, 1975; Waldichuk, 1974; Vernberg and Vernberg, 1974). Considering the thousands of co-existing, man-made chemicals and their transformation products which can reach the aquatic ecosystem and the number and diversity of the organisms involved, the problem is substantial. Is it reasonable to restrict the use of thousands of major industrial chemicals until extensive tests can assure complete safety?

Metabolic detoxification of these xenobiotics by aquatic organisms represents a significant mode of resistance (Crosby, 1973; Lu and Metcalf, 1975). The cells of plants and animals contain hundreds of distinctly different enzymes, which make it possible for an organism to carry out the chemical reactions of metabolism; the very survival and existence of any living organism is the result of its ability to modify chemically an almost unlimited range of compounds following their absorption or ingestion. Such transformations are usually accomplished by the metabolic processes of oxidation, reduction, hydrolysis, and conjugation, with distinct differences in rates and products for different species (Symposium on Comparative Pharmacology, 1967).

Prediction of the toxicity of a chemical to an organism based on known detoxification and activation processes has proven valuable when dealing with terrestrial animals (Parke, 1968). However, analogous procedures have not been applied to aquatic organisms due to inadequate methodology. In particular, the maintenance of intact, unstressed organisms and the isolation of extremely minute amounts of excreted metabolites from large volumes of water are substantial problems. It was the purpose of this work to develop a general system for the isolation and identification of metabolites of various chemicals from unstressed aquatic organisms. These date could eventually provide a basis for predicting those chemicals most likely to be detrimental to the aquatic ecosystem.

A pesticide was selected for this study because it represents typical environmental contaminants; the increasing production and usage of pesticidal materials (von Rumker, et al., 1974) is ample cause for concern by governmental agencies and researchers in this area. Surprisingly, little information exists on pesticides in the ocean (Crosby, 1975), although it is known that pesticides do occur in biological and physical components of the ocean (Cox, 1971; Duke and Dumas, 1974). In addition, pesticide metabolism can delineate many typical metabolic pathways (Menzie, 1969). The metabolism of parathion has been reviewed extensively (Menzie, 1969; O'Brien, 1967); Hollingworth (1971) has reviewed the comparative metabolism and selectivity of organophosphate insecticides.

For this study, animals were chosen as a representative cross-section of marine invertebrate phylogeny, in particular coelenterates, mollusks and echinoderms. It is obvious that even in the broadest studies only a tiny fraction of the numerous species are compared and that one cannot generalize safely on this basis. However, this study is only the beginning of research utilizing this system for aquatic metabolism determination. Future research with additional species, together with available *in vitro* and dose-response data, will allow more accurate generalization and predictions.

MATERIALS AND METHODS

Specimens

Organisms used in this study were collected from tidepools off the California coast and included the sea star (*Pisaster ochraceous*, Phylum Echinodermata), the California mussel (*Mytilus californianus*, Phylum Mollusca, Class Pelecypoda), the black turban snail (*Tegula funebralis*, Phylum Mollusca, Class Gastropoda), and the sea anemone (*Anthopleura artemesia*, *Anthopleura xanthogrammica*, Phylum Coelenterata). Specimens were keyed in the field according to the descriptions of Ricketts and Calvin (1968), and later in the laboratory with the keys of Smith *et al.*, (1970). All organisms were acclimated to experimental conditions in laboratory holding tanks containing circulating salt water maintained at 15^0C.

Apparatus

The system use for these studies is depicted in Figure 1. Water was delivered by siphon from a reservoir (A), through a constant level apparatus (B), and to the metabolism holding chamber (D). The chamber was a six liter, jacketed reaction flask (Kontes Glass Company) cooled to 15^0C, and fitted with a flat, perforated, plastic lid for these studies. This holding chamber was connected to a solid waste collector (E), which was a 300 ml jacketed reaction flask fitted with a #11 Neoprene stopper. Feces, skin, or shell settled into the collector, where cooling to 1^0C minimized degradation due to associated microorganisms. The collector was connected to a separatory funnel (F) containing 100 ml (wet volume) of Amberlite XAD-4 resin (Rohm and Haas Co.), where water flow through the system was regulated and chemicals used in the studies were concentrated for analysis.

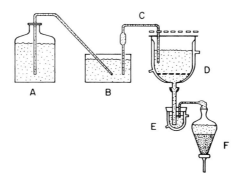

*FIGURE 1.   Aquatic metabolism system: reservoir (A); level-
er (B); siphon (C); chamber (D); separator (E); resin (F).*

Materials

Amberlite XAD-4 resin was cleaned by repeated rinsing with
water to remove salts and fines, sequential rinsing with 2N
sulfuric acid, distilled water, and acetone to remove the bulk
of impurities, and sequential Soxhlet extraction for 24 hours
with acetone and methanol. Following its use, the resin was
regenerated with this procedure.

All solvents were distilled in glass. Diazomethane was
prepared by distillation from Diazald reagent (Aldrich Chemical
Co.). Salt water was prepared from Instant Ocean Synthetic
Sea Salt (Aquarium Systems Inc.) according to the manufacturer's
directions to a specific gravity of 1.025 at 15^0C.

Ethyl parathion (o,o-diethyl-o-p-nitrophenyl phosphorothioate) was obtained with a 2,6-^{14}C ring label (99% purity,
1.8 mCi/mM, ICN Pharmaceuticals) and a ^{14}C-ethyl label (99%
purity, 15.9 mCi/mM, Amersham/Searle). 4-nitrophenol was obtained with a 2,6-^{14}C ring label (99% purity, 1.8 mCi/mM, ICN
Pharmaceuticals). Stock solutions were prepared by dilution
with ethanol, so that one microcurie could be delivered in 30
µl of carrier. Unlabeled chemicals included ethyl parathion
(Stauffer Chemical Co.), paraoxon, (o,o-diethyl-o-p-nitrophenyl
phosphate, Sigma Chemical Co.), amino parathion (o-p-aminophenyl
-o,o-diethyl phosphorothioate, Analabs), p-nitrophenol (MC/B
Manufacturing Chemists), 4-nitrophenyl-B-D-glucoside (Sigma
Chemical Co.), 4-nitrophenyl-B-D-glucuronide (Sigma Chemical
Co.), 4-nitrophenyl sulfate (Sigma Chemical Co.), o,o-diethyl
phosphate (DEP, Eastman Organic Chemicals), and o,o-diethyl
phosphorothioate (DETP, American Cyanamid). Stock solutions
(1 mg/ml) were prepared in acetone.

Experimental

Typical operating conditions for these studies included chamber temperature-15^{0}C, chamber aeration-100 ml/min, chamber salinity-1.025, chamber volume-2500 ml, flow rate-5000 ml in 24 hours, collector temperature-1^{0}C, and diurnal lighting (12 h/12 h) with white fluorescent lights.

To test the efficiency of XAD-4 resin, the metabolism chamber was inoculated with 25 μg of parathion, paraoxon, and amino parathion and 300 μg of 4-nitrophenol and 4-nitrophenyl sulfate. The system was operated with the conditions described above for 48 hours, when the water delivery flow was stopped and the system allowed to drain overnight through the resin column.

The resin column was aspirated to remove excess water and eluted with 500 ml of acetone. The acetone was removed from the eluate with a rotary evaporator and the residual water was extracted at pH 8 with three 50 ml portions of chloroform/hexane (1:1). The solvent was dried over anhydrous sodium sulfate, concentrated, and analyzed for parathion, paraoxon, and amino parathion on a Packard Model 417 gas chromatograph with a phosphorus thermionic detector (1.8 m x 2 mm i.d. glass, packed with equal parts 15% QF-1 and 10% DC-200 on 80/100 mesh Gas-Chrom Q at 215^{0}C).

The aqueous residual from the solvent extraction was acidified to pH 2 with 2N HCl and again extracted with the procedure described above. Aliquots of the extract were diluted with 1 N NaOH and 4-nitrophenol was analyzed as the yellow-colored phenate ion with a Bausch and Lomb Spectronic 20 at 400 nm.

The aqueous residual from above was concentrated to dryness by lyophilization (Kontes Glass Co.). The residual salts were rinsed with three 10 ml portions of methanol, which was analyzed for 4-nitrophenyl sulfate with a Beckman DK-2A Scanning Spectrophotometer at 300 nm.

For most animal studies, one microcurie (157 μg) of 2,6-ring labeled parathion was added to the chamber with the system off. The water was agitated to homogeneity and the animals of interest (one seastar, three mussels, twenty-four snails, or two sea anemones) were placed in the chamber and the water flow reestablished. For the first 8-12 hours, one ml aliquots were removed from the chamber, added to 15 ml of scintillation cocktail (PCS Solubilizer, Amersham/Searle), and analyzed with a Model 2425 Tri-Carb Liquid Scintillation Spectrometer (Packard Instrument Co.). Control systems were run without animals or with dead animals to serve as references for uptake studies.

The systems were operated for 96 hours; in some cases resin columns were changed and systems were operated for an additional 96 hours. At the conclusion, the water flow to the chamber was stopped; the animals were frozen in polyethylene bags; and the system was allowed to drain through the resin. All glassware

and excreta were rinsed with methanol and analyzed for radioactivity.

The resin was worked up according to the previous procedure,
except that the acidic and basic extracts were pooled for
analysis by thin layer chromatography (0.25 mm silica gel F_{254},
Brinkman Instruments, 3:1 diethyl ether/n-hexane) and autoradiography (Automatic TLC Radiochromatogram Scanner, Series
6000, Dot Recorder, Varian Aerograph). Areas of radioactivity
were scraped and quantitated by scintillation; confirmation
was by co-chromatography with standards on TLC and previously
described detectors. The aqueous residual was not lyophilized
since aliquots failed to contain appreciable radioactivity.

Entire frozen organisms were cut into 1 cc sections; mussels
were first separated from the shell. The sections were homogenized with dry ice in a 2-liter stainless steel blender
(Benville and Tindle, 1970). Replicate 50 gm subsamples were
mixed and dried with 100 gm of anhydrous sodium sulfate and
eluted in a glass extraction column (120 cm x 2.5 cm) with 400
ml of acetone (Hesselberg and Johnson, 1972). The eluate was
concentrated on a rotary evaporator and extracted and analyzed
according to the resin workup procedure. In some cases extracts
were cleaned by preparative TLC using previously described
conditions, followed by silica elution with 3:1 diethyl ether/
n-hexane.

For systems where DETP and DEP formation were studied, the
chamber was fortified with 150 µg of unlabeled parathion and
one microcurie (2 µg) of ^{14}C-ethyl labeled parathion. Throughout the study, all water was collected from the resin column
in 4-liter flasks, where the salinity was increased to 1.035
with sodium chloride and the pH was adjusted to 1.15 with concentrated HCl. All water was passed through another column of
XAD-4 resin at a flow rate of one liter per hour; this column
of resin was worked up and analyzed for DETP and DEP according
to the procedure of Daughton *et al.* (1976).

For studies of 4-nitrophenol metabolism, the chamber was
fortified with one microcurie (77 µg) of 2,6-^{14}C ring labeled
4-nitrophenol. Operating conditions were the same as those
previously mentioned.

RESULTS

Resin Recovery

The data in Table 1 demonstrate the efficiency of XAD-4
resin for the continuous removal of parathion and metabolites
from large volumes of salt water. After 48 hours of system
operation with typical operating conditions, a 99% replacement

TABLE 1.   Parathion and metabolites: Recovery[a,b] from marine metabolism system

Compound	Recovery %
Parathion	100
Paraoxon	100
Amino Parathion	94
4-Nitrophenol	102
4-Nitrophenyl sulfate	80

[a]Amount recovered compared to original fortification
[b]Data from single experiment

of the original salt water in the chamber is achieved. Recoveries from the resin were quantitative with little evidence of breakdown, volatilization, or sorption to glass. Although 4-nitrophenol existed as the more polar phenate ion in the system (pH 8.5), it was adsorbed quantitatively by the resin. Since conjugates are not extracted from water with organic solvents, the recovery of 4-nitrophenyl sulfate was a major achievement of the resin. Other studies with smaller volumes of salt water showed similar recoveries for 4-nitrophenyl glucoside (94%) and 4-nitrophenyl glucuronide (74%).

Parathion Uptake

Since the chamber volume (2500 ml) and flow rate (5000 ml in 24 hours) were constant, the gradient dilution of the chamber contents decreased the radioactivity exponentially (Figure 2); by comparing this rate of loss with any organism to an appropriate system control where no organism was present, the rate of uptake of parathion by the species studied was established. Dead animal studies and glassware solvent rinse verified that adsorption was not a significant factor (<4%). Uptake was linear for the first eight hours after dosage; fluctuations increased beyond this point due to the lower levels sampled and to other undefined variables. In particular, comparisons between the various species were facilitated by examining the time (T) necessary for half of the radioactive parathion to be removed from the chamber water; this value was similar for duplicate studies, with species variation probably resulting from differences in respiration rate, translocation processes, or fat content (Crosby, 1975).

Table 2 indicates the T-values calculated from Figure 2 (Tcontrol = 8 h); the mussels displayed the lowest T-value or most rapid rate of radioactive parathion uptake from the chamber water, while the two species of sea anemone had similar rates of uptake.

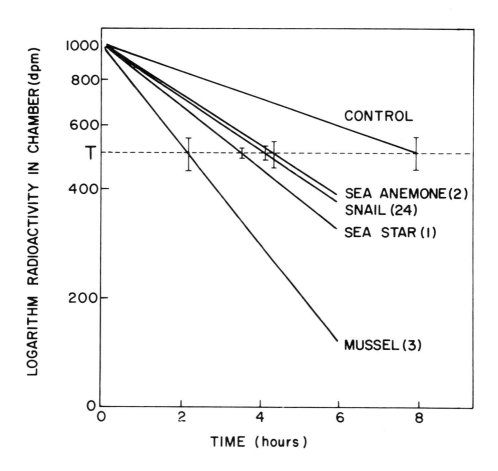

FIGURE 2. Parathion uptake by marine invertebrates: duplicate experiments; typical marine conditions; specific activity, 1.8 mCi/mM (157 µg); control represents gradient dilution of radioactivity in chamber, flow rate 5000 ml/24 hours, volume 2500 ml (63 ppb); T is time required for half of the radioactivity to be lost from the water in the chamber, from dilution, volatilization, uptake, adsorption.

TABLE 2. Parathion[a]: Radioactivity distribution and
analysis 100 x (dpm fraction/dpm fortification)

Fraction	Pisaster	Mytilus	Tegula	Anthopleura
T (hours)[b]	3.5/3.7	1.9/2.5	4.0/4.4	3.8/4.8
RESIN[b]				
parathion	23.8/28.8	39.2/46.6	25.1/29.3	40.0/45.8
p-nitrophenol	31.4/37.8	5.0/5.8	25.6/29.2	4.8/6.0
amino parathion	bkgd	bkgd	bkgd	bkgd
paraoxon	18.2/22.6	0.6/0.8	21.6/22.6	2.0/2.4
conjugates	bkgd	bkgd	bkgd	bkgd
TISSUE[c,d]				
parathion	4.0	31.5	4.2	25.3
OTHER[c,e]	9.2	6.3	10.7	14.5
TOTAL[f]	94.5	86.8	91.6	90.3

[a]Chamber temperature - 15°C, chamber aeration - 100 ml/min,
chamber salinity - 1.025, chamber volume - 2500 ml, flow rate -
5000 ml in 24 hours, collector temperature - 1°C diurnal
lighting, 1.8 mCi/mM, 1 μCi fortification (157μg),[14]C-ring
label.
[b]Values from duplicate experiments.
[c]Values from single experiment.
[d]Dry weight: sea star (1) - 122g; mussel (3) - 23g; snail
(24) - 43g; sea anemone (2) - 126g.
[e]Includes radioactivity in glassware rinse, resin soxhlet
extraction, rotary evaporator cold finger; GLC analysis indicated only parent compound.
[f]Sum of means plus values from single experiments.

Parathion Metabolism

At a ten fold chamber fortification above normal (1.5 mg, 630 ppb), the sea star displayed symptoms of intoxification including unusual arm and podia movements, followed by paralysis and death; these reactions were never observed at normal fortification levels (157 µg, 63 ppb). At similar high dosage levels, snails retracted and sealed their opercula, and mussels closed and discontinued filtration; after 24 hours of system operation, these species were observed to be functioning normally. These organisms did not display this avoidance response at normal dosage levels. Neither species of sea anemone deviated from normal behavioral patterns at any dosage employed.

Parathion can be metabolized according to the transformation scheme in Figure 3. Product analysis of radioactive fractions (Table 2) demonstrated some interesting differences and similarities between the various species examined. The sea star and snails displayed vigorous metabolism of parathion (PS) to paraoxon (PO) and p-nitrophenol (PNP); these species retained little PS after 96 hours. The two species of sea anemone metabolized only a small amount of parent compound to PO and PNP, while the mussels were only capable of minimal hydrolysis of PS to PNP; these species retained higher levels of radioactivity after 96 hours; extended analysis in the chamber revealed a gradual excretion of this compound (< 10% for

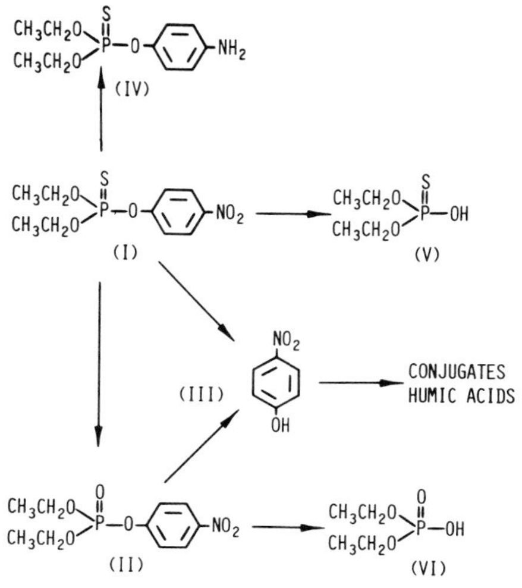

FIGURE 3: *Transformation of ethyl parathion: parathion (I), paraoxon (II), p-nitrophenol (III), amino parathion (IV), DETP (V), DEP (VI).*

additional 96 hours). Analysis of the lyophilized fractions demonstrated the lack of conjugated or other polar products for all species.

If the chamber was not aerated during system operation, the sea star died within 24 hours, while little effect was observed with the snails in behavior or metabolism. The sea anemones retracted within 48 hours, and PO was never observed as an excreted product or residue; the mussels displayed no behavioral changes but produced amino parathion as an excretion product (<3%).

Analysis of ^{14}C-ethyl parathion metabolism by the sea star and snails resulted in the isolation of DEP as the only excreted dialkylphosphate metabolite. Due to the low levels of PS metabolized, analysis for DETP and DEP were below the limit of detection for both sea anemone species; at the limit of detection, a trace of DETP was observed with the mussels. A system fortified with 200 µg each of DETP and DEP gave respective recoveries of 72% and 24% after 48 hours of normal operating conditions. A sea star, dosed by immersion in the chamber with 200 µg of DETP, displayed no intoxication and did not oxidize the compound to DEP after 48 hours of normal system operation.

^{14}C-4-nitrophenol displayed little uptake (T>7 h) and was not metabolized appreciably by any of the species examined; in particular, analysis of the lyophilized fractions revealed the absence of conjugated products. The compound was not accumulated in the tissues, and recoveries from the resin columns were quantitative (>90%).

DISCUSSION

While information is available on some physiological effects of exposure to parathion (Coppage and Matthews, 1974), behavioral toxicology has hardly been studied (Kleerekoper, 1974; Tripp, 1974). The snails and mussels displayed definite avoidance reactions to high fortification levels; recovery was the result of chamber dilution during system operation to lower concentrations. The high concentration of parathion was toxic to the sea star; symptoms resembled nervous disorders charactertistic of organophosphate poisoning (Koning and van der Meer, 1975).

The marine invertebrates in this study utilized typical pathways for parathion metabolism. Similar to reported terrestrial data (Lichtenstein, 1973; Neal, 1976), the sea star and snails converted parathion to the more toxic paraoxon, which was subsequently hydrolyzed to p-nitrophenol and diethyl

phosphoric acid; the ability to isolate small amounts of DETP
and DEP from the large volumes of salt water was a major
achievement for this metabolism system.

In this study mussels lacked the ability to oxidize parathion and in less aerobic conditions reduced it to amino parathion; other researchers also found the quahog, *Mercenaria
mercenaria,* to be deficient in oxidative pathways but capable
of nitroreductase activity (Carlson, 1972). Similar to *in vivo*
metabolism of parathion in the rat and housefly (Nakatsugawa,
et al., 1969), the mussels hydrolyzed parathion to p-nitrophenol
and diethyl phosphorothionic acid. Under certain conditions,
the reduction of parent compound may assume a more important
role as a detoxification mechanism than hydrolysis (Hitchcock
and Murphy, 1967). Both sea anemone species were able to form
paraoxon and p-nitrophenol.

The absence of conjugates of p-nitrophenol in all studies
is contrary to reported terrestrial metabolism (Menzie, 1969).
Dutton (1966) demonstrated that the mollusks *Helix pomatia* and
Arion ater were capable of detoxification of simple phenols by
conjugation with glucose to form glucosides; Allsop (1965)
determined the comparative metabolism of phenols by marine
invertebrates and found conjugation mechanisms common to mollusks and echinoderms. While p-nitrophenol was appreciably
ionized in Instant Ocean (pH 8.5) and displayed little uptake
by all species, bonding in the parathion molecule facilitated
lipid membrane penetration, with hydrolysis liberating p-nitrophenol *in situ*; conjugation of p-nitrophenol was either lacking in these species or not competitive with excretion mechanisms.

SUMMARY

This system was used to study the comparative metabolism
of parathion by marine intertidal invertebrates. Organisms
were chosen to present a representative cross-section of marine
invertebrate phylogeny; as with most comparative studies,
accurate generalization is difficult due to the limited number
of species studied. However, broad conclusions were apparent;
metabolism of this pesticide by the invertebrates studied
involved typical metabolic pathways of oxidation, reduction,
and hydrolysis. These invertebrates never metabolized parathion to the same extent or displayed the diversity of metabolic pathways characteristic of terrestrial animal studies.

Prediction of the toxicity of a chemical to an organism
based on known detoxification and activation processes has
proven valuable when dealing with terrestrial species; by

utilizing this aquatic metabolism system, analogous procedures can now be applied to aquatic organisms. This type of data eventually could provide a basis for predicting those chemicals most likely to be detrimental to the aquatic ecosystem.

LITERATURE CITED

Allsop, T. F. 1965. Comparative detoxification of phenols by some marine invertebrates. M. Sc. in Biochemistry. Victoria University of Wellington.

Benville, P. C. and R. C. Tindle. 1970. Dry ice homogenization procedure for fish samples in pesticide residue analysis. J. Agr. Food Chem., 18: 948.

Carlson, G. P. 1972. Detoxification of foreign organic compounds by the quahog, *Mercenaria mercenaria*. Comp. Biochem. Physiol. 43B: 295.

Coppage, D. L. and E. Matthews. 1974. Short-term effects of organophosphate pesticides on cholinesterases of estuarine fishes and pink shrimp. Bull. Environ. Contam. Toxicol., 11: 483.

Cox, J. L. 1971. DDT residues in seawater and particulate matter in the California current system. Fish. Bull., 69: 443.

Crosby, D. G. 1973. The fate of pesticides in the environment. Ann. Rev. Plant Physiol., 24: 467.

Crosby, D. G. 1975. The toxicant-wildlife complex. Pure Appl. Chem., 42: 23.

Daughton, C. G., D. G. Crosby, R. L. Garnas and D. P. H. Hsieh. 1976. Analysis of phosphorous-containing hydrolytic products of organophosphorus insecticides in water. J. Agr. Food Chem., 24: 236.

Duke, T. W. and D. P. Dumas. 1974. Implications of pesticide residues in the coastal environment. In Pollution and Physiology of Marine Organisms, p. 137, ed. F. J. Vernberg and W. B. Vernberg. Academic Press: New York.

Dutton, G. J. 1966. Uridine diphosphate glucose and the synthesis of phenolic glycosides by mollusks. Archs. Biochem. Biophys., 116: 399.

Hesselberg, R. J. and J. L. Johnson. 1972. Column extraction of pesticides from fish, fish food, and mud. Bull. Environ. Contam. Toxicol., 7: 115.

Hitchcock, M. and S. D. Murphy. 1967. Enzymatic reduction of o,o-diethyl-o-(4-nitrophenyl) phosphate, and o-ethyl-o-(4-nitrophenyl) benzene thiophosphonate by tissues from mammals, birds, and fishes. Biochem. Pharmacol., 16: 1801.

Hollingworth, R. M. 1971. Comparative metabolism and selectivity of organophosphate and carbamate insecticides. Bull. Org. Mond. Sante Bull. Wld. Hlth. Org., 44: 155.

Kleerekoper, H. 1974. Effects of exposure to a subacute concentration of parathion on the interaction between chemoreception and water flow in fish. In: Pollution and Physiology of Marine Organisms, p. 237, ed. by F. J. Vernberg and W. B. Vernberg. Academic Press: New York.

Koning, L. J. L. W. and C. van der Meer. 1975. The cause of death of *Palaemonetes varians* (Leach, 1814) treated with cholinesterase inhibitors. Comp. Biochem. Physiol., 51C: 73.

Lichtenstein, E. P., T. W. Fuhremann, A. A. Hochberg, R. N. Zahlten and F. W. Stratman. 1973. Metabolism of ^{14}C-Parathion and ^{14}C-Paraoxon with fractions and subfractions of rat liver cells. J. Agr. Food Chem., 21: 416.

Lu, P.-Y. and R. L. Metcalf. 1975. Environmental fate and biodegradability of benzene derivatives as studied in a model aquatic ecosystem. Environ. Health Perspect., 10: 269.

Menzie, C. M. 1969. Metabolism of Pesticides. Special Scientific Report, Wildlife No. 127, Bureau of Sport Fisheries and Wildlife, Washington, D. C.

Nakatsugawa, T., N. M. Tolman, and P. A. Dahm. 1969. Metabolism of S^{35}-Parathion in the house fly. J. Econ. Entomol., 62: 408.

Neal, R. A. 1967. Studies on the metabolism of diethyl 4-nitrophenyl phosphorothionate (parathion) *in vitro*. Biochem. J., 103: 183.

O'Brien, R. D. 1967. Insecticides: Action and Metabolism. Academic Press: New York.

Park, D. V. 1968. The Biochemistry of Foreign Compounds. Pergamon Press: Oxford.

Portmann, J. E. 1974. Marine pollution. Effluent Water Treat. J., 14: 655.

Ricketts, E. F. and J. Calvin. 1968. Between Pacific Tides. 4th Edition revised by J. W. Hedgpeth, Standford University Press.

Smith, R. I., F. A. Pitelka, D. P. Abbott and F. M. Weesner. 1970. Intertidal Invertebrates of the Central California Coast. From S. F. Light's Laboratory and Field Text in Invertebrate Zoology. University of California Press.

Symposium on Comparative Pharmacology. 1967. Fed. Proc. 26.

Tripp, M. R. 1974. Effects of organophosphate pesticides on adult oysters *(Crassostrea virginica)*. In: Pollution and Physiology of Marine Organisms, p. 225, ed. by F. J. Vernberg and W. B. Vernberg. Academic Press: New York.

Vernberg, F. J. and W. B. Vernberg (eds.). 1974. Pollution and Physiology of Marine Organisms. Academic Press: New York.

von Rumker, R., E. W. Lawless and A. F. Meiners.   1974.
    Production, Distribution, Use, and Environmental Impact
    of Selected Pesticides.   EPA 540/1-74-001, Washington, D.C.
Waldichuk, M.   1974.   Coastal marine pollution and fish.
    Ocean Management, 2: 1.

PHYSIOLOGICAL AND BIOCHEMICAL INVESTIGATIONS OF
THE TOXICITY OF PENTACHLOROPHENOL TO CRUSTACEANS

K. Ranga Rao, Ferris R. Fox, Philip J. Conklin,
Angela C. Cantelmo[1] and Anita C. Brannon

Faculty of Biology
University of West Florida
Pensacola, Florida

INTRODUCTION

Pentachlorophenol (PCP) and its water soluble salt, sodium
pentachlorophenate (Na-PCP), are broad spectrum biocides used
in a variety of pesticide formulations. Worldwide production
of PCP has been estimated to be near 200 million pounds per
year (Detrick, 1977). PCP production in the United States
amounted to 52.4 million pounds during 1974 and it was expected
to be near 80 million pounds during 1977 (Cirelli, 1978).
Although nearly 80% of the PCP produced appears to be used for
wood preservation and treatment, the antimicrobial, antifungal,
herbicidal, insecticidal and molluscicidal properties of this
chlorophenol led to a widespread application of PCP formulations. Bevenue and Beckman (1967) reviewed the early literature on chemistry, toxicity and environmental residues of PCP.
A recent volume on pentachlorophenol (Rao, 1978) contains articles on chemistry, pharmacology and environmental toxicology
of PCP and related compounds. The occurrence of PCP as a
residue in human and animal tissues (Bevenue and Beckman, 1967;
Dougherty, 1978) and its presence in 85% or more of the urine
samples of people exposed non-occupationally to PCP (Kutz et
al., 1978) suggest a ubiquitous distribution of this compound.
Pentachlorophenol often finds its way into the aquatic
environment, especially in runoff waters and effluents from

[1]Present address: Department of Biology, Ramapo College of
New Jersey, Mohwah, New Jersey.

307

wood-treatment plants. Usage of PCP-treated wood pilings and
jetties, and antifouling paints containing PCP are additional
sources of contamination of the aquatic environment. Pentachlorophenol and its sodium salt are used in drilling muds
and well completion fluids to control bacterial growth
(Robichaux, 1975; Shaw, 1975) during oil and gas exploration.
There have been reports of toxic concentrations of PCP in
aquatic environments (Fountaine et al., 1976) and fish kills
associated with PCP spills (Pierce et al., 1977; Kobayashi,
1978). Although there have been numerous reports on the toxicity of PCP and related compounds to freshwater organisms,
relatively little is known about their effects on estuarine
and marine organisms (Rao, 1978).

Previous investigations on the toxicity of PCP to aquatic
organisms indicate that crustaceans are less sensitive than
fish to PCP and related compounds (Goodnight, 1942; Tomiyama
and Kawabe, 1962; Kaila and Saarikoski, 1977). These investigations on crustaceans involved short term toxicity tests,
and there have been no attempts to elucidate the physiological
and biochemical basis for the toxicity of PCP to crustaceans.
Based on the effects of PCP on rat liver mitochondria and
snail albumen gland mitochondria, the biocidal effects of PCP
are thought to be due to uncoupling of oxidative phosphorylation (Weinbach, 1957). However, subsequent investigations of
the effects of PCP on fish (Boström and Johansson, 1972;
Holmberg et al., 1972) revealed that this phenol has variable
effects on several enzymes, leading to alterations in carbohydrate and lipid metabolism.

In view of the general paucity of information on the toxicity of pentachlorophenol to crustaceans we have conducted a
series of physiological and biochemical investigations. The
aims of our studies were to determine (1) the toxicity of Na-
PCP to grass shrimp at different stages of the molt cycle, (2)
the effects of Na-PCP on limb regeneration in grass shrimp,
(3) the effects of Na-PCP and 2,4-dinitrophenol (DNP) on
oxygen consumption by grass shrimp and tissues from blue crabs
and (4) the effects of Na-PCP and DNP on blue crab hepatopancreatic enzymes. Tests with DNP, which is also a known
uncoupler, permitted a comparison of the effects of both
phenols.

EXPERIMENTAL ANIMALS AND METHODS

Grass shrimp, *Palaemonetes pugio,* were collected from grass
beds in Santa Rosa Sound, Gulf Breeze, Florida. They were
used in experiments within two weeks after collection. The
average length of the shrimp used was 25 mm, with a range of

22 to 28 mm. Shrimp were maintained in seawater (300 milliosmoles) at 20 ± 1°C under 12 hr light: 12 hr dark conditions.
They were not fed during 96-hr toxicity tests. During long-
term experiments the shrimp were fed live brine shrimp larvae
on alternate days. Shrimp were kept individually in glass
jars containing 250 ml of seawater or experimental medium.
Analytical grade Na-PCP (City Chemical Corp., New York;
supplied by the Gulf Breeze Environmental Research Laboratory)
was dissolved in seawater. Solutions of Na-PCP were prepared
daily, immediately prior to use, and diluted to the desired
concentrations. The media of control and experimental shrimp
were replaced daily with fresh solutions.

Mature blue crabs, *Callinectes sapidus,* were purchased
from the Rollo Seafood Company, Milton, Florida. The crabs
were kept in individual containers in seawater (300 or 940
milliosmoles).

Additional details of experimental methods are given at
appropriate places in the text.

TOXICITY OF SODIUM PENTACHLOROPHENATE
TO THE GRASS SHRIMP, *PALAEMONETES PUGIO,*
IN RELATION TO THE MOLT CYCLE

The periodic shedding of exoskeleton is an important event
associated with the growth of crustaceans. The permeability
of crustacean cuticle to water and salts varies in relation to
the cyclic shedding, secretion and calcification of the exoskeleton (Passano, 1960). The cuticle of a newly molted
crustacean is relatively thin and less protective than the
calcified cuticle of an intermolt (hard shell stage) crustacean.
Although there have been previous reports of increased sensitivity of newly molted crustaceans to pollutants (Duke *et al.,*
1970; Nimmo *et al.,* 1971; Hubschman, 1967; Armstrong *et al.,*
1976) the relative toxicity of a given pollutant at specific
stages of the molt cycle was not determined. Furthermore,
whether the increased sensitivity of newly molted animals was
due to an abrupt increase in the uptake of toxicants at this
stage of the molt cycle remained to be evaluated.

The aim of our first series of experiments was to determine the toxicity of Na-PCP to grass shrimp at known stages
of the molt cycle. The edges of uropods of each test shrimp
were examined microscopically for incidence and extent of
apolysis (detachment of the epidermis from the exoskeleton)
and progress in neosetogenesis (formation of new setae) to

identify the stage in the molt cycle (Drach and Tchernigovtzeff, 1967; Rao *et al.*, 1973; Conklin and Rao, 1978a). The toxicity of Na-PCP was determined using standard 96-hr bioassays and long-term (66 days) exposures (Conklin and Rao, 1978a,b).

The average duration of the molt cycle for representative grass shrimp (rostrum-telson length: 25mm) was 14 days. Although the duration of stages C (intermolt) and D_O (early proecdysis) varied among the individuals, the duration of the later proecdysial stages (D_1-D_4) appeared relatively constant (Fig. 1). As the shrimp completed the later proecdysial stages it was possible to predict how soon ecdysis would occur.

In standard 96-hr toxicity tests (American Public Health Association, 1975) the median lethal concentrations (LC_{50}; computed using probit analysis, Finney, 1971) of Na-PCP for intermolt (stage C shrimp were not significantly different from

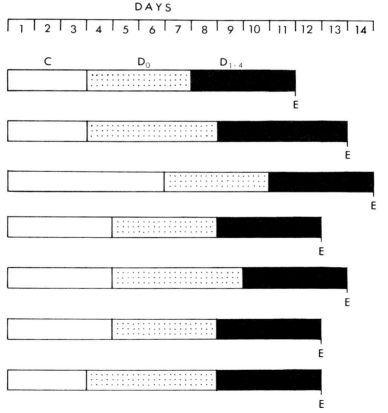

FIGURE 1. Relative durations of intermolt (C), early proecdysial (D_O) and later proecdysial (D_1-D_4) stages of the molt cycle of grass shrimp, Palaemonetes pugio. E: ecdysis. (from Conklin and Rao, 1978a).

those for shrimp in early proecdysis (stage D_O, Fig. 2).
However, shrimp in late proecdysial stages D_3-D_4 exhibited
substantially greater senstivity to Na-PCP as indicated by the
lower LC_{50} values. The shrimp that were in D_3-D_4 stages of
the molt cycle at the beginning of the 96-hr test usually
completed ecdysis (molt) within the first 48 hours of the test
period. Mortalities occurred shortly after ecdysis. Therefore, the observed increase in the sensitivity of shrimp in
D_3-D_4 stages of the molt cycle was due to the incidence of
ecdysis (molt) during the test period.

   A long-term experiment (66 days) was conducted to further
evaluate the toxicity of Na-PCP to grass shrimp in relation to
the molt cycle. Experimental shrimp were exposed to nominal
concentrations of 0.1, 0.5 and 1.0 ppm Na-PCP while controls
were maintained in seawater. Exposure to Na-PCP did not alter
the relative durations of stages in the molt cycle, but it
affected the survival of shrimp after ecdysis. A majority
(63%) of the experimental shrimp died within 24 to 48 hours
after ecdysis (Fig. 3). Whether these deaths occurred
following first or second ecdysis depended on the concentration

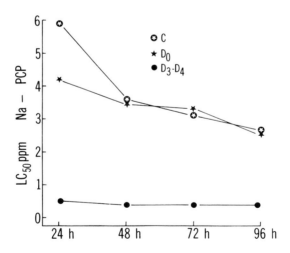

FIGURE 2.   Short-term (96-hr) toxicity of sodium penta-
chlorophenate (Na-PCP) to Palaemonetes pugio *at different*
stages of the molt cycle. The median lethal concentrations
(LC_{50}) were computed based on the observed mortalities at 24,
48, 72 and 96 hours after exposure. It should be noted that
the shrimp which were in stages D_3-D_4 of the molt cycle at
the beginning of the test underwent ecdysis during the first
24-48 hours of the test period. (from Conklin and Rao, 1978a).

of Na-PCP in the medium.   At the highest concentration of Na-PCP tested (1.0 ppm), a majority (76%) of deaths occurred after the first ecdysis, whereas among the shrimp exposed to 0.5 ppm Na-PCP a majority (60%) died after the second ecdysis. Exposure to 0.1 ppm Na-PCP did not affect the survival of shrimp.

To determine whether the observed mortality of newly molted shrimp may be related to an increase in the uptake of PCP from the medium, the relative uptake of ^{14}C-PCP (Pathfinder Lab.) by shrimp at different stages of the molt cycle was studied. Shrimp were exposed for one hour to a medium containing 2 ppm PCP.   At the end of this exposure the shrimp were allowed to swim in PCP-free water for two minutes and then processed for radioactivity measurements.   It can be seen from figure 4 that intermolt and premolt shrimp accumulated very little PCP, the bioconcentration factor (concentration of PCP in the shrimp divided by the concentration of PCP in the medium) being 2 to 3X.   However, within the same period of exposure (1 hr) to media containing PCP, newly molted shrimp accumulated PCP to a substantially greater extent (bioconcentration factor of approximately 120 X).   Although a progressive decline in the

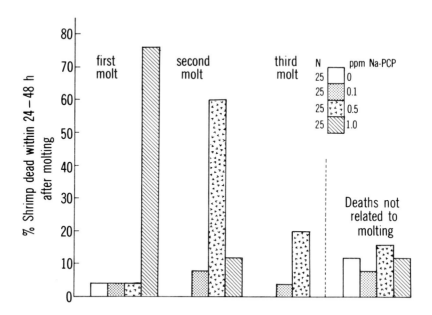

FIGURE 3.   Results of a long-term (66 days) exposure of the grass shrimp, Palaemonetes pugio, to sodium pentachloro-phenate (Na-PCP).   The incidence of mortalities in relation to ecdysis (molt) is shown.   (from Conklin and Rao, 1978a).

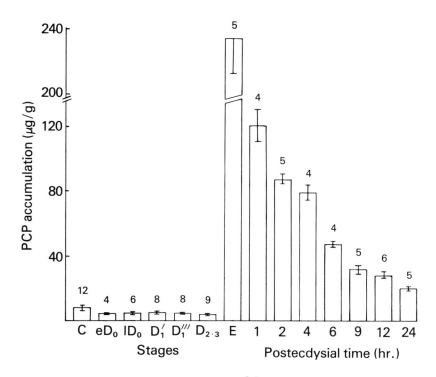

FIGURE 4. *Relative uptake of* ^{14}C-*PCP by the grass shrimp,* Palaemonetes pugio, *at different stages of the molt cycle and at known intervals after ecdysis. Shrimp were exposed for one hour to a medium containing 2 μg PCP.ml.*

uptake of PCP was noted during the postecdysial period, the relative uptake of PCP by early postecdysial shrimp (up to 24 hours after ecdysis) was still greater than that by inter-molt and premolt shrimp. Therefore, the increased sensitivity of newly molted (postecdysial) shrimp appear to be due to an increased uptake of PCP from the medium at this period of the molt cycle. Ultrastructural examination of gills, hepato-pancreas and gut tissues of shrimp exposed to Na-PCP revealed that deterioration of these tissues begins soon after ecdysis and proceeds rapidly during the postecdysial period (Doughtie and Rao, 1978). These results and a recent report of variations in microsomal aryl hydrocarbon hydroxylase activities of crus-tacean tissues in relation to the molt cycle (Singer and Lee, 1977) suggest that the dynamics of bioaccumulation, metabolism and elimination of pollutants may vary in relation to the molt cycle.

COMPARATIVE TOXICITY OF PENTACHLOROPHENOL
AND SODIUM PENTACHLOROPHENATE TO CRUSTACEANS

The available toxicological data are summarized in Table 1.
The 96-hr LC_{50} value (0.436 ppm; 95% C.I., 0.361-0.498) for
late proecdysial grass shrimp that molted during the test
period is the lowest of all LC_{50} values reported for adult
crustaceans. This LC_{50} value for adult *Palaemonetes pugio* is
lower than that reported by Borthwick and Schimmel (1978)
for larvae of this species (LC_{50}: 0.649 ppm; 95% C.I., 0.494-
0.915 ppm), but closer to that for the larvae of *Palaemonetes
varians* (LC_{50}: 0.363 ppm; 95% C.I., 0.200-0.680; van Dijk *et
al.*, 1977).

Although a majority of crustaceans are known to be less
sensitive than fishes to pentachlorophenol, the LC_{50} value
obtained for adult and larval grass shrimp, *Palaemonetes pugio*
(0.436 and 0.649 ppm respectively), and other decapod crustacean larvae (0.084 to 0.363 ppm; van Dijk *et al.*, 1977) are
well within the range of LC_{50} values reported for larval and
adult fish (0.037 to 0.516 ppm; Davis and Hoos, 1975; Adelman
et al., 1976; Borthwick and Schimmel, 1978; Schimmel *et al.*,
1978).

INHIBITORY EFFECTS OF SODIUM PENTACHLOROPHENATE
ON LIMB REGENERATION IN THE GRASS SHRIMP,
PALAEMONETES PUGIO

Environmental pollutants such as insecticides (Weis and
Mantel, 1976) and heavy metals (Weis, 1976, 1977) have been
shown to have marked effects on molting and limb regeneration
in fiddler crabs. As a part of an investigation of sublethal
effects of Na-PCP, Rao *et al.* (1978) studied the patterns of
limb regeneration in grass shrimp. The left fifth pereiopod
was removed from each shrimp with a fine-tipped jeweler's
forceps. The shrimp were microscopically examined at two day
intervals to assess the progress in limb regeneration and
advances in the molt cycle. The regeneration index, R value,
was determined following the method of Bliss (1956):

$$R \text{ value} = \frac{\text{length of limb bud}}{\text{carapace length}} \times 100 \qquad (1)$$

The incidence of limb regeneration, the rate of growth
of limb buds and the size of the new limb after ecdysis
depended on (a) stage of molt cycle at which limb removal
occurred and (b) the interval between limb removal and ecdysis

TABLE 1. Toxicity of Pentachlorophenol to Crustaceans

		Toxicity	
Animal	Test	ppm	Source
ADULTS			
Asellus communis	72 hour toxicity	>5.0	Goodnight, 1942
Astacus fluviatilis	192 hour LC_{50}; pH 6.5	9.0	Kaila and Saarikoski, 1977
Astacus fluviatilis	192 hour LC_{50}; pH 7.5	53.0	Kaila and Saarikoski, 1977
Cambarus (Orconectes) virilis	72 hour toxicity	>5.0	Goodnight, 1942
Crangon crangon	96 hour LC_{50}	1.79	Van Dijk et al.,1977
Daphnia magna	24 hour toxicity	<1.0	Weber, 1965
Daphnia pulex	72 hour toxicity	>5.0	Goodnight, 1942
Hyelella knickerbockerii	72 hour toxicity	>5.0	Goodnight, 1942
Palaemon elegans	96 hour LC_{50}	10.39	Van Dijk et al., 1977
Palaemon (Leander) japonicus[a]	48 hour TLM	2.3	Tomiyama and Kawabe, 1962
Palaemonetes pugio	96 hour LC_{50}	0.436	Conklin and Rao, 1978a
Palaemonetes varians	96 hour LC_{50}	5.09	Van Dijk et al., 1977
LARVAE			
Crangon crangon	96 hour LC_{50}	0.112	Van Dijk et al., 1977
Palaemon elegans	96 hour LC_{50}	0.084	Van Dijk et al., 1977
Palaemonetes pugio	96 hour LC_{50}	0.649	Borthwick and Schimmel, 1978
Palaemonetes varians	96 hour LC_{50}	0.363	Van Dijk et al., 1977

[a] Shrimp in later stages (D_3-D_4) of the molt cycle at the beginning of the test.

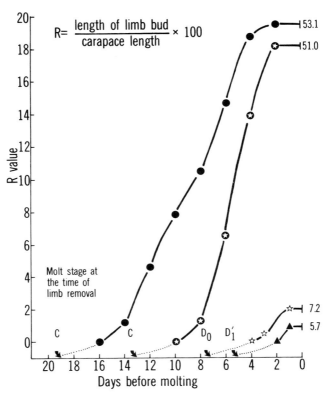

$$R= \frac{\text{length of limb bud}}{\text{carapace length}} \times 100$$

FIGURE 5. Representative patterns of limb regeneration in the grass shrimp, Palaemonetes pugio. *The left fifth pereiopod was removed (arrow) from shrimp in different stages of the molt cycle. The vertical line at the end of each curve indicates the incidence of ecdysis (molting) and the values given next the vertical bar represent postecdysial R values. (from Rao* et al.*, 1978).*

(Fig. 5). A successful completion of limb regeneration was noted only in animals subjected to limb removal during the intermolt (Stage C) period. At the time of ecdysis the limb bud stretched to a new dimension (2.5 to 3 times longer) due to the uptake of water. The postecdysial size of the new limb was dependent on the degree of limb bud development during the proecdysial period.

As noted earlier, exposure of shrimp to 0.1, 0.5 and 1.0 ppm Na-PCP did not alter the relative duration of the inter-molt cycle. However, limb regeneration was inhibited in a dose-dependent manner among the shrimp exposed to Na-PCP. An examination of the R values on the 7th, 9th and 11th days after limb removal shows the inhibitory effects of Na-PCP on

limb regeneration (Fig. 6). This inhibition of limb bud
growth during the interval between limb removal and ecdysis
led to proportional reductions in the length of new limbs
stretching out at ecdysis (Fig. 7). The nutritional state
of the shrimp did not influence the inhibitory effects of
Na-PCP on limb regeneration.

In order to determine which phases of limb regeneration were
affected by Na-PCP, shrimp were exposed to NA-PCP beginning at
varying times after limb removal. The extent of inhibition
decreased with increase in the interval between limb removal
and the initiation of exposure to Na-PCP (Fig. 8). Therefore,
the early phases of regeneration (wound healing, mitosis and
early dedifferentiation) appeared to be more sensitive to Na-
PCP than the later phases of limb regeneration. The inhibitory
effect of PCP on mitotic activity has been previously observed
in experiments with root cells and flower buds *Vicia fabra*
(Amer and Ali, 1960).

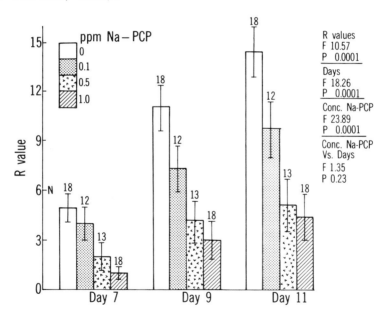

FIGURE 6. A comparison of the R values noted on the 7th,
9th and 11th day after removal of the left fifth pereiopod
from the grass shrimp, Palaemonetes pugio. Control shrimp
were maintained in seawater while the experimental shrimp
were exposed to sodium pentachlorophenate beginning the day
of limb removal until the completion of ecdysis. The shrimp
were not fed during the experiment. The figure is based on
mean values and standard errors of the mean. (from Rao et al.,
1978).

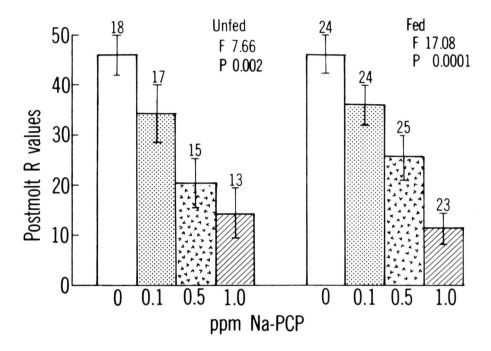

FIGURE 7.   A comparison of the postecdysial (post molt) R
values of unfed and fed grass shrimp, Palaemonetes pugio, exposed to sodium pentachlorophenate.   The R values are for the
left fifth pereiopod.   The figure is based on mean values and
standard errors of the mean.   (from Rao et al., 1978).

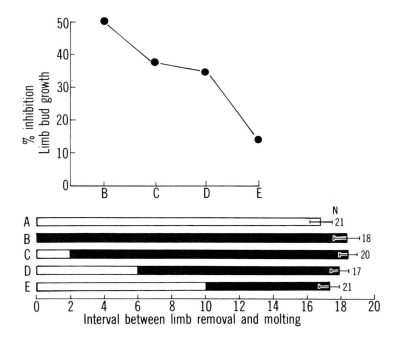

FIGURE 8. The relationship between duration of exposure
to sodium pentachlorophenate and the extent of inhibition
(relative to controls) of regeneration of the left fifth
pereiopod in the grass shrimp, Palaemonetes pugio. The control
shrimp (A) were exposed to seawater throughout the experiment.
B, shrimp exposed to 1.0 ppm Na-PCP beginning the day of limb
removal through the completion of ecdysis. C, shrimp exposed
to 1.0 ppm Na-PCP beginning two days after limb removal. D,
shrimp exposed to 1.0 ppm Na-PCP beginning six days after
limb removal. E, shrimp exposed to 1.0 ppm Na-PCP beginning
ten days after limb removal. (from Rao et al., 1978).

EFFECTS OF SODIUM PENTACHLOROPHENATE
AND 2,4-DINITROPHENOL ON OXYGEN CONSUMPTION

    Measurement of oxygen consumption not only indicates meta-
bolic rate but also provides an index of stress to organisms.
This series of experiments was designed to determine the effects of Na-PCP on in vivo and in vitro oxygen consumption in
crustaceans. For comparative purposes the effects of 2,4-dini-
trophenol, a commonly used uncoupler of oxidative phosphorylation, were also determined.
    Measurements of oxygen consumption were made using a

Gilson Differential Respirometer. Each respirometer flask
contained one grass shrimp in 5 ml of seawater (300 milliosmoles) with or without the test chemicals. Readings were
made every 15 minutes and the respirometers were re-equili-
brated every four or five hours. Oxygen consumption was
measured at the same temperature (20 ± 1°C) and light conditions (12 hr light: 12 hr dark) as in laboratory, and the
animals were not fed during or two days prior to experimentation. Because of variability in oxygen consumption associated with the movement of shrimp in flasks, a basal rate of
oxygen consumption was calculated based on data collected over
a 18 to 24 hour period. Since the flasks contained only 5 ml
of water, concentrations of Na-PCP used in these experiments
had to be higher than those employed in standard 96-hour
toxicity tests or long-term exposures discussed earlier.

The effects of Na-PCP on oxygen consumption varied depending on the stage of the molt cycle, concentration of Na-PCP
and extent of pre-exposure of shrimp to Na-PCP (Cantelmo *et al.*,
1978). At concentrations of 1.5 and 5.0 ppm Na-PCP did not
alter the oxygen consumption of intermolt and proecdysial
shrimp. Late proecdysial shrimp (stages D_3-D_4) exposed to 5.0
ppm, exhibited an increase in oxygen consumption in relation
to ecdysis to the same level as that of control shrimp in seawater (Fig. 9). However, following ecdysis, the shrimp
exposed to 5.0 ppm Na-PCP exhibited decline in oxygen consumption and died within three hours.

A decline in oxygen consumption as noted above could be
induced in intermolt (Stage C) shrimp by higher concentrations
of Na-PCP. Exposure of Shrimp to 10 or 20 ppm Na-PCP, or to
5 ppm followed by 20 ppm Na-PCP caused an initial increase in
oxygen consumption and a subsequent decline leading to death
(Cantelmo *et al.*, 1978). Since the respirometer flasks were
reequilibrated every four to five hours the death of shrimp
could not have been due to an insufficient oxygen supply.
The mean basal rate of oxygen consumption of grass shrimp in
seawater was 325.8 ± 25.2 µl O_2/g wet weight/hr. When these
shrimp were exposed to 20 ppm Na-PCP their oxygen consumption
increased significantly (P < 0.025) to 406.8 ± 22.8 µl O_2/g
wet weight/hr during the initial few hours of exposure but
this was followed by a rapid decline in oxygen consumption
(Cantelmo *et al.*, 1978.

To determine whether the observed inhibition of oxygen
consumption was due to an inhibition of tissue respiration by
Na-PCP rather than a secondary effect associated with the
death of the organism the following *in vitro* experiments were
conducted. In view of the difficulties involved in handling
tissues of a small animal such as the grass shrimp, this
series of experiments was conducted on tissues from the blue
crab, *Callinectes sapidus*.

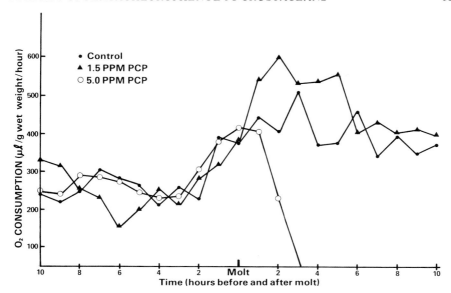

FIGURE 9. *Effect of Na-PCP on the oxygen consumption in*
Palaemonetes pugio *preceding, during and after ecdysis. Late*
proecdysial shrimp (stages D_3-D_4) were used and a basal rate
was determined prior to the addition of sodium pentachloro-
phenate to the medium. Each curve is based on data from a
representative shrimp from each group. The oxygen consump-
tion during a period 10 hours prior to ecdysis and 10 hours
after ecdysis in control shrimp (closed circles), shrimp
exposed to 1.5 ppm Na-PCP (triangles) and 5 ppm Na-PCP (open
circles) is shown. (from Cantelmo et al., 1978.)

Estuarine organisms inhabit an environment that exhibits
wide fluctuations in abiotic factors such as salinity. In
order to live in this environment organisms such as *Callinectes*
possess osmoregulatory abilities. However, a rapid alteration
of the salinity still presents an osmotic stress to the organism. The regulatory responses to an osmotic stress require
energy expenditure. We examined the effects of Na-PCP and DNP
on crab tissues subjected to normal and acutely stressful
osmotic conditions.

Blue crabs were acclimated to either 940 milliosmole (30%)
or 300 milliosmole (10%) seawater for two weeks prior to experimentation. Crabs were sacrificed and the gills, muscle and
hepatopancreas were removed. Na-PCP and DNP were tested on
tissues from crabs acclimated to 300 and 940 milliosmole seawater and on tissues from crabs acclimated to 940 milliosmole
seawater but subjected to acute hypoosmotic stress *in vitro*.

Oxygen consumption was measured using a Gilson Differential
Respirometer. The osmotic concentrations of the medium for

in vitro tests on gills corresponded with osmotic concentrations
of the media in which the whole crabs were acclimated. However,
for *in vitro* tests on the hepatopancreas and muscle the osmotic
concentrations of the test medium were adjusted to correspond to
the hemolymph osmotic concentrations of crabs acclimated to
either 940 or 300 milliosmole medium for two weeks. The osmotic
concentration of hemolymph of crabs in 300 milliosmole medium
was 670 milliosmoles while the hemolymph of crabs in 940 milliosmole medium was isoosmotic with the medium. Thus, to induce
hypoosmotic stress in muscle or hepatopancreas *in vitro*, these
tissues were dissected from crabs acclimated to 940 milliosmole
medium and placed in respirometer flasks containing 670 milliosmole medium.

The oxygen consumption of gill and muscle tissues from
crabs acclimated to 300 milliosmoles was higher than tissues
from crabs acclimated to 940 milliosmole medium. Hypoosmotic
stress *in vitro* caused an elevation of oxygen consumption in
the gill and muscle tissues. The oxygen consumption of hepatopancreatic tissue did not vary in response to acclimation to
low salinity or hypoosmotic stress *in vitro* (Cantelmo and Rao,
1978). At concentrations of 1 X 10^{-6}M (0.3 ppm Na-PCP or 0.2
ppm DNP) and 5 X 10^{-5}M (14.5 ppm Na-PCP or 9.2 ppm DNP) the
phenols did not alter the oxygen consumption of the tissues
of blue crabs. However, at a concentration of 5 X 10^{-3}M (1445
ppm Na-PCP or 920 ppm DNP) both compounds caused a reduction
in oxygen consumption of all tissues. The osmotic condition of
the crab and the type of tissue examined did not appear to influence the inhibitory effects of Na-PCP on oxygen consumption.
Even though the oxygen consumption was higher in the gill
(Fig. 10) and muscle tissues under hypoosmotic conditions, the
extent of inhibition of oxygen consumption exerted by Na-PCP
remained the same as observed under isoosmotic conditions.
This observation is consistant with the finding that although
the oxygen consumption of gill tissues from *Cancer irroratus*
and *Carcinus maenas* varied inversely with salinity, the amount
of inhibition of oxygen consumption by cadmium was independent
of the test salinity (Thurberg *et al.,* 1973).

Pentachlorophenol is known to bind more tightly than dini-
trophenol to mitochondrial proteins and is thought to be a
more effective uncoupler of oxidative phosphorylation (Weinbach
and Garbus, 1965). However, under most of the *in vitro* conditions tested in the present study DNP and Na-PCP caused comparable inhibition of oxygen consumption (Fig. 11). The only
exception was noted in tests with gills from crabs acclimated
to 300 milliosmole medium in which DNP caused significantly
less (P < 0.05) inhibition of oxygen consumption than Na-PCP.
The reason for this observed difference remains unclear.

Weinbach (1956b) found that at concentrations below 2.5 X 10^{-5}M (6.7 ppm PCP) the oxygen consumption of mitochondria from the pond snail, *Lymnaea stagnalis*, was mildly stimulated

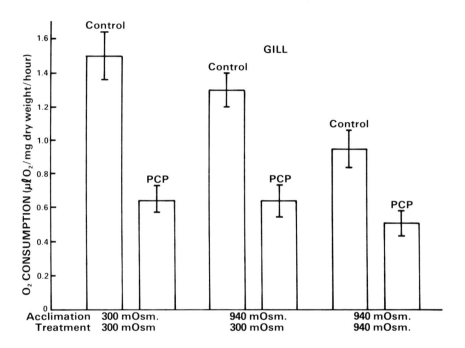

FIGURE 10. *Effect of sodium pentachlorophenate in vitro on the oxygen consumption of the gill tissue from the blue crab,* Callinectes sapidus. *All experimental flasks contained 5 X 10^{-3}M Na-PCP. The gills were removed from crabs acclimated to either 300 or 940 milliosmole seawater, and placed in 300 or 940 milliosmole seawater. The vertical line on each bar represents the standard error of the mean (From Cantelmo et al., 1978).*

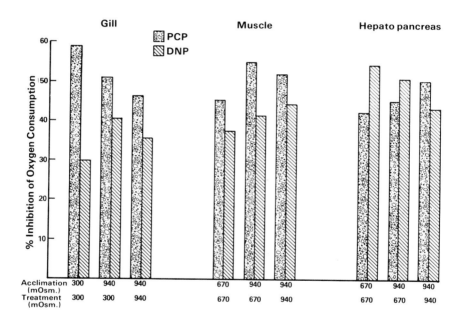

FIGURE 11. *Comparative effects of sodium pentachlorophe-nate and 2,4-dinitrophenol on the* in vitro *oxygen consumption of muscle, gill and hepatopancreatic tissues from the blue crab,* Callinectes sapidus. *The concentration of Na-PCP and DNP was 5 X 10⁻³M. (from Cantelmo et al., 1978).*

while phosphorylation was completely suppressed. At higher concentrations (5×10^{-4}M: 134 ppm PCP) phosphorylation was still completely suppressed and in addition, oxygen consumption was inhibited. Similar effects were noted in rat liver mitochondria (Weinbach, 1954). When exposed to 5×10^{-5} and 5×10^{-4} PCP (13.4 and 134 ppm) the oxygen consumption of tissue homogenates of snail, *Biomphalaria alexandrina,*was inhibited (Ishak *et al.*, 1972), the degree of inhibition being comparable to that observed by us in blue crab tissues. The decline in oxygen consumption as noted above has been attributed to the buildup of oxaloacetate resulting from an inhibition of succinate dehydrogenase activity by PCP (Chance and Williams, 1956; Chappell, 1964; Crestea and Gurban, 1964; Ishak *et al.*, 1972). In order to determine whether a similar inhibition of succinate dehydrogenase occurs in crustacean tissues and to assess the overall effects of PCP and DNP on crustacean intermediary metabolism the following experiments were conducted.

EFFECTS OF SODIUM PENTACHLOROPHENATE
AND 2,4-DINITROPHENOL ON BLUE CRAB
HEPATOPANCREATIC ENZYMES

For *in vivo* experiments, Na-PCP or DNP was injected (6 µg/g body weight: 6 ppm) into blue crabs. Three hours after injection the crabs were dissected and their hepatopancreatic tissues were homogenized and subjected to differential centrifugation (Fox and Rao, 1978b). The microsomal, mitochondrial and soluble (cytoplasmic) fractions were assayed for enzyme activity. For *in vitro* experiments the hepatopancreas was removed from intermolt animals and washed in 0.25M sucrose. The tissue was homogenized as above and fractionated. Each fraction was incubated with an appropriate concentration of either Na-PCP or DNP for 5 minutes and then subjected to enzyme bioassays. The assay conditions and other details are described elsewhere (Fox and Rao, 1978a,b).

Tricarboxylic Acid Cycle Enzymes

The enzymes studied were: fumarase, isocitrate dehydrogenase, malate dehydrogenase and succinate dehydrogenase. Fumarase, malate dehydrogenase and succinate dehydrogenase from blue crab hepatopancreatic fractions were inhibited by Na-PCP and DNP under both *in vitro* and *in vivo* conditions (Fig. 12, 13). Isocitrate dehydrogenase was stimulated at a

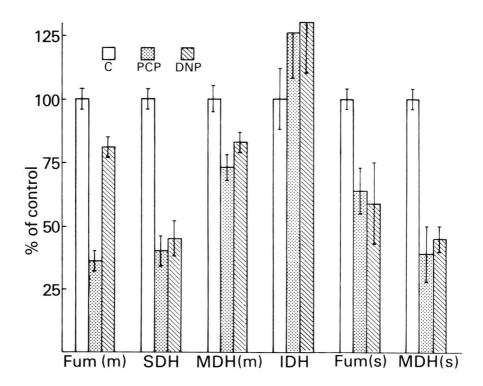

FIGURE 12.   In vivo *effects of sodium pentachlorophenate and 2,4-dinitrophenol on certain tricarboxylic acid (TCA) cycle enzymes in the hepatopancreas of the blue crab,* Callinectes sapidus.   *The crabs were injected with 6 μg Na-PCP or DNP per gram body weight.   After three hours the hepatopancreatic tissue was dissected, homogenized and sub-jected to differential centrifugation.   The resultant frac-tions were assayed for enzyme activity.   The values are mean ± S. D. of 5 experiments.   (from Fox and Rao, 1978a).*

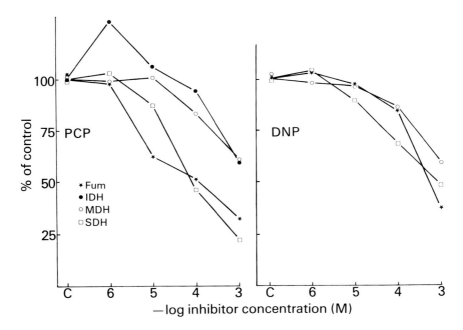

FIGURE 13. In vitro *effects of sodium pentachlorophenate and 2,4-dinitrophenol on certain tricarboxylic acid (TCA) cycle enzymes in the hepatopancreas of the blue crab,* Callinectes sapidus. *For in vitro studies the hepatopancreatic fraction was preincubated with the phenol for 5 minutes. Each data point represents the mean value derived from 5 experiments. (from Fox and Rao, 1978a).*

concentration of 10^{-6}M Na-PCP while higher concentrations inhibited it (Fig. 13) under *in vitro* conditions. Both Na-PCP and DNP stimulated this enzyme *in vivo* in blue crabs. Krueger *et al.* (1966) reported that PCP treatment causes stimulation of isocitrate dehydrogenase in the fish, *Cichlasoma bimaculatum*.

Boström and Johansson (1972) found that fumarase in the liver of eels was stimulated by Na-PCP *in vivo* and inhibited *in vitro*. This is in contrast with our observations on blue crabs in which this enzyme was inhibited by Na-PCP under *in vivo* and *in vitro* conditions. The differential effects as seen in eels may indicate that the phenol may not be directly affecting the enzyme or that the membrane protein-phenol interactions may be somewhat different under those experimental conditions. The protein-phenol interaction is thought to play an important role in the uncoupling phemomenon as well as inhibition of enzymes associated with oxidative phosphorylation (Weinbach and Garbus, 1965).

Glucose Metabolism

The blue crab hepatopancreatic pyruvate kinase and lactic dehydrogenase were inhibited by Na-PCP and DNP under *in vivo* and *in vitro* conditions (Figs. 14 and 15). Glucose-6-phosphate dehydrogenase, a pentose pathway enzyme was inhibited by the phenols *in vivo,* but was fairly stable under *in vitro* conditions when exposed to Na-PCP concentrations ranging from 10^{-6} to 10^{-3}M (0.3 to 289 ppm).

In eels exposed to a medium containing Na-PCP the two glycolytic enzymes, pyruvate kinase and lactic dehydrogenase were inhibited while the glucose-6-phosphate dehydrogenase was stimulated (Boström and Johansson, 1972). Liu (1969) reported that in cichlid fish the pentose pathway was inhibited in the presence of PCP. The carbohydrate metabolism shifted entirely to glycolysis when cichlid fish were kept in media containing PCP, but even this pathway was inhibited to some extent (Cheng, 1965). Working on snail tissues, Weinbach (1956b) found that 5 X 10^{-4}M PCP had no effect *in vitro* on glycolytic phosphorylation while higher concentrations of this phenol inhibited glycolysis completely. Boström and Johansson (1972) found eel liver pyruvate kinase to be inhibited by Na-PCP at 3 X 10^{-5} and 10^{-3}M Na-PCP. The lactic dehydrogenase of this animal was more resistant than pyruvate kinase showing 59% activity at 10^{-3}M Na-PCP while the latter enzyme was inhibited completely at this concentration. As observed in the blue crabs the eel hepatic glucose-6-phosphate dehydrogenase was the most stable enzyme, resistant to Na-PCP under *in vitro* conditions.

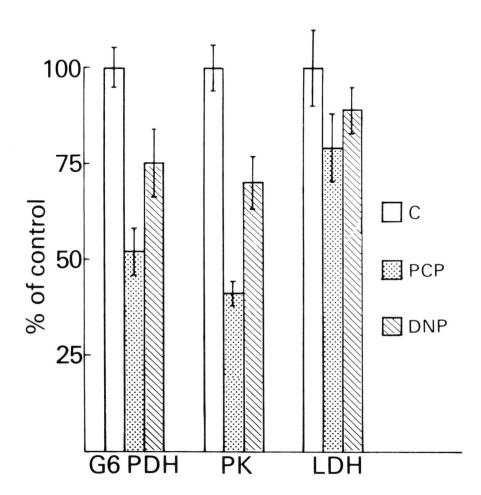

FIGURE 14. In vivo *effects of sodium pentachlorophenate and 2,4-dinitrophenol on enzymes involved in carbohydrate metabolism in the hepatopancreas of the blue crab,* Callinectes sapidus. *The values are mean ± S.D. of 5 experiments. (from Fox and Rao, 1978a).*

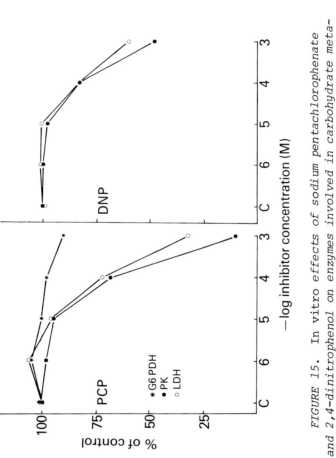

FIGURE 15. In vitro effects of sodium pentachlorophenate and 2,4-dinitrophenol on enzymes involved in carbohydrate metabolism in the hepatopancreas of the blue crab, Callinectes sapidus. Each data point represents the mean value derived from 5 experiments. (from Fox and Rao, 1078a).

Other Metabolic Enzymes

   Glutamate-pyruvate transaminase from the blue crab hepatopancreas was inhibited both *in vivo* and *in vitro* by Na-PCP
and DNP (Fox and Rao, 1978a). This transaminase in the liver
and muscle of eels was shown to be less active in the presence
of Na-PCP (Holmberg et al., 1972).
   PCP has been shown to cause an increase in lipid catabolism
in eels (Holmberg et al., 1972) and steelhead trout alevins
(Chapman and Shumway, 1978). The limited work on lipid metabolism in blue crabs injected with Na-PCP provided evidence
for increased lipid catabolism (Bose and Fujiwara, 1978).

d) ATPases:

   Adenosine triphosphate phosphohydrolases (APTases) are
vital for regulating oxidative phosphorylation, ionic transport, muscle function and several other transport-dependent
phenomena. In view of the known uncoupling effects of PCP
and DNP, their action on mitochondrial ATPases has been studied
by previous investigators. Weinbach (1956a) reported that the
mitochondrial Mg^{++} - ATPase from snail tissues was not stimulated by low concentrations of PCP, but was inhibited at
higher concentrations. The albumen gland from *Lymnaea stagnalis*
yielded a mitochondrial ATPase that had characteristics associated with oxidative phosphorylation. This enzyme was stimulated considerably at $5 \times 10^{-5}M$ PCP. A biphasic response to
PCP was exhibited by rat liver ATPase (Weinbach, 1954) and
by oligomycin-sensitive (mitochondrial) Mg^{++} - ATPase from
rat liver and kidney tissues (Desaiah, 1978). At low concentrations PCP stimulated these **enz**ymes while higher concentrations ($10^{-4}M$ and greater) inhibited them. Unlike PCP, dini-
trophenol seemed to stimulate mitochondrial ATPase at higher
concentrations (Lardy and Wellman, 1953).
   Very little is known of the effects of PCP on other cation-
transporting ATPases. The Na^+, K^+ - ATPase from rat kidney
was inhibited by PCP while the Na^+, K^+ - ATPase from rat brain
exhibited a biphasic response to PCP. This phenol had a sti-
mulatory effect on a muscle Ca^{++} = ATPase (Kameyama et al.,
1974) and an inhibitory effect on a bacterial Ca^{++} - APTase
(Ishida and Mizushima, 1969). Fox and Rao (1978b) studied the
characteristics of a microsomal Ca^{++} - ATPase from blue crab
hepatopancreas. This enzyme was inhibited by Na-PCP and DNP
under *in vivo* and *in vitro* conditions. DNP was less effective
than Na-PCP in causing inhibition. The results shown in Figure
16 reveal that the enzyme activity decreased with increase of
phenol concentration. Interference with Ca^{++} - ATPase activity
may affect active transport of calcium. Knudsen et al. (1974)
noted a striking dose-dependent decrease of calcium deposits

K. RANGA RAO *ET AL.*

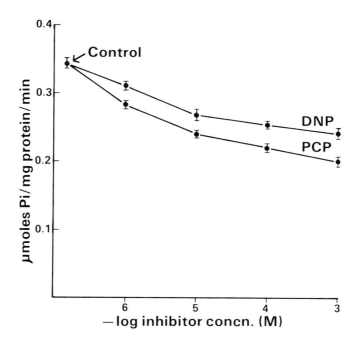

FIGURE 16. In vitro *effects of sodium pentachlorophenate and 2,4-dinitrophenol on a calcium-activated ATPase in the microsomal fraction of the hepatopancreas from the blue crab,* Callinectes sapidus. *The values are mean ± S. D. of 5 experiments. (from Fox and Rao, 1978a).*

in the kidneys of rats fed pentachlorophenol. Exposure of grass shrimp (*Palaemonetes pugio*) to media containing Na-PCP led to an apparent increase in the dry weight of exuvia and the total quantity of calcium in the exuvia (Brannon and Conklin, 1978). Whether the observed changes in exuvia of shrimp exposed to Na-PCP are due to a decrease in the resorption of calcium from the old exoskeleton remains to be established.

SUMMARY AND CONCLUSIONS

The toxicity of sodium pentachlorophenate (Na-PCP) to the grass shrimp, *Palaemonetes pugio*, varied with the stage in the molt cycle. In both short-term (96 hr bioassay) and long-term (66 days) studies newly molted shrimp exhibited the greatest sensitivity to Na-PCP. The increased mortalities noted during the early postecdysial period appeared to be due to an abrupt

increase in the uptake of PCP from the medium during this
period of the molt cycle. The LC_{50} values obtained for adult
Palaemonetes and the recent reports on the toxicity of PCP
to crustacean larvae indicate that at certain phases of life
history crustaceans are as sensitive as fishes to PCP and its
salts.

Crustaceans are capable of regenerating lost limbs. The
incidence of limb regeneration, the rate of growth of the limb
bud and the relative size of the new limb after ecdysis depend
on (a) the stage of the molt cycle at which limb removal occurs
and (b) the interval between limb removal and ecdysis. Exposure to media containing Na-PCP caused a dose-related inhibition of limb regeneration in grass shrimp. The early phases
of regeneration (wound healing, cell division and dedifferentiation) appeared to be more sensitive to Na-PCP than later
phases of regeneration.

The effects of Na-PCP on oxygen consumption by the grass
shrimp varied depending on the stage of the molt cycle and
the concentration of Na-PCP. Intermolt shrimp exposed to high
concentrations of Na-PCP (10 or 20 ppm) exhibited an initial
increase in oxygen consumption followed by a rapid decline,
leading to death. A decline in oxygen consumption followed by
death can be induced in newly molted shrimp by a lower concentration (5 ppm) of Na-PCP. Tests on isolated tissues (muscle,
gill and hepatopancreas) from the blue crab, *Callinectes
sapidus,* revealed that Na-PCP and 2,4-dinitrophenol (DNP)
inhibit oxygen consumption *in vitro*.

Na-PCP and DNP had inhibitory effects on several hepatopancreatic enzymes in blue crabs. The enzymes affected were:
fumarase, succinate dehydrogenase, malate dehydrogenase,
glucose-6-phosphate dehydrogenase, pyruvate kinase, lactic
dehydrogenase, glutamate-pyruvate transaminase and a microsomal calcium-activated ATPase. Under *in vitro* conditions,
isocitrate dehydrogenase was stimulated by a low concentration
of Na-PCP (10^{-6}M) while higher concentrations inhibited it.

The present studies on crustaceans and the previous studies
on fish, molluscs and rats seem to indicate that PCP affects
carbohydrate metabolism, lipid metabolism, ion transport and
possibly protein metabolism. The inhibitory effects on a
wide variety of enzymes suggest that the actions may be due
to non-specific interactions of this phenol with membrane proteins. In addition to the proven uncoupling effects on oxidative phosphorylation, the overall inhibitory effects of PCP
on a variety of enzymes as noted in the present study may account for the broad-spectrum biocidal effects of pentachlorophenol and its salts.

ACKNOWLEDGMENTS

   This investigation was supported by Grant R-804541-01
from the U. S. Environmental Protection Agency. We are
thankful to Dr. Norman L. Richards, Associate Director for
Extramural Activities, Gulf Breeze Environmental Research
Laboratory, for his suggestions and encouragement during this
investigation.

LITERATURE CITED

Adelman, I. R., L. L. Smith, Jr., and G. D. Siesennop. 1976.
    Effect of size or age of goldfish and fathead minnows on
    use of pentachlorophenol as a reference toxicant. Water
    Res., 10: 685-687.
Amer, S., and E. M. Ali. 1960. Cytological effects of pesti-
    cides. IV. mitotic effects of some phenols. Cytologia,
    34: 533-540.
American Public Health Association. 1975. Standard methods
    for the examination of water and wastewater including
    bottom sediments and sludges. 14th edition, 1197 p.,
    American Public Health Association, Washington, D. C.
Armstrong, D. A., D. V. Buchanan, M. H. Mallon, R. S. Caldwell
    and R. E. Millemann. 1976. Toxicity of the insecticide
    methoxychlor to the Dungeness crab, *Cancer magister*.
    Marine Biol., 38: 239-252.
Bevenue, A., and H. Beckman. 1967. Pentachlorophenol: A
    discussion of its properties and its occurrence as a
    residue in human and animal tissues. Residue Rev., 19:
    83-134.
Bliss, D. E. 1956. Neurosecretion and the control of growth
    in a decapod crustacean. In: Bertil Hanström, Zoological
    Papers in Honor of his Birthday, November 20, 1956 (K. G.
    Wingstrand, ed.), pp. 56-75, Zoological Institute, Lund,
    Sweden.
Borthwick, P. W., and S. C. Schimmel. 1978. Toxicity of
    pentachlorophenol and related compounds to early life
    stages of selected estuarine animals. In: Pentachloro-
    phenol: Chemistry, Pharmacology and Environmental Toxico-
    logy (K. R. Rao, ed.), pp. 141-146, Plenum Press, New York.
Bose, A. K., and H. Fujiwara. 1978. Fate of pentachloro-
    phenol in the blue crab, *Callinectes sapidus*. In:
    Pentachlorophenol: Chemistry, Pharmacology and Environmental
    Toxicology (K. R. Rao, ed.), pp. 83-88, Plenum Press, New
    York.

Boström, S. L., and R. G. Johansson. 1972. Effects of pentachlorophenol on enzymes involved in energy metabolism in the liver of the eel. Comp. Biochem. Physiol., 41B: 359-369.

Brannon, A. C., and P. J. Conklin. 1978. Effect of sodium pentachlorophenate on exoskeletal calcium in the grass shrimp, *Palaemonetes pugio*. In: Pentachlorophenol: Chemistry, Pharmacology and Environmental Toxicology (K. R. Rao, ed.) pp. 205-211, Plenum Press, New York.

Cantelmo, A. C., P. J. Conklin, F. R. Fox and K. R. Rao. 1978. Effects of sodium pentachlorophenate and 2,4-dinitrophenol on respiration in crustaceans. In: Pentachlorophenol: Chemistry, Pharmacology and Environmental Toxicology (K. R. Rao, ed.), pp. 251-263, Plenum Press, New York.

Cantelmo, A. C., and K. R. Rao. 1978. The effects of pentachlorophenol (PCP) and 2,4-dinitrophenol on the oxygen consumption of tissues from the blue crab, *Callinectes sapidus*. Comp. Biochem. Physiol., In press.

Chance, B., and C. R. Williams. 1956. The respiratory chain and oxidative phosphorylation. Adv. Enzymol., 17: 65-134.

Chapman, G. A., and D. L. Shumway. 1978. Effects of sodium pentachlorophenate on survival and energy metabolism of embryonic and larval steelhead trout. In: Pentachlorophenol: Chemistry, Pharmacology and Environmental Toxicology (K. R. Rao, ed.), pp. 285-299, Plenum Press, New York.

Chappell, J. B. 1964. Effects of 2,4-dinitrophenol on mitochondrial oxidations. Biochem. J., 90: 237-248.

Cheng, J. T. 1965. The effects of potassium pentachlorophenate on selected enzymes in fish. Master's thesis, Oregon State University, Corvallis.

Cirelli, D. P. 1978. Patterns of pentachlorophenol usage in the United States of America--an overview. In: Pentachlorophenol: Chemistry, Pharmacology and Environmental Toxicology (K. R. Rao, ed.), pp. 13-;7, Plenum Press, New York.

Conklin, P. J., and K. R. Rao. 1978a. Toxicity of sodium pentachlorophenate to the grass shrimp, *Palamonetes pugio*, in relation to the molt cycle. In: Pentachlorophenol: Chemistry, Pharmacology and Environmental Toxicology (K.R. Rao, ed.), pp. 181-192, Plenum Press, New York.

Conklin, P. J., and K. R. Rao. 1978b. Toxicity of sodium pentachlorophenate to the grass shrimp, *Palaemonetes pugio*, at different stages of the molt cycle. Bull. Environ. Contam. Toxicol., in press.

Crestea, E., and C. Gurban. 1964. Experimental evidence of a metabolic control mechanism in mitochondrial respiration. Biochim. Biophy. Acta, 98: 176-179.

Davis, J. C., and R. A. W. Hoos. 1975. Use of sodium pen-
    tachlorophenate and dehydroabietic acids as reference
    toxicants for salmonid bioassays. J. Fish. Res. Bd.
    Canada, 32: 411-416.
Desaiah, D. 1978. Effect of pentachlorophenol on the ATPases
    in rat tissues. In: Pentachlorophenol: Chemistry,
    Pharmacology and Environmental Toxicology (K.R. Rao, ed.),
    pp. 277-283, Plenum Press, New York.
Detrick, R. S. 1977. Pentachlorophenol: possible sources of
    human exposure. Forest Prod. J., 27: 13-16.
Dougherty, R. C. 1978. Human exposure to pentachlorophenol.
    In: Pentachlorophenol: Chemistry, Pharmacology and Environ-
    mental Toxicology (K. R. Rao, ed.), pp. 251-361, Plenum
    Press, New York.
Doughtie, D. G., and K. R. Rao. 1978. Ultrastructural changes
    induced by sodium pentachlorophenate in the grass shrimp,
    Palaemonetes pugio, in relation to the molt cycle. In:
    Pentachlorophenol: Chemistry, Pharmacology and Environ-
    mental Toxicology (K. R. Rao, ed.), pp. 213-250, Plenum
    Press, New York.
Drach, P., and C. Tchernigovtzeff. 1967. Sur la methode
    de détermination des stages d'intermue et son application
    generale aux Crustaces. Vie Millieu A18: 595-610.
Duke, T. W., J. I. Lowe, and A. J. Wilson, Jr. 1970. A
    polychlorinated biphenyl (Aroclor 1254^R) in the water,
    sediment, and biota of Escambia Bay, Florida. Bull.
    Environ. Contam. Toxicol., 5: 171-180.
Finney, D. J. 1971. Probit Analysis. 3rd edition, 333 p.,
    Cambridge University Press, London.
Fountaine, J. E., P. B. Joshipura, and P. N. Keliher. 1976.
    Some observations regarding pentachlorophenol levels in
    Haverford Township, Pennsylvania. Water Res., 10: 185-188.
Fox, F. R., and K. R. Rao. 1978a. Effects of sodium penta-
    chlorophenate and 2,4-dinitrophenol on hepatopancreatic
    enzymes in the blue crab, *Callinectes sapidus.* In:
    Pentachlorophenol: Chemistry, Pharmacology and Environ-
    mental Toxicology (K. R. Rao, ed.), pp. 265-275, Plenum
    Press, New York.
Fox, F. R., and K. R. Rao. 1978b. Characteristics of a
    Ca^{2+}-activated ATPase from the hepatopancreas of the blue
    crab, *Callinectes sapidus.* Comp. Biochem. Physiol., 59B:
    327-331.
Goodnight, C. J. 1942. Toxicity of sodium pentachlorophenate
    and pentachlorophenol to fish. Industr. Eng. Chem., 34:
    868-872.
Holmberg, B., S. Jensen, A. Larsson, K. Lewander, and M. Olsson,
    1972. Metabolic effects of technical pentachlorophenol
    (PCP) on the eel *Anguilla anguilla* L. Comp. Biochem.
    Physiol., 43B: 171-183.

Hubschman, J. H. 1967. Effects of copper on the crayfish *Orconectes rusticus* (Girard). I. Acute toxicity. Crustaceana, 12: 33-42.

Ishak, M. M., A. A. Sharaf, and A. H. Mohamed. 1972. Studies on the mode of action of some molluscicides on the snail, *Biomphalaria alexandrina*. II. Inhibition of succinate oxidation by Bayluscide, sodium pentachlorophenate and copper sulphate. Comp. Gen. Pharmacol., 3: 385-390.

Ishida, M., and S. Mizushima. 1969. Membrane ATPase of *Bacillus megaterium*. I. Properties of membrane ATPase and its solubilized form. J. Biochem., 66: 33-43.

Kaila, K., and J. Saarikoski. 1977. Toxicity of pentachlorophenol and 2,3,6-trichlorophenol to the crayfish *(Astacus fluviatilis* L.). Environ. Pollut., 12: 119-123.

Kameyama, T., S. Hayakawa, and T. Sekine. 1974. Effect of phenol derivatives and chemical modification on the adenosine triphosphatase activities of heavy meromyosin and subfragment I. J. Biochem., 75: 381·387.

Knudsen, I., H. G. Verschuuren, E. M. Den Tonkelaar, R. Kroes, and P. F. W. Helleman. 1974. Short-term toxicity of pentachlorophenol in rats. Toxicology, 2: 141-152.

Kobayashi, K. 1978. Metabolism of pentachlorophenol in fishes. In: Pentachlorophenol: Chemistry, Pharmacology and Environmental Toxicology (K. R. Rao, ed.), pp. 89-105, Plenum Press, New York.

Krueger, H., S. D. Lu, G. Chapman, and J. T. Cheng. 1966. Effects of pentachlorophenol on the fish, *Cichlasoma bimaculatum*. Abstracts from the 3rd Int. Pharm. Cong. S. Paulo, Brazil, 24-30 July 1966.

Kutz, F. W., R. S. Murphy and S. C. Strassman. 1978. Survey of pesticide residues and their metabolites in urine from the general population. In: Pentachlorophenol: Chemistry, Pharmacology and Environmental Toxicology (K. R. Rao, ed.), pp. 363-369, Plenum Press, New York.

Lardy, H. A., and H. Wellmann. 1953. The catalytic effect of 2,4-dinitrophenol on adenosinetriphosphate hydrolysis by cell particles and soluble enzymes. J. Biol. Chem., 201: 357-370.

Liu, D. H. W. 1969. Alterations of Carbohydrate Metabolism by Pentachlorophenol in Cichlid Fish. Ph.D. dissertation, Oregon State University, Corvallis.

Nimmo, D. R., R. R. Blackman, A. J. Wilson, and J. Forester. 1971. Toxicity and distribution of Aroclor[R] 1254 in the pink shrimp *Penaeus duorarum*. Marine Biol., 11: 191-197.

Passano, L. M. 1960. Molting and its control. In: The Physiology of Crustacea (T. H. Waterman ed.), Vol. 1, pp. 473-536, Academic Press, New York.

Pierce, R. H., Jr., C. R. Brent, H. P. Williams, and S. G.
    Reeves. 1977. Pentachlorophenol distribution in a fresh-
    water ecosystem. Bull. Environ. Contam. Toxicol., 18:
    251-258.
Rao, K. R., (ed.). 1978. Pentachlorophenol: Chemistry,
    Pharmacology and Environmental Toxicology, 416 p., Plenum
    Press, New York.
Rao, K. R., P. J. Conklin and A. C. Brannon. 1978. Inhibition
    of limb regeneration in the grass shrimp, *Palaemonetes
    pugio,* by sodium pentachlorophenate. In: Pentachlorophenol:
    Chemistry, pharmacology and Environmental Toxicology (K. R.
    Rao, ed.), pp. 193-203, Plenum Press, New York.
Rao, K. R., S. W. Fingerman, and M. Fingerman. 1973. Effects
    of exogenous ecdysones on the molt cycles of fourth and
    fifth stage American lobsters, *Homarus americanus.*
    Comp. Biochem. Physiol., 44A: 1105-1120.
Robichaux, T. J. 1975. Bactericides used in drilling and
    completion operations. Conference on Environmental
    Aspects of Chemical Use in Well-Drilling Operations.
    Houston, Texas, May 21-23, 1975. EPA-560-1-75-004,
    pp. 182-198. Office of Toxic Substances, United States
    Environmental Protection Agency, Washington, D. C.
Schimmel, S. C., J. M. Patrick, Jr., and L. F. Faas. 1978.
    Effects of sodium pentachlorophenate on several estuarine
    animals: toxicity, uptake and depuration. In: Pentachloro-
    phenol: Chemistry, Pharmacology and Environmental Toxico-
    logy (K. R. Rao, ed.), pp. 147-155, Plenum Press, New York.
Shaw, D. R. 1975. The toxicity of drilling fluids, their
    testing and disposal. Conference on Environmental Aspects
    of Chemical Use in Well-Drilling Operations. Houston,
    Texas, May 21-23, 1975. EPA-560-1-75-004, pp. 463-471,
    Office of Toxic Substances, United States Environmental
    Protection Agency, Washington, D. C.
Singer, S. C., and R. F. Lee. 1977. Mixed function oxygenase
    activity in blue crab, *Callinectes sapidus:* tissue distri-
    bution and correlation with changes during molting and
    development. Biol. Bull., 153: 377-386.
Thurberg, F. P., M. A. Dawson, and R. S. Collier. 1973.
    Effects of copper and cadmium on osmoregulation and oxygen
    consumption in two species of estuarine crabs. Marine Biol.
    23: 171-175.
Tomiyama, T., and K. Kawabe. 1962. The toxic effect of
    pentachlorophenate, a herbicide, on fishery organisms in
    coastal waters. I. the effect on certain fishes and a
    shrimp. Bull. Jap. Soc. Sci. Fish., 28: 379-382.
van Dijk, J. J., C. van der Meer, and M. Wijnans. 1977.
    The toxicity of sodium pentachlorophenolate for three
    species of decapod crustaceans and their larvae. Bull.
    Environ. Contam. Toxicol., 17: 622-630.

Weber, E., Einwirkung von Pentachlorophenolnatrium aur Fische
    und Fischnahrtiers. Biol. Zentralbl., 84: 81-93.
Weinbach, E. C. 1954. The effect of pentachlorophenol on
    oxidative phosphorylation. J. Biol. Chem., 210: 545-550.
Weinbach, E. C. 1956a. Pentachlorophenol and mitochondrial
    adenosine triphosphatase. J. Biol. Chem., 221: 609-618.
Weinbach, E. C. 1956b. The influence of pentachlorophenol
    on oxidative and glycolytic phosphorylation in snail tissue.
    Arch. Biochem. Biophys., 64: 129-143.
Weinbach, E. C. 1957. Biochemical basis for the toxicity of
    pentachlorophenol. Proc. National Acad. Sci., 43: 393-397.
Weinbach, E. C., and J. Garbus. 1965. The interaction of
    uncoupling phenols with mitochondria and with mitochondrial
    protein. J. Biol. Chem., 240: 1811-1819.
Weis, J. S. 1976. Effects of mercury, cadmium, and lead salts
    on limb regeneration and ecdysis in the fiddler crab, *Uca
    pugilator*. U. S. Fish Wildl. Serv., Fish Bull., 74: 464-
    467.
Weis, J. S. 1977. Limb regeneration in fiddler crabs: species
    differences and effects of methyl mercury. Biol. Bull.,
    152: 263-274.
Weis, J. S., and L. H. Mantel. 1976. DDT as an accelerator
    of regeneration and molting in fiddler crabs. Estuarine
    Coastal Mar. Sci., 4: 461-466.

A BENTHIC BIOASSAY USING TIME-LAPSE PHOTOGRAPHY
TO MEASURE THE EFFECT OF TOXICANTS ON THE FEEDING BEHAVIOR
OF LUGWORMS (POLYCHAETA:ARENICOLIDAE)

Norman I. Rubinstein

Faculty of Biology
The University of West Florida
Pensacola, Florida

INTRODUCTION

Federal legislation (i.e. the Marine Protection, Research
and Sanctuaries Act of 1972) requires that permits for the discharge of materials into coastal waters be evaluated on the
basis of their ecological impact on the marine environment.
Under this legislative mandate the U. S. Environmental Protection Agency (EPA) has been delegated the responsibility to
establish guidelines for conducting bioassays used to define
the types and amounts of materials that may be released into
the marine environment. The bioassay, therefore, serves as a
regulatory tool used by federal agencies and private industry
to assess the ecological impact of pollutants on the marine
environment.

Bioassay procedures recommended by EPA (1976) for conducting
toxicity evaluations utilize a variety of sensitive epibenthic
and pelagic species but do not include representative infaunal
organisms. This is due, in part, to the relative lack of
sensitivity displayed by many infaunal species and the difficulty in observing biological effects while organisms are
buried in sediment. However, many macrofaunal invertebrates
are deposit feeders that have a great effect on the benthic
community as a result of their substrate reworking activity.
These organisms have been shown to influence benthic community
trophic structure and sediment stability (Rhoads and Young, 1971).
In addition, sediment processing organisms provide a pathway
for cycling organic material, nutrients and pollutants between
the sediment and the water column (Rhoads, 1973; Meyers, 1977).

341

Therefore, meaningful evaluations of the impact of a pollutant on the marine environment must include information regarding effects on representative members of the infaunal community. The objective of this study was to develop a sensitive and practical method that could be used to assess biological effects of pollutants on estuarine and marine infaunal organisms.

Various benthic organisms including holothurians, crustaceans, pelecypods and polychaetes, produce distinct, characteristic topographical features on the substrate surface as a result of their normal activity (i.e. feeding, burrowing and excretion). These features in some cases may serve as *in situ* indicators of the organisms' activity (Rowe *et al.*, 1974). With the aid of time-lapse photography, surface features can be monitored and analyzed statistically to determine the relative effects of xenobiotics on organisms selected for study.

The benthic bioassay described here utilizes time-lapse photography to measure the formation of feeding funnels produced by the lugworm, *Arenicola cristata* Stimpson. Comparisons of rates of feeding funnel formation between exposed and control lugworms serve as the test criterion.

Ecological Significance

Lugworms are sedentate polychaetes distributed throughout the world in most littoral habitats. Their activities, which are somewhat analogous to those of the earthworm, are responsible for bioturbations of the sediment to depths as great as 50 cm. Populations of the European species, *Arenicola marina*, have been observed to turn over nearly 500 tons of sand (dry weight) per acre per year (Blegvad, 1914).

Lugworms form u-shaped burrows in a variety of substrates ranging from silt and mud to coarse gravel and mud. *Arenicola cristata* normally builds its burrows in muddy sand (particle size 200-700 μm) to depths of 20-25 cm and in densities as great as 20 per square meter (D'Asaro, 1976). The burrow consists of a tail shaft, a horizontal gallery and a head shaft. Periodically, the lugworm moves forward in the gallery and ingests sand along with associated organic material (living and dead). The resulting displacement of the overlying sediment produces a subsiding column of sand marked by a funnel shapped depression on the substrate surface. When the organic content in the region of the head shaft is depleted, *A. cristata* forms new feeding funnels in adjacent areas. Figure 1 illustrates the progressive formation of feeding funnels produced by one lugworm at 12-hour intervals.

Feeding and the consequent formation of feeding funnels is an integral part of an activity sequence which also incorporates excretion and peristaltic pumping of water through the burrow

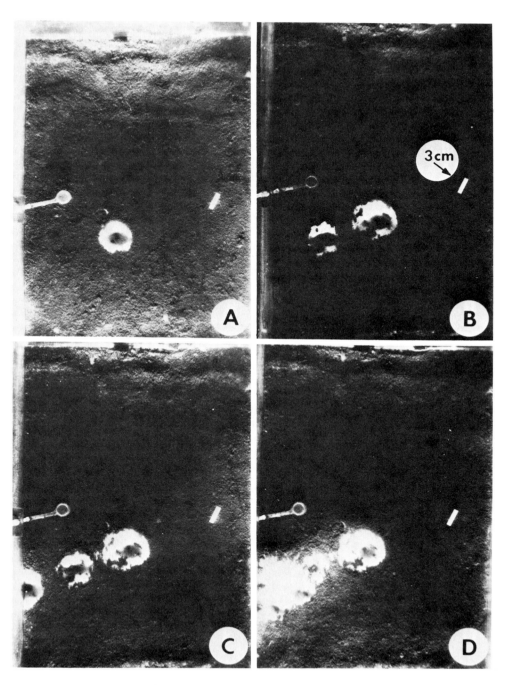

FIGURE 1. Feeding funnels produced by one lugworm at 12-hour intervals.

for respiration and ventilation. These combined activities comprise the "normal cyclical pattern" which is believed to be controlled by internal pacemakers and is therefore independent of normal environmental variables (Wells, 1966). A decrease in the formation of feeding funnels would indicate an interruption in this activity cycle.

As the lugworm feeds and pumps water through its burrow, it mixes organic material and oxygenated water into the substrate (especially in the vicinity of the head shaft). This process is undoubtedly beneficial to other infaunal organisms living in association with the lugworm, for it provides them additional sources of food and oxygen. However, during periods of environmental perturbation, contaminating agents are also transported into the substrate. Garnas *et al.* (1977) demonstrated that lugworm activity affected movement of the pesticide methyl parathion from the water column into the sediment. The fate of such compounds as they interact with infaunal organisms raises questions of great complexity. The role of the lugworm in this regard is not yet fully understood. However, the lugworm has been observed to suspend substrate modifying activity when environmental stress reaches a threshold level (Rubinstein, 1976). This type of behavioral response to environmental contaminants could serve as a sensitive indicator for various polluting agents.

Feeding Activity of Unstressed Lugworms

To insure that comparisons between exposed and non-exposed lugworms could be made, preliminary tests were conducted to determine if lugworms of similar size rework the substrate at equivalent rates when subjected to the same conditions. Due to a limitation in available time-lapse photography equipment, only two groups of lugworms could be compared at a given time.

METHODS

Bioassay Procedure

Two 125 ℓ aquaria were used as test habitats. Both aquaria contained 0.25 m^2 of sand (particle size 200-700 μm) to a depth of 25 cm and 72 ℓ of 20 μm filtered seawater. All tests were conducted at salinities between 20 and 22o/oo and temperatures between 22 and 25^{0}C. Water was aerated by airstones except when automatically turned off prior to taking photographs.

Lugworms were obtained from stocks cultured at the University of West Florida Marine Laboratory, Sabine Island, Pensacola, Florida by the method of D'Asaro (1976). Six worms 8.5 to 9.5

cm long (measured fully contracted) were placed in each
aquarium and allowed 48 hours to establish burrows and acclimate to test conditions. Following acclimation, 50 grams of
ground seagrass (predominantly *Thallasia testudinum)* were added
to both aquaria. The seagrass formed a dark mat on the sediment surface and served as the detrital component of the benthic
system. It was used as food by lugworms and also provided
photographic contrast against the white underlying sand when
turned under by feeding animals.

A 35-mm single lens reflex camera was positioned above each
aquarium (Figure 2). The cameras were equipped with an automatic advance mechanism, 24-hour timers and an automatic lighting system consisting of four strobes and two floodlights.
Photographs of the substrate surface were taken at 12-hour
intervals for 72 hours and then analyzed to determine the surface area disturbed by feeding lugworms. Surface area was
calculated by using a system of point counting in a coherent
grid modified from Hyatt (1973). The outline of feeding
funnels were traced onto a 0.25 cm grid overlay, all points
on the 5 mm intersections were counted and then converted to
actual surface area (cm^2). Total surface area turned under
by feeding lugworms was plotted against time to determine the
substrate reworking rate of lugworms in both aquaria. Rates
were subjected to linear regression analysis and the calculated
regression lines were compared (Netter and Wasserman. 1975);
differences due to treatment were considered significant at
∞ = 0.01.

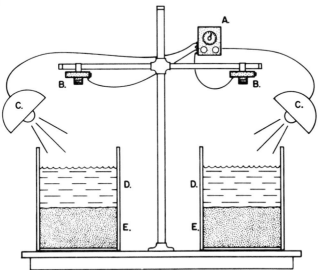

FIGURE 2. Photo-Bioassay System (A-24-hour timer; B-35
mm camera with automatic advance; D, E-aquaria with 25 cm of
sand and 75 ℓ of seawater).

Six replicate tests were conducted with a different group
of lugworms for each test (Figure 3). Variability in the rate
and magnitude of surface area disturbed between tests can be
attributed to slight variations in size, age and conditions
of lugworms. For this reason comparisons of substrate reworking
rates were made within a test and not between separate tests.
Rates of feeding funnel formation were not significantly different (t{97.5; 8}) between aquaria within each of the six tests.
Therefore, the use of two aquaria, one experimental and one
control, is valid because the rates of substrate reworking between lugworms of similar size and condition are comparable.

Application of Feeding Activity to Toxicity Testing

We tested the sensitivity of lugworm feeding activity to
the insecticide Kepone (dodecachlorooctahydro-1, 3, 4-metheno-
2H cyclobuta {cd} pentalene) using methods perfected in preliminary tests with unstressed lugworms. Kepone was selected
because of the extreme hazard it poses to aquatic life in the
James River Estuary and Chesapeake Bay (Hansen *et al.*, 1976),
areas in which the lugworm is endemic. Although the acute
toxicity of Kepone to several estuarine fishes and invertebrates has been investigated (Hansen *et al.*, 1976; Schimmel
and Wilson, 1977; Nimmo *et al.*, 1977) the sub-lethal effects
of this compound on infaunal organisms have not been evaluated.

A series of six tests using diminishing concentrations
of Kepone was conducted using the bioassay procedure previously
described. A stock solution of Kepone (88% pure) in nanograde
acetone was dispersed in the water of one aquarium; the second
aquarium received an equivalent amount of the acetone carrier
and served as the control. Aquaria were dosed following lugworm acclimation and approximately one hour after the ground
seagrass was added. From this point on photographs of the
substrate surface were taken at 12-hour intervals for 144 hours.
One liter water samples taken from aquaria one hour after introduction of the test compound were analyzed by gas chromatography using the method of Schimmel and Wilson (1977). Measured concentrations of Kepone tested were 29.5, 7.4, 6.6, 4.5
and 2.8 µg/ℓ. A non-detectable level (<0.02 µg/ℓ) was also
tested.

RESULTS AND DISCUSSION

Kepone was acutely toxic to lugworms at the highest concentration tested. All lugworms exposed to 29.5 µg/ℓ died
while no mortalities occurred at the lower concentrations

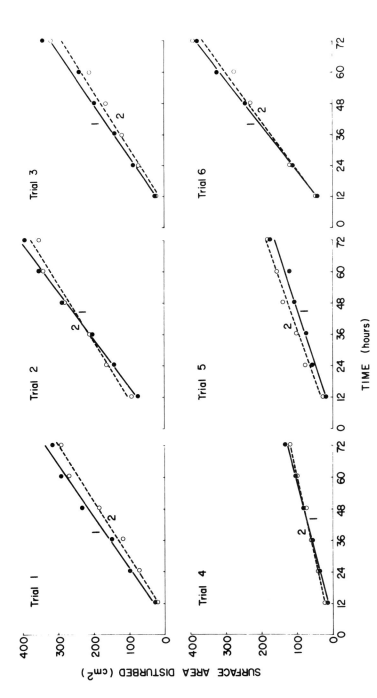

FIGURE 3. Comparison of the rates of sediment turned under by groups of similar size. A different group of lugworms was used for the six replicate tests.

during the 144 hour period. Significant inhibition of lugworm feeding activity in both magnitude and rate was observed
at all detectable levels of Kepone tested (F $\geq$ 91, d.f. 2, 10;
∞ = 0.01). Comparisons of substrate reworking rates between
exposed and control lugworms for selected tests are shown in
Figure 4.

Lugworms were sensitive to Kepone at sub-lethal levels as
low as 2.8 µg/ℓ. During the first 48 hours of exposure to
concentrations ranging from 7.4 to 2.8 µg/ℓ exposed and control
lugworms displayed similar rates of feeding funnel formation.
However, between 60 and 144 hours a significant reduction in
the amount of surface area disturbed by exposed lugworms was
observed. This latent effect suggests that lugworms may
gradually accumulate Kepone until a threshold level is reached
which then interrupts the "normal cyclical pattern."

The relative toxicity of Kepone to several estuarine species
has been determined (Schimmel and Wilson, 1977; Nimmo et al.,
1977). The species examined, and their 96-hour LC$_{50}$ values
were: grass shrimp (Palaemonetes pugio), 121 µg/ℓ; blue crab
(Callinectes sapidus), 210 µg/ℓ; mysid (Mysidopsis bahia), 10.1
µg/ℓ; sheepshead minnow (Cyprinodon variegatus), 69.5 µg/ℓ;
and spot (Leiostomas xanthurus), 6.6 µg/ℓ. Although the 96-
hour LC$_{50}$ value for the lugworm was not determined, complete
mortality was observed within 48 hours at 29.5 µg/ℓ. Based
on these findings it is apparent that the lugworm is sensitive
to Kepone when compared with other estuarine species.

A bioassay system has been presented for quantifying the
effects of a pollutant on a marine infaunal polychaete. The
toxicity tests conducted with Kepone serve to illustrate applicability of the bioassay technique. At sublethal concentrations
(7.4 to 2.8 µg/ℓ) the "normal cyclical pattern" of the lugworm
was interrupted, resulting in a decrease in substrate reworking
activity. Although the consequences of reduced lugworm activity
are speculative at this time, it is possible that a suspension
of substrate reworking by the lugworm and other infaunal organisms with similar deposit feeding habits could reduce the exchange of pollutants between the sediment and water column
thereby prolonging the residence time of a pollutant in the
water. Long-term effects of reduced lugworm feeding activity
could eventually result in the depauperation of lugworm populations. Such an event would affect the overall transport of
nutrients and pollutants through the benthic system as well as
alter food chains of which the lugworm forms a part.

Whatever the environmental significance of this deviation
from normal activity caused by pollutant stress, it is clear
that this technique can demonstrate a behavioral effect on an
important infaunal organism. Such a test, which demonstrates
biological effects of low levels of pollutants on an important

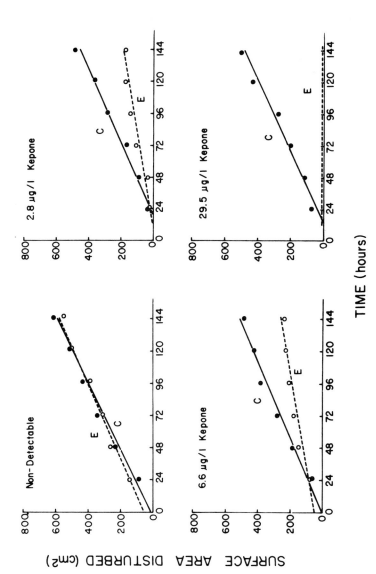

FIGURE 4. *Comparison of the rates of sediment turned under by lugworms.*
C: control; E: experimental group exposed to Kepone. Each
group consisted of six lugworms.

ecological process such as sediment reworking, will be of
value in determining the potential impact of a contaminant on
the marine environment.

SUMMARY

1.   A benthic bioassay was developed to ellucidate the
effect of pollutants on the substrate reworking activity of an
infaunal polychaete (*Arenicola cristata*).
2.   Groups of lugworms of similar size and conditioning
rework sediments at comparable rates.
3.   Lugworms exposed to the pesticide Kepone (at water concentrations as low as 2.8 $\mu g/\ell$) showed a significant reduction
in sediment processing activity when compared with non-exposed
animals.
4.   Reduction of substrate reworking activity by lugworms
and other organisms with similar deposit feeding habits could
affect sediment-water column dynamics and ultimately alter
benthic food chains.
5.   Toxicity tests using sub-lethal responses to pollutants
provide a sensitive and realistic way to evaluate the ecological
impact of pollutants on the benthic component of the marine
environment.

ACKNOWLEDGMENTS

This study was supported by an EPA grant (R804458) to the
University of West Florida. I would like to thank Dr. C. N.
D'Asaro (University of West Florida) for giving me the oppor-
tunity to carry out his original suggestions, Dr. N. R. Cooley
and Mr. D. J. Hansen (EPA) for editorial comments, Mr. Lowell
Bahner (EPA) for help with the statistical analysis and Ms.
Lynn Faas (EPA) for the chemical analysis. Facilities to com-
plete the research were made available by Dr. T. Duke of the
Gulf Breeze Environmental Research Laboratory.

LITERATURE CITED

Blegvad, H.   1914.   Food and conditions of nourishment among
     the communities of invertebrate animals found on or in
     the sea bottom in Danish waters.   Rep. Danish Biol. Sta.
     22: 41-78.

D'Asaro, C. N. 1976. Lugworm aquaculture Part I. A preliminary plan for a commercial bait-worm hatchery to produce
the lugworm *Arenicola cristata* Stimpson. Report 16; State
University System of Florida. A Sea Grant College Program.
EPA. 1976. Bioassay procedure for the ocean disposal permit program EPA-600/9-76-010,May 1976. US.Environmental Research
Laboratory, Gulf Breeze, Florida.
Garnas, R. L., C. N. D'Asaro, N. I. Rubinstein and R. A. Dime.
1977. The fate of methyl parathion in a marine benthic
microcosm. Paper #44 in Pesticide Chemistry Division,
173rd ACS meeting, New Orleans, Louisiana, March 20-25, 1977.
Hansen, D. J., A. J. Wilson, D. R. Nimmo, S. C. Schimmel, L. H.
Bahner and R. Huggett. 1976. Kepone: Hazard to aquatic
organisms. Science. 193: 528.
Hyatt, M. H. 1973. Principles and techniques of electron
microscopy. Vol. 3. van Nostrand Rheinhold Co., New York,
pp. 239-289.
Meyers, A. C. 1977. Sediment processing in a marine subtidal
sandy bottom community: I Physical aspects; II Biological
consequences. J. Mar. Res. 35 (3): 609-647.
Netter, J. **and** W. Wasserman. 1975. Applied linear statistical
models. p. 160-167. Richards D. Irwin, Inc., Homewood,
Ill. 841 pp.
Nimmo, D. R., L. H. Bahner, R. A. Rigby, J. M. Sheppard and
A. J. Wilson. 1977. *Mysidopsis bahia* - an estuarine
species suitable for life-cycle toxicity tests to determine
the effects of a pollutant. Aquatic Toxicology and Hazard
Evaluation, ASTM STP 634, F. L. Mayer and J. L. Hamelink
Eds. 109-116.
Rhoads, D. C. 1973. The influence of deposit-feeding benthos
on water turbidity and nutrient recycling. Am. J. of Sci.
273: 1-22.
Rhoads, D. C. and D. K. Young. 1971. Animal sediment relations
in Cape Cod Bay, Massachusetts. II. Reworking by *Molpadia
oolitica* (Holothuroidea).
Rowe, G. T., G. Keller, H. Staresinic and N. MacIlvaine. 1974.
Time-lapse photography of the biological reworking of
sediments in Hudson Bay submarine canyon. J. Sed. Petrol.
2: 549-552.
Rubinstein, N. I. 1976. Thermal and haline optima and lethal
limits affecting the culture of *Arenicola cristata* (Polychaeta:
Arenicolidae). Masters Thesis, University of West Florida.
Schimmel, S. C. and A. J. Wilson, Jr. 1977. Acute toxicity
of Kepone to four estuarine animals. Cheas. Sci. 18(2)
224-227.
Wells, G. P. 1966. The lugworm *(Arenicola)* a study in adaptation. Neth. J. Sea Res. 3: 294-313.

PART IV

Multiple Factor Interactions

EFFECT OF DIMILIN[R] ON DEVELOPMENT OF LARVAE OF THE STONE CRAB *MENIPPE MERCENARIA*, AND THE BLUE CRAB, *CALLINECTES SAPIDUS*

John D. Costlow

Duke University Marine Laboratory
Beaufort, North Carolina

INTRODUCTION

With the trend toward development of more specialized and less persistent chemicals for the control of a variety of insects, a number of studies have appeared in recent years on how these compounds may affect the development and reproduction of nontarget organisms, especially in the closely-related Crustacea. Initially, the studies concentrated on the effects of juvenile hormone mimics, largely to determine if their presence would inhibit metamorphosis of marine crustaceans in the same way that it has been demonstrated to prevent metamorphosis in insects. Gomez et al. (1973) demonstrated that hydroprene (Altozar[R]) actually accelerated metamorphosis of the barnacle *Balanus galeatus* and Ramenofsky, Faulkner, and Ireland (1974) subsequently investigated the effect of hydroprene on cyprid larvae of *B. galeatus*. Studies to date on the effect of both methoprene (Altosid[R]) and hydroprene (Altozar[R]) on larvae of the Brachyura (true crabs) have indicated that these compounds will not inhibit metamorphosis (Bookhout and Costlow, 1974; Costlow, 1977; Christiansen, Costlow, and Monroe, 1977a, b; Costlow and Bookhout, 1978). There were, however, levels of both of these compounds which were demonstrated to be toxic to crab larvae, depending in part upon stage of development and the natural environmental conditions of salinity and temperature during the larval stages. Payen and Costlow (1977), working with adult female *Rhithropanopeus harrisii*, reported a number of abnormalities in gametogenesis, including the progressive inhibition of vitellogenesis and spermatogenesis in concentrations of 1-3 ppm of methoprene.

More recently, a number of studies have been directed toward Dimilin[R] (TH-6040), a compound classified as an "insect growth regulator" and the way in which its inhibition of the enzyme chitinase may affect development and reproduction in such nontarget organisms as crab larvae and the brine shrimp, *Artemia salina*. Christiansen, Costlow, and Monroe (in press) observed that concentrations of Dimilin[R] ranging from 0.5 to 7.0 ppb had significant effects in reducing survival of zoeae and megalopae of two species of estuarine crabs, *Rhithropanopeus harrisii* and *Sesarma reticulatum*, and that levels of 10 ppb were lethal. An earlier study by Cunningham (1976) indicated that *Artemia* nauplii exposed to concentrations of Dimilin[R] greater than 10 ppb would not survive beyond three days, but that in lower concentrations, survival was similar to that observed for nauplii of the control series. Cunningham (1976) also reported that the hatchability of cysts (encapsulated overwintering zygotes) was less than 5% at Dimilin[R] exposures of 1.0, 5.0, and 10.0 ppb to parents as compared to 23% hatchability of the controls.

Inasmuch as Dimilin[R] is highly effective in the control of salt marsh mosquitoes, it seemed appropriate to examine further what effects this compound might have on two additional species of marine Crustacea that are found as larvae, juveniles, or adults in estuarine and coastal waters adjacent to those salt marsh areas that might be selected for spraying. Two species were selected, in part because they are known to occupy estuarine and coastal waters adjacent to salt marshes during at least some portion of their life history and in part because of their commercial significance to sizable portions of the east and Gulf coasts of the United States. *Menippe mercenaria* Say, the stone crab, is a xanthid crab and would thus offer a comparison with the results reported for larvae of *Sesarma reticulatum* Say by Christiansen *et al.* (in press). *Sesarma reticulatum* is a grapsid crab but, during its adult life, it is normally found within the salt marsh as opposed to being sublittoral as is *M. mercenaria*. *Callinectes sapidus* Rathbun, the common blue crab, normally occupies waters of higher salinity during its zoeal stages with the final larval stage, the megalopa, and the numerous juvenile stages which precede the adult found in the shallow estuarine waters of lower salinity adjacent to the salt marshes along the east and Gulf coast estuaries.

The primary objective of this study was to determine if the concentrations of Dimilin[R] which have been shown to significantly affect survival in the larvae of *S. reticulatum* and *R. harrisii* would also produce significant reductions in survival for the larvae of *M. mercenaria* and *C. sapidus* and if morphological abnormalities and delays in development would occur. The effects of Dimilin[R] on all larval stages of

M. mercenaria were considered while with *C. sapidus,* which previous studies have demonstrated to have high variability in survival during the 7-8 zoeal stages, only the megalopa stage was studied.

MATERIAL AND METHODS

The larvae of both species under consideration were reared in the laboratory, from hatching to the first crab, following the techniques which have previously been described (Costlow and Bookhout, 1959; Costlow, 1967; Ong and Costlow, 1970). In order to determine if salinity would alter the effect of varying concentrations of Dimilin^R on the larvae, *Menippe mercenaria* larvae were hatched and maintained in four salinities (20, 25, 30 and 35 ppt) and 25°C. The megalopa of *Callinectes sapidus,* reared from hatching until the final larval stage at the combination of temperature and salinity which has shown to be optimum (30 ppt, 25°C), were maintained in six combinations of temperature and salinity: 20°C - 20 ppt, 20°C - 30 ppt; 25°C- 20 ppt, 25°C - 30 ppt; 30°C- 20 ppt, and 30°C - 30 ppt. All larvae were changed daily to freshly filtered seawater of the appropriate temperature and salinity, provided with recently hatched *Artemia* nauplii, and, at that time, the mortality and/or presence of abnormalities were noted. Larvae of both species were maintained in culture cabinets during the entire period of the experiment at a photoperiod of 12 hrs light and 12 hrs dark.

A stock solution of 1 ppt Dimilin^R was prepared in pesticide grade acetone and maintained in a refrigerator throughout the experiment. Dilutions of the compound were prepared daily in filtered seawater of the appropriate salinity and included 0.5 ppb, 1.0 ppb, 3.0 ppb, and 6.0 ppb. A control series, without Dimilin^R or acetone, was provided for each of the temperature/salinity combinations and, inasmuch as previous studies have indicated that acetone alone at a concentration of 1 ppt has not deleterious effect on survival or development of crustacean larvae, an acetone control series was not included.

RESULTS

As indicated in Figure 1, all experimental concentrations of Dimilin^R used were lethal to the larvae of the stone crab, *Menippe mercenaria.* Although larvae of the control series did complete the five zoeal stages, the one megalopa stage, and complete metamorphosis to the first crab in salinities

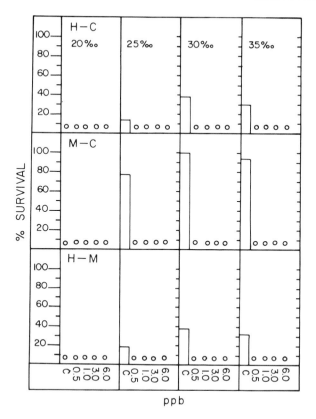

Menippe mercenaria

FIGURE 1.   Percent survival of larvae of the stone crab,
Menippe mercenaria, from hatch to megalopa (H-M), megalopa to
crab (M-C), and throughout development from hatch to crab
(H-C).   Larvae were maintained at 25°C and four salinities
(20, 25, 30 and 35 ppt) combined with four concentrations of
Dimilin[R] (0.5, 1.0, 3.0, 6.0 ppb and C, control).

ranging from 25 ppt to 35 ppt, none of the larvae in the experimental condition containing Dimilin[R] survived beyond the
first zoeal stage.   During the first three to four days following hatching, first stage zoeae appeared normal in all respects
but, at the time of the molt from the first zoeae to the second
zoeae, all died.
     Survival of megalopae of Callinectes sapidus showed con-
siderable variation, both in relation to the concentration of
Dimilin[R] and relative to the salinity/temperature combination

(Fig. 2). At 20°C, 20 ppt and 20°C, 30 ppt, a percent survival similar to that observed for the control series was
observed in concentrations of Dimilin[R] as high as 1 ppb.   In
the same temperature/salinity combinations, concentrations of
3 ppb and 6 ppb resulted either in total mortality or less
than 5 percent survival. In the higher temperature, 25°C,
there was a general increase in survival of the megalopa in
control and experimental conditions with total mortality, or
less than 5 percent survival, at 3 ppb and 6 ppb. As the
temperature was increased further to 30°C, regardless of the
salinity, the concentrations at which total mortality occured
was 0.5 ppb - 1.0 ppb.

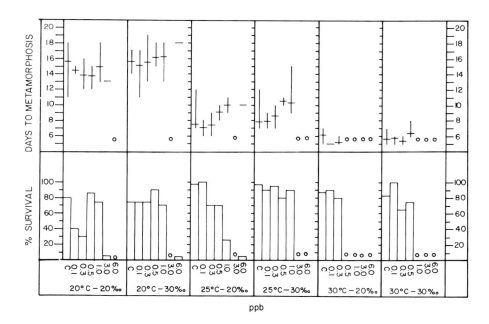

ppb

Callinectes sapidus

FIGURE 2. Effect of six concentrations of Dimilin[R]
combined with six combinations of temperature and salinity
(20°C - 20 ppt; 20°C - 30 ppt; 25°C - 20 ppt; 25°C - 30 ppt;
30°C - 20 ppt; 30°C - 30 ppt) on survival and rates of
development of megalopa of the blue crab, Callinectes sapidus.

DISCUSSION

From the limited observation on the effect of Dimilin^R on developmental stages of Brachyura (Christiansen, Costlow, and Monroe, in press) it would appear that the levels of the compound which are toxic to insects, especially saltwater mosquitoes, are also toxic to zoeal stages. The degree of sensitivity, however, appears to vary with the species. Larvae of the xanthid crab, *Rhithropanopeus harrisii*, demonstrated decreased survivorship at 1 ppb, larvae of the grapsid crab, *Sesarma reticulatum*, exhibited similar reduction in survival at 3 ppb, and zoeae of another xanthid, *Menippe mercenaria*, could not survive 0.5 ppb. Tolerance of the megalopa of *Callinectes sapidus* was slightly higher, dependent upon temperature and salinity, with a somewhat higher tolerance at 20°C than at 30°C.

As in the previous studies on larvae of *R. harrisii* and *S. reticulatum* (Christiansen *et al.*, in press) the larvae of *M. mercenaria* and the megalopa of *C. sapidus* appear perfectly normal during intermolt periods and the effects of Dimilin^R, usually expressed in mortality or morphological abnormalities, are not apparent until molting. In certain instances the morphologically aberrant first crab stages of *Callinectes sapidus*, which appeared following metamorphosis from the megalopa did "correct" the limb abnormalities during the molt to the second crab or the third crab. Cunningham (1976), working with *Artemia* exposed to Dimilin^R, reported an increased mortality in adult females which molt prior to mating every 4-6 days but did not observe increased mortality in males.

As opposed to the effects of a variety of other pollutants, which at non-lethal levels cause an extension of the larval period (Bookhout and Costlow, 1976; Bookhout, Costlow and Monroe, 1977; Rosenberg and Costlow, 1977) or modifications in the behavioral response of the larvae to light (Forward and Costlow, 1974; Forward and Costlow, 1976), no significant delay in larval development was observed in the presence of sublethal concentrations of Dimilin^R.

Evidence for synergistic effects of Dimilin^R and temperature or salinity is not clear-cut from the results of this experiment, although it was apparent that the levels of Dimilin^R which could be tolerated by the larvae did change with the change in temperature (Fig. 2). Synergistic effects of a variety of pollutants and extremes of natural environmental conditions have been reported for the larvae of a number of species (Bookhout and Costlow, 1977; Rosenberg and Costlow, 1977). Although there are species differences, in general, tolerance of the larvae to pollutants is reduced as the salinity/temperature regimes depart from the "optimum" and may be quite pronounced in those extremes of temperature and/or salinity which would otherwise permit some survival in the absence of the pollutant.

From the results of this work, as well as the limited observations on other non-target organisms, it is apparent that Dimilin[R], while a highly effective mechanism for the control of a variety of insects, must be used with extreme caution in areas which serve as a watershed to estuarine and coastal waters. The four species which have been studied thus far are widely distributed and are dependent upon the salt marsh-estuarine water for their larval or adult habitats. Although two of the species could not possibly be construed as having commercial value (*Sesarma reticulatum* and *Rhithropanopeus harrisii*), they do contribute to the established balance within the estuarine ecosystems, both in the marshes themselves and in the water column in terms of the planktonic food web. In the case of the two species which have significant commercial value, *Menippe mercenaria* and *Callinectes sapidus*, the use of Dimilin[R] in salt marshes adjacent to estuarine waters where these two species abound could only be described as potentially disastrous.

At the present time there is insufficient information to determine what effects Dimilin[R] might be expected to have on a large number of other estuarine and marine Crustacea, some of which have planktonic larvae and some of which do not. It might be expected, however, that although the levels of tolerance would differ, the compound could be expected to interfere with the molting process and the formation of new integument, regardless of the time at which this physiological process may occur within the life cycle of the individual species. Because of the specificity in action of the compound, it will be important that any future studies are designed in such a way as to provide for observations over at least one molting period and not be restricted to an intermolt period when, as exhibited by zoeal and megalopa stages, there is not outward manifestation of the effects of the compound. It is equally important to arrive at an understanding of the rate at which the compound deteriorates within the estuarine waters and marsh substrates, as well as what levels of toxicity might be expected from the breakdown products themselves. Although data are available for the rates of deterioration of Dimilin[R] in a variety of terrestrial and freshwater environments, there does not appear to be any similar information published for estuarine and coastal environments.

As with other "second generation" pesticides, such as the juvenile hormone mimics, it is obvious that they do offer great potential as replacements for the more persistent pesticide, but extreme care must be taken to be certain that we do not replace the persistent pesticides with something which can be far more harmful to the estuarine and coastal environments.

ACKNOWLEDGMENTS

    These studies were partially supported by a contract between the U.S. Environmental Protection Agency (R803838-01-1) and Duke University, Durham, North Carolina.

LITERATURE CITED

Bookhout, C. G. and J. D. Costlow, Jr. 1974. Crab develop-
    ment and effects of pollutants. Thalassia. Yugoslav.,
    10:77-87.
Bookhout, C. G. and J. D. Costlow, Jr. 1976. Effects of
    mirex on the larval development of blue crab. Water,
    Air and Soil Pollution, 4:113-126.
Bookhout, C. G., J. D. Costlow, Jr. and R. Monroe. 1976. Ef-
    fects of methoxychlor on larval development of mud-crab
    and blue crab. Water, Air and Soil Pollution, 5:349-
    365.
Christiansen, M. E., J. D. Costlow, Jr. and R. J. Monroe.
    1977a. Effects of the juvenile hormone mimic ZR-515
    (Altosid) on larval development of the mud-crab
    Rhithropanopeus harrisii in various salinities and cyclic
    temperatures. Mar. Biol., 39:269-279.
Christiansen, M. E., J. D. Costlow, Jr. and R. J. Monroe.
    1977b. Effects of the juvenile hormone mimic ZR-512
    (Altozar) on larval development of the mud-crab
    Rhithropanopeus harrisii at various cyclic temperatures.
    Mar. Biol., 39:281-288.
Christiansen, M. E. and J. D. Costlow, Jr. and R. J. Monroe.
    1978. Effects of the insect growth regulator Dimilin[R]
    (TH-6040) on larval development of two estuarine crabs.
    (in press).
Costlow, J. D., Jr. 1967. The effect of salinity and
    temperature on megalops of the blue crab, Callinectes
    sapidus Rathbun. Helgolander wiss. Meeresunters, 15:
    84-97.
Costlow, J. D., Jr. 1977. The effect of juvenile hormone
    mimics on development of the mud crab, Rhithropanopeus
    harrisii (Gould). Physiol. Responses Mar. Biota to
    Pollutants. Academic Press. pp. 439-457.

Costlow, J. D., Jr. and C. G. Bookhout. 1959. The larval
    development of *Callinectes sapidus* Rathbun reared in the
    laboratory. Biol. Bull., 116:373-396.
Costlow, J. D. and C. G. Bookhout. 1978. Second generation
    pesticides and crab development. Paper presented at
    symposium "The State of Marine Environmental Research,"
    EPA Environmental Research Laboratory, Narragansett,
    Rhode Island, May 1977.
Cunningham, P. A. 1976. Effects of Dimilin (TH 6040) on
    reproduction in the brine shrimp, *Artemia salina*.
    Environ. Entomol., 5:701-706.
Forward, R. B., Jr. and J. D. Costlow, Jr. 1974. The
    ontogeny of phototaxis by larvae of the crab
    Rhithropanopeus harrisii (Gould). Mar. Biol. 26:27-33.
Forward, R. B., Jr. and J. D. Costlow, Jr., 1976. Crustacean
    larval behavior as an indicator of sublethal effects of
    an insect juvenile hormone mimic. In: Symposium on
    Behavior as a Measure of Sublethal Stress on Marine
    Organisms (B. Olla, ed.), Third International Estuarine
    Research Federation Conference, Galveston, TX. New York:
    Academic Press, Inc. pp. 279-289.
Gomez, E. D., D. J. Faulkner, W. A. Newman and C. Ireland.
    1973. Juvenile Hormone Mimics: effect on cirriped
    crustacean metamorphosis. Science, NY, 179:813-814.
Ong, Kah-Sin and J. D. Costlow, Jr. 1970. The effect of
    salinity and temperature on the larval development of the
    stone crab, *Menippe mercenaria* Say, reared in the
    laboratory. Ches. Sci., 11:16-29.
Payen, G. and J. D. Costlow, Jr. 1977. Effects of a juvenile
    hormone mimic on male and female gametogenesis of the
    mudcrab, *Rhithropanopeus harrissi* (Gould) (Brachyura:
    Xanthidae). Biol. Bull., 152:199-208.
Ramenofsky, M., D. John Faulkner and C. Ireland. 1974.
    Effect of juvenile hormone on cirriped metamorphosis.
    Biochem. Biophys. Res. Comm., 60:172-178.
Rosenberg, R. and J. D. Costlow, Jr. 1977. Synergistic
    effects of cadmium and salinity combined with constant
    and cyclic temperatures on the larval development of two
    estuarine crab species. Mar. Biol., 38:291-303.

SEASONAL MODULATION OF THERMAL ACCLIMATION AND BEHAVIORAL THERMOREGULATION IN AQUATIC ANIMALS

John L. Roberts

Department of Zoology
University of Massachusetts
Amherst, Massachusetts

INTRODUCTION

No one seriously attempts now to find a universal indicator organism or its counterpart in application--a universal test procedure--for studies on effects of pollutants and gross climatic changes on aquatic ecosystems. The reason is simply that the resulting stresses, sometimes caused by human manipulation of environments, often are novel to organisms or selective in effects they produce. Consequently, continuing expansion is needed of our catalogue of ways and means organisms use to detect stresses, initiate protective mechanisms and employ these mechanisms to permit survival and reproductive success. Despite the informational deficiency, pollution biologists must be ready to identify and quantify overt stress responses as early warning signals that all is not well in a particular ecosystem if their concerns are to be appreciated and effective action taken.

Procedures used by biologists who have presented studies at this symposium illustrate that there is hearty growth in the variety and sophistication of assays for contaminant effects on organisms. My own experience has centered on induction of adaptive responses[1] to thermal stress in respiratory, circulatory and locomotory systems of aquatic ectotherms and

[1]Adaptation is used as equivalent to acclimation, compensation, adjustment, etc. The author chooses to avoid scientific sterotyping of common English terms and phases particularly when their meanings might vary in intent from commom usage.

the modifying factor of daily photoperiod length. Because a
number of detailed reviews on direct thermal responses of
organisms have appeared since Precht differentiated between
resistance and capacity adaptations in 1950 (Precht *et al.*,
1973) (translated from *Resistanzadaptation* and *Leistungsadapta-
tion*), my discussion will be limited as the title indicates.
Some of the reviews that might be consulted for background
information about resistance and capacity adaptations appear
in a convenient fashion in several monographs (Whittow, 1970;
Vernberg and Vernberg, 1972; Precht *et al.*, 1973; Prosser,
1973; Wieser, 1973; Newell, 1976). A separate paper by Hazel
and Prosser (1974) is a useful review of molecular mechanisms
of thermal compensation because of its comprehensive coverage
and extensive referencing of the literature.

Two assumptions underlie my selection of examples to
reinforce an argument made recently by Olla and Studholme
(1976) that behavioral responses offer, "sensitive and eco-
logically pertinent measures of pollutant effects on aquatic
organisms." Both are suggested as "rules of thumb" or precautions to be considered in the design of tests for effects
of pollutant agents when applied to test organisms as syner-
gists or secondary stress agents with temperature change as
the primary stress.

The first is precautionary, for adoption of a 12:12 LD
photoperiod (12 h light, 12 h dark) cannot be relied upon as a
control photoperiod or as a short-cut to simplify methodology.
Depending upon the season at which test organisms are collected, photoperiods near 12 h may serve as triggers for seasonal
events in life cycles which can effect rate functions like
metabolism which are often used as contaminant indicators.

The second is to emphasize changing ideas on what thermal
preferenda of aquatic animals represent. Newer evidence suggests that this term or its equivalents--eccritic temperature,
thermal optimum, preferred temperature (Fraenkel and Gunn,
1961; Hutchison, 1976)--actually define ranges in temperature
at which an organism ceases to receive aversive thermal feedback from the enviornment, shown for example by the remarkable
experiments of Rozin and Meyer (1961). That is, behavioral
thermoregulation by many ectothermal organism consists only of
avoidance activity in the absence of other than thermal cues
(most simply, ortho- and klinokineses; Fraenkel and Gunn,
1961). Furthermore, the argument will be taken here that the
analogue of a single, variable set-point thermostat is not an
adequate control model for behavioral thermoregulation of
fishes above and below thermal preferenda. Models of proportioning control are a better match with more recent studies,
especially those done with "aquatic" versions of an electronic
shuttlebox.

Evidence to support the first of these assumptions is extensive with regard to photoperiodic effects on reproductive events and migration (e.g., salinity preference changes by anadromous fishes), but still is meager regarding the role of photoperiod in resistance and capacity adaptation (Fry and Hochachka, 1970). Yet this assumption, to be considered first with primary emphasis on capacity adaptations, does link to the second assumption in important ways.

CAPACITY ADAPTATIONS TO TEMPERATURE
AND PHOTOPERIOD CHANGE

Routine Metabolism

Studies with sunfish, *Lepomis gibbosus* (Burns and Roberts, personal observations), reiterates the common experience that changes in temperature and photoperiod length are closely tied in timing of seasonally appropriate activities during life cycles of organisms (Roberts, 1964, 1967; Burns, 1975, 1976). Withrow (1959) has pointed out the precision of photo-period change as a potential trigger of seasonal events. Seasonal fluctuations of temperature from year to year even in aquatic environments lack this precision for reasons related to detection sensitivity. Yet, another factor our experiments turned up was that temperature is critical for the expression of photoperiodic effects between about 10 to 25°C. Two of our experimental series are combined in Figure 1 to show the hysteresis loop of seasonal changes in the routine metabolism (defined by Fry, 1957) of sunfish adapted to natural conditions (Burns, 1975) compared to similar data from sunfish adapted to "steady-state" temperatures (5 to 25°C) and photoperiods (9 and 15 h LD, 24 h daily cycle) in the laboratory.

The quality of the seasonal adaptations by sunfish was demonstrated in another way by Burns (1975). His monthly fish collections were brought to a common temperature of 17.5°C over a period of 20 to 30 h followed by a second measurement of routine rates of oxygen uptake. The seasonally adapted rates at 17.5°C showed as expected, an inverse mirror image of the monthly collection temperatures except early during spawning (May and June) when day lengths and temperatures rose from about 13 to 15 h and about 8 to 17°C (Type 3 acclimation, Precht *et al.*, 1973). At this time, the routine metabolism of the sunfish rose and fell dramatically (Fig. 2).

Periodic enhancement of routine rates of metabolism with progression of the seasons has been documented for a variety

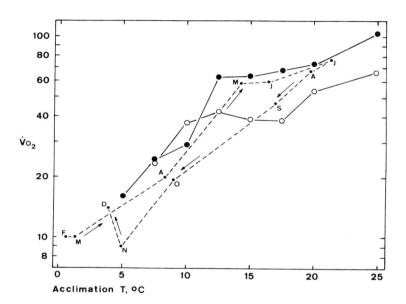

FIGURE 1. Routine oxygen consumption, $\dot{V}_{O_2}$ $(ml \cdot kg^{-1} \cdot h^{-1})$ of sunfish, Lepomis gibbosus, acclimated to steady-state temperatures at photoperiods of 9:16LD (●-●) and 16:9 LD (O-O), compared to monthly routine metabolic rates of sunfish determined the day following collection from a pond exposed to natural conditions (●--●). Statistical variances deleted from data after Roberts (1964) and Burns (1975) for clarity. Monthly rates from May through August (M,J,J,A) do not differ significantly (P>0.05) from laboratory acclimated sunfish concurrently exposed to a 9-h photoperiod. Rates for the September (S) collection differ significantly (P<0.05) from the summer sunfish but not from rates of the laboratory animals adapted to a 16-h photoperiod between acclimation and test temperatures 10 to 20°C. Arrows indicate monthly sequence.

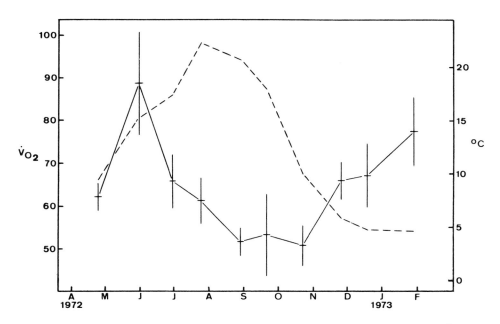

FIGURE 2.   Seasonal changes in the routine $\dot{V}_{O_2}$ $(ml \cdot kg^{-1} \cdot h^{-1})$
of sunfish, Lepomis gibbosus, measured at $17.5^\circ C$ after 20-23 h
at $17.5^\circ C$ (solid line).   The dashed line gives temperatures
at times of seasonal collections (2 days previous).   Vertical
bars $\pm$ 2 standard errors of mean values.   (From Burns, 1975.)

of fishes (Beamish, 1964; Burns, 1975).   In most cases these
have been related to reproductive events, and seem to be subject to hormonal mediation via some sort of photoperiod time
measurement system (Burns, 1976).

Similar seasonal effects on rates of routine oxygen uptake
have been reported as well for a variety of invertebrates.   For
example, closely related species of fiddler crabs, Uca sp.,
show seasonal changes in their routine oxygen consumption
(Vernberg, 1969).   Rates of winter U. pugnax from the Carolinas
have been found to be higher when tested between 7 and 28°C
than the rates of summer animals.   In contrast, tropical U.
rapax (Florida and Central America) gave no hints of a seasonal
effect upon their routine metabolism.   Grapsoid shore crabs of
the western coastal states show seasonal changes in routine
metabolism that relate mainly to winter conditions.   The three
most commonly found species [Pachygrapsus crassipes (Roberts,
1957); Hemigrapsus oregonensis and H. nudus (Dehnel, 1958, 1960)]

show only slight thermal acclimation which is most evident
during winter months. The more terrestrial, and more active
P. crassipes seems to avoid the heat stress of summer behaviorally.

The case for *Hemigrapsus* species is complexed by the factor
of salinity to the extent that clear separation of thermal and
salinity adaptations on routine metabolism has not been accomplished. However, Dehnel (1958) also has reported that routine
rates of oxygen uptake by summer animals of both species of
Hemigrapsus (salinity 11 ^{o}/oo) can be significantly enhanced
by adaptation of the crabs to short photoperiods, 8:16LD.

Routine Versus Standard and Active
Levels of Metabolism

That levels of routine metabolism seem so labile seasonally
is obvious from the examples cited and raises the question:
What does determination of the rate of routine oxygen consumption actually measure? Seasonal changes of temperature, photoperiod, salinity, and perhaps other physical factors in particular environments often interact thus requiring careful
application of routine metabolism as a response indicator.
Although measurement of routine metabolism, unlike standard
(resting or basal) and active (steady-state work load), is
relatively simple from standpoints of time and apparatus
requirements, routine measures represent a compromise. Routine
metabolism includes a quantitatively unpredictable combination
of energy expenditures for spontaneous movements, muscular
tonus and body maintenance activities that has been well documented (Fry, 1957; Beamish, 1964b; Beamish and Mookhergii,
1964). For example, routine rates of metabolism of laboratory
adapted sunfish measured at their acclimation temperatures
(Fig. 1), suggest the components of spontaneous movements (fin
sculling, etc.) and tonus were greatest in those fish adapted
to the 9-h photoperiod, and acclimated to the temperature of
12.5°C. Neither of these components were actually measured as
they can be by using the more complex techniques of Spoor
(1946), and Beamish and Mookhergii (1964) to obtain standard
rates (see also, Fry, 1957).

Heart and opercular rate measures were also recorded from
these sunfish, but revealed little. The explanation seems to
be that they show a large capacity for accommodation to work
load which is indirectly borne out by the fact that these
secondary indicators show less sensitivity to temperature
change (low Q_{10} values) than oxygen uptake, even in active
species like salmonids (Roberts, 1964, 1967; Heath and Hughes,
1973). Sutterlin (1969) found other evidence of accommodation
in fishes under the routine conditions shown in Figure 3 as a
waxing and waning in the frequencies of respiratory cycles of

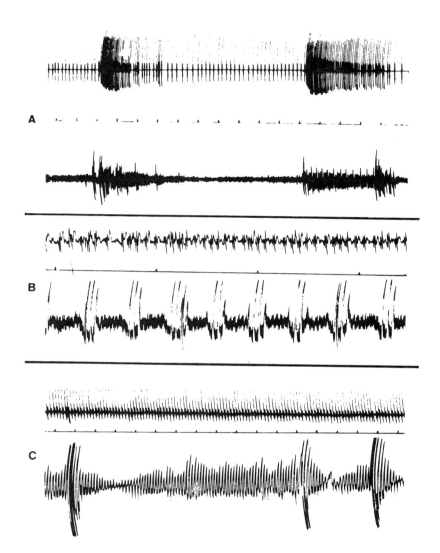

FIGURE 3. Routine ECG's and respiratory movements (upper and lower traces, respectively) of the brown bullhead, Ameiurus nebulosus *(A), the sunfish,* Lepomis gibbosus *(B), and the brown trout,* Salmo trutta *(C). Bradycardia parallels respiratory pauses [(bullhead, sunfish) or marked excursions by the branchial pumps (trout)]. Breathing movements of the bullhead are only clearly differentiated in the right hand part of the trace. Time marks at second intervals. (Modified from A. M. Sutterlin, 1966, Ph.D. Dissertation in Zoology, Univ. Massachusetts, Amherst.)*

fish. To be sure, such patterns are most obvious in sluggish
swimmers and least obvious in active species in which specia-
ized expansion of the opercular and branchiostegal baskets has
not been extensive (Roberts, 1975).

Standard, routine and activity determinations have their
specific uses. Standard and routine measures probably are more
sensitive and show subtle stress influences that would be masked during measurement of active, steady-state power output for
the reason that locomotion requires an ever increasing proportion of total body metabolism as velocity increases. Yet,
studies of active metabolism have proven to be highly useful
adjuncts in determining energy budgets and thermal requirements of fishes (Brett, 1964, 1970). By extrapolation,
standard metabolism also can be derived from a graded series
of active rates (Brett, 1964). However, the method of Beamish
and Mookhergii (1964), adapted from one used by Spoor (1946),
is more applicable for tests on effects of pollutant agents
because of its relative simplicity. Their technique utilizes
detection of spontaneous swimming movements in a small respiration vessel which enables extrapolation of standard rates from
measurements of routine metabolism at different levels of
spontaneous activity. Excellent correlation in the detection
of standard rates by both the Brett and Beamish methods has
been demonstrated by Brett (1964).

Quantification of standard metabolic rates or routine
metabolism minus standard metabolism in relation to spontaneous
activity has been less successful among invertebrates, other
than sedentary filter feeders or animals that have visible
periods of activity and quiescence (Mangum and Sassaman, 1969).
Perhaps best known is the interesting attempt by Newell (1966)
and his associates (Newell and Northcroft, 1967) to establish
standard and active rates of oxygen uptake (Y-intercept values
at unity size) from scatter data plots of individual respiration rates versus body size, a problem still without a generally
applicable technical solution when the presence or absence of
motor activity cannot be otherwise identified. They assumed
that the different activity levels, standard and active, could
be derived by the simple expedient of fitting regression lines
to minimal and maximal rates. Several aspects of their techniques (Newell and Northcroft, 1967; Newell, 1970) have been
criticized by Tribe and Bowler (1968) as faults of technique
and data analysis. Other comments by Mangum and Sassaman
(1969) and by Coyer and Mangum (1973) support these critical
views. These are worthwhile reconsidering as precautions in
attempts to differentiate between standard, routine and active
metabolism, and useful to later arguments relating spontaneous
locomotion to behavioral thermoregulation.

The derivative procedures used by Newell and Northcroft (1967) seem to have been originated by Courtney and Newell (1965), and were based on monitoring oxygen uptake rates in closed containers with membrane covered, polarographic oxygen electrodes (Kanwisher, 1959). In the case of *Cardium edule*, a lamellibranch mollusc, Newell and Northcroft state (1967; p. 287):

> No attempt was made to quantify the relation
> between oxygen uptake and volume of water
> pumped although it was noted that cockles spend
> a considerable proportion of the time with
> their siphons extruded and open but with no
> trace of a current of water flowing to and from
> them.

This was the "quiescent phase," also identified in an earlier study on the oxygen uptake of *C. edule* over the temperature range of 2.5 to 20°C (Newell, 1966). Tribe and Bowler (1969) point out the obvious question of identification of the "quiescent phase" (no branchial flow) when it is known that many intertidal lamellibranchs - contained within an oxygen-impermeable shell - can sustain long periods of anaerobiosis. In fact, pay-back of an oxygen debt may not occur at all or only very slowly during the "active phase" when powered flow of the respiratory stream returns. If *C. edule* and other invertebrates have the capacity for facultative anaerobiosis possessed by the American oyster, *Crassostrea virginica* (Hammen, 1969) quiescent phases could not be considered equivalent to standard metabolism. Thus the contention by Newell and Northcroft (1967) that the standard metabolism of many of the intertidal animals they examined is temperature independent, must remain equivocal until the aerobic-anaerobic abilities of the animals they studied are better known.

However, their related argument (Newell and Northcroft, 1967; Newell, 1970) that temperature change effects active metabolism more sharply than standard metabolism has been broadly based on studies of both invertebrates (Newell, 1970) and vertebrates (Fry and Hochachka, 1970). An instructive example is illustrated in Figure 4 from the study by Halcrow and Boyd (1967). It shows relative sensitivities of standard and routine rates of oxygen uptake by isopods, *Gammarus oceanicus*, measured at acclimation temperatures. A significant feature was the measurement of routine metabolic rates at spontaneous, not at "paced" activity levels. That is, the isopods were not subject to forced movement (upstream swimming) as has been usual in similar studies with fishes (Fry and Hochachka, 1970).

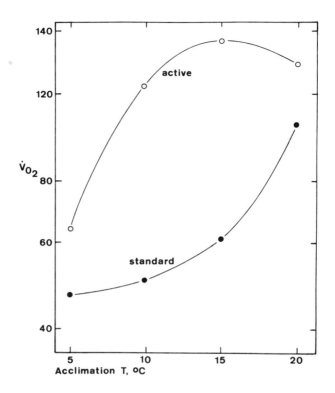

FIGURE 4. Acclimated $\dot{V}_{O_2}$ levels $(ml \cdot g^{-1} \cdot h^{-1}$, wet weight) of active and inactive isopods, Gammarus oceanicus. After Halcrow and Boyd, 1967.

The answer to why active metabolism seems to be more sensitive to temperature change than standard metabolism has been based on the supposition that a temperature change effects a proportionately greater change on levels of spontaneous motor activity than on levels of non-motor, maintenance functions. It is an assumption of particular relevance to understanding the role of spontaneous locomotion in the aggregation of animals at their thermal preferenda. But is is also necessary to recall that the physical properties of water are only slightly effected by temperature change. For those animals whose primary motor activity is swimming in water (e.g. fishes) or pumping it (filter feeders) the effect of temperature on drag forces opposing flow is negligible (Webb, 1975). Brett's (1964) elegant study of swimming energetics of young sockeye salmon (Onchorhynchus nerka) has demonstrated this fact explicitly as shown in Figure 5 for the oxygen uptake of young salmon given as metabolic scope (active minus standard metabolism). When the steady state power output of young sockeye

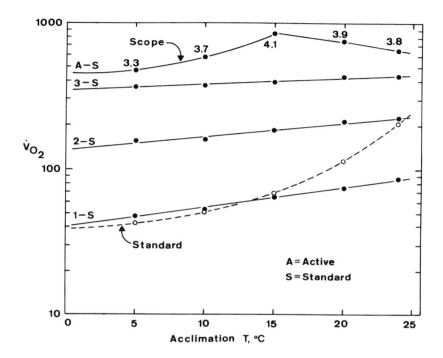

*FIGURE 5. Metabolic scopes (active minus standard) for
locomotion as $\dot{V}o_2$ values (mg·kg^{-1}·h^{-1}) of young sockeye salmon,
Onchorhyncus nerca, at acclimation temperatures for three,
steady-state swimming speeds (body lengths per sec) and for
maximal sustainable speeds (A-S, 60 min test period). Standard
or maintenance oxygen uptake shown as dashed line. From Brett,
1964.*

is increased from a swimming velocity of 1 body length per
sec (about 18 cm·sec^{-1}) to 3 body lengths, the slope of the
metabolic-scope relationship to temperature falls because the
proportion of the total body metabolism devoted to swimming
thrust increases relative to the level of standard metabolism.

The concept that Brett's studies has generated is significant not only because it agrees well with hydrodynamic conclusions (Webb, 1975). That is, the steady-state power output
required for swimming at a fixed velocity or the "inside-out"
cases of lamellibranchs, worms or crabs propelling water over
gills at steady volume-rates, has a negligible thermal sensitivity. It is also significant because it affirms the view
that thermal sensitivity of spontaneous motor activity is
primarily behavioral - a function of the tonic drive level of
the control system.

BEHAVIORAL THERMOREGULATION

The choice of techniques for determination of thermal preferenda has resulted in considerable argument and the generation of complicated hypotheses about how ectotherms, especially fishes, "orient" and aggregate in regions where temperature is favorable during particular seasons or stages in life cycles (Reynolds, 1977). Some investigators reject ortho- and klino-kineses as simplistic, favoring some grade of directive behavior or thermotaxis (Fraenkel and Gunn, 1961). Others reject both kineses and taxes and suggest a sort of "seeking" behavior (Reynolds, 1977). But it is difficult to understand how seeking responses can be interpreted as other than types of taxes following the descriptions outlined by Fraenkel and Gunn (1961). Discussions by Neill and Magnuson (1974), Norris (1963) and Reynolds (1977) should be consulted to gain an overview of the current status of the argument.

Thus to pursue the implications of the second introductory assumption, a kinetic model has been adopted to explain thermal selection by many fishes. No doubt my comments will not settle the argument but they are intended to give reasons why kinetic arguments still have validity.

Sensory Detection and Control

Except for sharp discontinuities such as power-plant thermal plumes, lake or oceanic thermoclines, and sun-warmed surfaces, temperature gradients met by aquatic organisms most of the time are gentle. Detection sensitivity of fishes happens to be of a high order ($\pm$ 0.1°C or better) although neither specific receptors or their locations have been identified (Murray, 1971). Other evidence suggests that fish can differentiate directionally, small thermogradients along the body (Dizon *et al.*, 1974; Crawshaw, 1976). Because water is essentially opaque to infra-red radiation, directional orientation to radiant heat sources, which is so important in the thermal orientation (taxes) of almost all terrestrial animals, simply is not available as a sensory cue in aquatic environments. A fish's thermosensory world must be very primitive; detection has to occur literally at the body surface; and given the unpredictability of small-scale current patterns existing in open waters, the integration system also has to be "confusion proof." Consequently these factors must have severely restricted the range of sensory means modern aquatic organisms could have evolved to directionally orient in their thermal environments. If thermoreceptors of aquatic organisms do indeed have poor directional specificity, the "choices" probably are limited much of the time to behaviorally negative or aversive responses in order to escape a stress temperature change unless non-thermal cues also are available. For highly motile animals like fishes, my choice here is to define direct thermal responses as klino-kinetic; which according to the classification of Fraenkel and Gunn (1961), requires an increase in both the rate of spontaneous swimming and the frequency of turning when the prevailing temperature becomes stressful.

A part of the confusion in arguments on how fishes select favorable thermal environments seems to rest with the notion that a thermal preferendum of a fish at a particular time of the year is an expression of the action of a central, variable set-point "thermostat," albeit a sloppy one, located in the preoptic hypothalamus (Hammel et al., 1973). Identification of the thermostat as having a variable set point relates to extensive evidence that thermal preferenda of fishes acclimate to seasonal temperatures, usually higher during summer than in winter (Fry and Hochachka, 1970; Coutant, 1977; Reynolds, 1977) and to local conditions (Norris, 1963). Part of the confusion relates as well to the impression that statistical variance in temperature selection by animals in linear gradients (cf. Norris, 1963) often is less than the variance found in experiments done with the newer, electronic shuttleboxes (Neill et al., 1972; Reynolds and Casterlin, 1976).

Satinoff (1978) has just challenged the prevailing view that a single hypothalamic thermostat (Hammel et al., 1973) controls mammalian thermogenesis and thermoregulatory behavior. In a sense she questions the suitability of the thermostat analogy to describe thermoregulation. She suggests a rather less restrictive approach which does not demand that lower vertebrates should have to behaviorally regulate like mammalian endotherms. Her remarks offer another reminder of the direction that evolution has proceeded by suggesting again that functional revision of many reflexes "invented" during the evolution of lower vertebrates has occurred with time on the geological scale (Roberts, 1974). As an example, vaso-motor control plays many roles in cardio-vascular accommodations in all vertebrates (Burnstock, 1969), but with the possible exception of some unusual cases among fishes (tunas, lamnid sharks, etc.), there are no real hints that vasomotor control has a role in thermoregulatory processes of lower vertebrates as it does in reptiles, birds and mammals.

Thermal Preferenda and Spontaneous Motor Activity

Satinoff's article (1978) appeared during the writing of this paper and reinforces an argument stated in the secondary introductory assumption - that a thermal preferendum represents a range in temperature at which an organism ceases to receive

aversive thermal feedback. Performance of fishes in gradients
of the shuttlebox type seem to verify this assumption for the
turn-about temperatures of carp (*Cyprinus carpio*) and large-
mouth bass (*Micropterus salmoides*) illustrated in Figure 6
from the experiments of Neill *et al.* (1972). Similar results
were obtained by Reynolds and Casterlin (1976) using a variant
design of the same apparatus with some of the same species.

Turn-abouts of a fish in the experiments of Neill *et al.*
(1972) and Neill and Magnuson (1974), detected when it swam
through the tunnel connecting warming and cooling tanks,
occurred at least once an hour at warming and cooling rates of
3° to $5^O C$. Special circuits controlled the temperature between
the two tanks at a differential of $2^O C$, and reversed heating or
cooling of the tank the fish just left so that the fish performed as its own "thermostat." Upper and lower turn-about
temperatures showed considerable individual variation in the
3-day runs employed. Variance was also large in the magnitude
of temperature differences between pairs of turn-about excursions from the warming to the cooling tank or the reverse.

Reynolds and Casterlin (1976) have given their shuttlebox
data as "modal preferenda" or values lying between upper and
lower turn-about temperatures and corrected for skewness in the
frequency distributions of times spent by fish in the warming
and cooling tanks. Neill and Magnuson (1974) have chosen to
present their data as mean-turn around temperatures or as
ranges of preferred temperature which turned out to be about

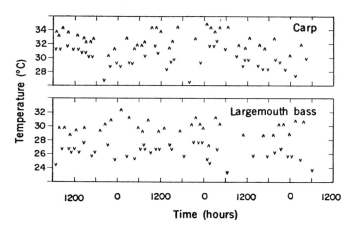

FIGURE 6. Self-programmed temperature selection by carp,
Micropterus salmoides, *during 3 day test periods in an elec-*
tronic, shuttlebox aquarium. Turn-around temperatures, upper
(∧) and lower (∨) are shown for movements between right and
left tanks (2^O C differential, right tank higher). From Neill
et al., *1972.*

3.5°C for five of the six species tested. Thus the statistical treatment of Neill and Magnuson satisfies the kinetic argument (e.g., klinokinesis). Data used by Reynolds and Casterlin emphasizes a more complex behavior, possibly of a bimodal nature (see also, Roberts, 1974). However, all variants of laboratory gradient devices that have been used introduce some behavioral conditioning to other than thermal cues and influence a fish's initial or basic thermal orientation. Therefore, a requirement for increasingly complex behavior is likely to increase the variance in measuring the role of spontaneous locomotion in behavioral thermoregulation. In fact, Norris (1963) has demonstrated that prior thermal history (natural versus laboratory acclimation) can produce differences in gradient performance that are not readily explainable.

Perhaps these problems can be best illustrated for gradient performance by goldfish. Their final thermal preferendum is about 28°C (Fry, 1947; Roy and Johansen, 1970; Reynolds et al., 1978). In one form of a linear gradient, Roy and Johansen established a selection range from 29° to 31°C (2° differential) for young goldifsh acclimated to 20°C. Somewhat larger goldfish tested in an electronic shuttlebox showed a mean hourly preferendum of 26° to 29.7°C (3.7° differential). When goldfish were trained by Rozin and Meyer (1961) to turn down the temperature of a heated aquarium (near the upper thermal limit) by paddle bumping to let in cold water, they ceased working the paddle when the temperature dropped to about 35°C. Rozin and Meyer (1961) comment by way of explanation that the temperature of 35° or 7°C above the published thermal preferendum, may represent the lowest temperature at which the paddle-pushing can be induced. These examples for goldfish not only illustrate that different levels of behavior can be induced to reinforce basic thermal responses, by involvement of other sensory systems (e.g., visual), but also the inadequacy of the single, variable set-point thermostat as an analogue model.

Few experiments have been designed to eliminate non-thermal sensory signals in studies on the role of spontaneous movements in aversive behavior. Nevertheless, the potentially great value of electronic shuttleboxes should not be discounted as a technique in pollution studies for the expressions of more complex thermal behavior can be a highly sensitive indicator as Reynolds et al. (1976) have shown in their unique experiments on induction of "behavioral fever" in fishes.

Several authors explain residence in regions of preferred temperature as a range in temperatures where spontaneous movements are minimal, that is, the kinetic argument (Sullivan, 1954; Norris, 1963). The Sullivan article (1954) unfortunately contains what appears to be an editorial error which follows the statement of this generalization by another denying it.

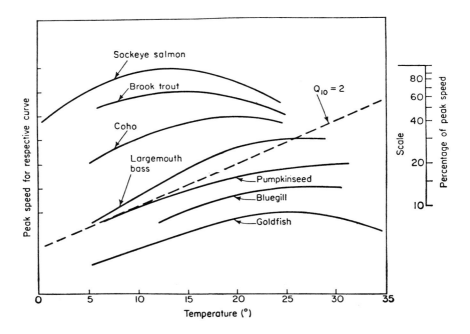

FIGURE 7. *Sustainable cruising velocities (various test periods) of fishes over temperature ranges relative to peak speeds of young sockeye salmon. In some cases test and acclimation temperatures were the same. The dashed line is the Q_{10} isopleth. From Fry, 1967 (various sources).*

An often quoted summary figure (Fig. 7) of curves for cruising velocities, established over a range of acclimation temperatures for a variety of fishes, indicates that maximal velocities coincide with final thermal preferenda (Fry, 1967). However, Fry and Hochachka later have cautioned (1970) that the data does not constitute an appropriate assessment of spontaneous locomotion because of operant conditioning. That is, the swimming speeds were obtained by forced or "paced" activity which for several of the fishes shown in Figure 7, are maximal cruising velocities. Another contrasting experimental series completed by Olla and Studholme (1976) probably is the best available case to illustrate the role of spontaneous locomotion in thermoregulatory behavior because operant conditioning was kept minimal by design (Fig. 8).

Experimental procedures used by Olla and Studholme were simple but required a large ellipsoid tank, 10.6m x 4.5m x 3m depth (121 kiloliter), to allow unimpeded spontaneous swimming. Small schools of adult bluefish (*Pomatomus saltatrix*) and Atlantic mackerel (*Scomber scomber*) were timed in this tank

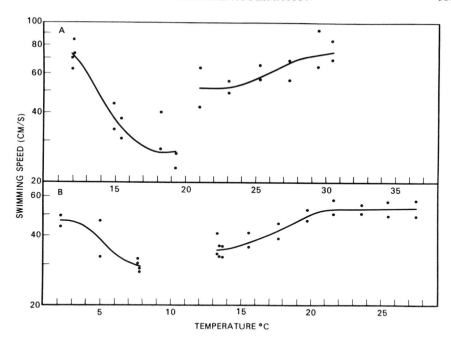

FIGURE 8. Spontaneous high and low swimming velocities for small schools of adult bluefish, Pomatomus saltarix *(A), and Atlantic mackerel,* Scomber scomber *(B), during slow warming (about 0.5°C per day) from acclimation temperatures of 19.9°C and 13.3°C. The fish swam steadily in a large ellipsoid aquarium where it was possible to reduce operant conditioning to a minimum. From Olla and Studholme, 1976.*

during 4 to 5 day intervals for their paired mean high and low swimming speeds at the temperature means for the periods. Because temperature change was gradual or about 0.5°C per day above and below starting acclimation temperatures (19.9°C, bluefish; 13.3°C, mackerel), all of the velocities in Figure 8 can be considered acclimated rates. Although thermal preferenda of these migratory species are unknown, Olla and Studholme (1976) state the peak abundance temperatures for bluefish off our eastern coasts is 18° to 20°C, and that the optimal thermal range for Atlantic mackerel is 12° to 14°C.

Partial reliance on a variety of sensory information as well as on detection of seasonal temperature changes likely serves for guidance during migration of nearshore pelagic fishes. That thermoregulatory behavior of pelagic fishes can be of a simple sort, namely a kinesis, seems convincing from effects of temperature change on swimming speeds of Atlantic

mackerel and bluefish observed by Olla and Studholme (1976).
Closer identification of their kinetic behavior, as ortho- or
klinokinesis, would be difficult because random turning
frequency, as the most likely determinant, is highly sensitive
to unsuspected operant disturbances. If spontaneous movements
are responsible for expression of thermal preferenda, then
the kinetic argument in its simplest form means that a pelagic
fish will be found most often in a thermal environment where
it is least active and least often in an environment where it
is most active.

Spontaneous thermoregulatory movements of less motile and
territorial fishes may be dominated by behavioral "programs"
that mask or supplant kinetic locomotion. For example, both
young and adult tautogs (*Tautoga onitis*) show a marked stress
orientation toward their adopted shelters in large temperature-
controlled tanks (Olla and Studholme, 1975; Olla *et al.*, 1978).
When the tautogs were warmed slowly (1 - 1.3°C/hr) from acclimation temperatures about 20°C, their spontaneous swimming,
feeding and aggression toward subordinate fish did not change
much until the tank temperature approached tolerance limits
(about 28°C). Their activity then dropped nearly to nighttime
low levels and most of the fish tended to remain close to or
in their shelters. Olla and Studholme have suggested (1975)
that young tautogs have limited abilities to move from a zone
of stress which may relate to a very different behavior, that
of predator avoidance. Apparently their shelter orientation
is so dominant over kinetic responses they will remain shelter-
ed even if the thermal stress reaches lethal limits.

Just how species, which like tautogs become seclusive when
exposed to high thermal stress, would behave in a thermal
gradient apparatus might be a revealing experiment. The contrasting behavior shown by the pelagic and shelter-seeking
fishes studied by Olla and his associates (1975, 1978) affirms
the utility of overt behavioral responses as highly sensitive
stress indicators. Yet an investigator also must recognize
the evolutionary consequences of niche selection. For tautogs,
one consequence seems to have been a change in the priority of
behavioral means used to meet thermal stress. As Olla and his
associates have stated regarding their behavioral studies
(personal communication):

> The only caveat before extrapolating these
> results to the natural environment is that we
> understand the precise role that the behavior
> plays in the life habits and ecology of the
> animal.

SUMMARY COMMENTS

1. One of the commonly used indicators of pollutant effects
on metabolic processes of nearshore and estuarine animals is
oxygen consumption. Secondary metabolic indicators such as
cyclic heart and respiratory movements (or branchial pumping)
can be useful because their rates are easily measured. However, these often show low sensitivity as indicators of stress
at standard or routine levels of metabolism (e.g., fishes and
large crustaceans). As maintenance functions, heart and
respiratory movements both accommodate for large changes in
energy expenditure for motor activity (5-10 times standard
levels), but accommodations represents not only changes in
rate but also in volume output per cycle (stroke volume). The
latter usually proves to be a difficult parameter to measure.
2. Special emphasis has been placed here on the potential of
behavioral responses for indication of contaminant effects on
aquatic organisms. This potential has been illustrated by the
utility of shuttlebox aquariums for determination of thermal
preferenda. Thermal preferenda have been identified as ranges
in temperature at which aquatic organisms do not receive
thermal feedback from the environment that is inductive of
aversive or avoidance responses. The thermal preferendum of a
species is labile to modification by seasonal changes in
temperature and probably as well by photoperiod change. Therefore a preferendum has been expressed in kinetic terms as a
seasonally variable range in temperatures where a particular
species is likely to aggregate because locomotory activity is
least (Fraenkel and Gunn, 1961). These views have been discussed in relation to the role of spontaneous motor activity
in behavioral thermoregulation and its link to standard,
routine and active levels of metabolism of a variety of animals
as defined by Fry (1957).
3. Levels of spontaneous activity often show high sensitivity
to incremental temperature increases, not because steady-state
power output is influenced *per se* by temperature, but because
spontaneous motor activity reflects marked effects of temperature change on levels of both maintenance metabolism (standard)
and motor output from the central nervous system (tonic drive).
The fact that locomotory activity of aquatic ectotherms can
increase in some proportion with the magnitude of temperature
change above and below a seasonal thermal preferendum (aversive responses) suggests that behavioral thermoregulation in
subject to a sort of central proportioning control. That is,
a variable set-point thermostat does not seem to provide an
appropriate analogue for the control system.
4. Behavioral thermoregulation of aquatic ectotherms probably
involves integration of other, non-thermal sensory modalities

for with some exceptions (e.g., thermal plumes, thermoclines), directional thermal orientation is limited due to the thermal properties of water. For example, the performance of fishes in linear thermal gradients, thermal selection shuttlebox aquariums, and in devices used to measure sustained cruising speeds relative to temperature, is subject to differing degrees of operant conditioning of the test fish to non-thermal cues (visual, lateral-line current detection, etc.).

LITERATURE CITED

Beamish, F. W. H.   1964.   Seasonal changes in the standard rate of oxygen consumption of fishes.   Can. J. Zool. 42: 189-194.

Beamish, F. W. H. and P. S. Mookherjii.   1964.   Respiration of fishes with special emphasis on standard oxygen consumption.   I. Influence of weight and temperature on respiration of goldfish, *Carassius auratus* L.   Can. J. Zool. 42: 161-175.

Brett, J. R.   1964.   The respiratory metabolism and swimming performance of young sockeye salmon.   J. Fish. Res. Board Can. 21: 1183-1226.

Brett, J. R.   1970.   Fish - The energy cost of living, pp. 37-52.   In: (W. J. McNeill, ed.) Marine Aquiculture. Oregon State University Press.

Burns, J. R.   1975.   Seasonal changes in the respiration of pumpkinseed, *Lepomis gibbosus*, correlated with temperature, day length, and stage of reproductive development. Physiol. Zool. 48: 142-149.

Burns, J. R.   1976.   The reproductive cycle and its environmental control in the pumpkinseed, *Lepomis gibbosus* (Pices:Centrarchidae).   Copeia, 1976, 3: 449-455.

Burnstock, G.   1969.   Evolution of the autonomic innervation of visceral and cardiovascular systems in vertebrates. Pharmacol. Rev. 21: 247-324.

Courtney, W. A. M. and R. C. Newell.   1965.   Ciliary activity and oxygen uptake in *Branchiostoma lanceolatum* (Pallas). J. Exp. Biol. 43: 1-12.

Coutant, C. C.   1977.   Compilation of temperature preference data.   J. Fish. Res. Board Can. 34: 739-745.

Coyer, P. E. and C. P. Mangum.   1973.   Effects of temperature on active and resting metabolism in polychaetes, pp. 173-180.   In: (W. Wieser, ed.) Effects of Temperature on Ectothermic Organisms.   Springer-Verlag, Berlin.

Crawshaw, L. I.   1976.   Effect of rapid temperature change on mean body temperature and gill ventilation in carp.   Am. J. Physiol. 231: 837-841.

Dehnel, P. A. 1958. Effect of photoperiod on the oxygen consumption of two species of intertidal crabs. Nature 181: 1415-1417.

Dehnel, P. A. 1960. Effect of temperature and salinity on the oxygen consumption of two intertidal crabs. Biol. Bull. 126: 354-372.

Dizon, A. E., E. D. Stevens, W. H. Neill, and J. J. Magnuson. 1974. Sensitivity of restrained skipjack tuna (*Katsuwonus pelamis*) to abrupt increases in temperature. Comp. Biochem. Physiol. 49A: 291-299.

Fraenkel, G. S. and D. L. Gunn. 1961. The Orientation of Animals. Dover Publications, Inc., New York. 376p.

Fry, F. E. J. 1947. Effects of the environment on animals activity. Univ. Toronto Stud. Biol. Ser. 55, Publ. Ont. Fish. Res. Lab. 68: 1-62.

Fry, F. E. J. 1957. The aquatic respiration of fish, pp. 1-63. In: (M. E. Brown, ed.) The Physiology of Fishes, Vol. I. Academic Press, Inc., New York.

Fry, F. E. J. 1967. Responses of vertebrate poikilotherms to temperature, pp. 375-409. In: (A. H. Rose, ed.) Thermobiology. Academic Press, Inc., New York.

Fry, F. E. J. and P. W. Hochachka. 1970. Fish, pp. 79-134. In: (G. C. Whittow, ed.) Comparative Physiology of Thermoregulation, Vol. 1, Invertebrates and non-mammalian vertebrates. Academic Press, Inc., New York.

Halcrow, K. and C. M. Boyd. 1967. The oxygen consumption and swimming activity of the amphipod *Gammarus oceanicus* at different temperatures. Comp. Biochem. Physiol. 23: 233-242.

Hammel, H. T., L. I. Crawshaw, and H. P. Cabanac. 1973. The activation of behavioral responses in the regulation of body temperature in vertebrates, pp. 124-141. In: (P. Lomax and E. Schönbaum, eds.) The Pharmacology of Thermoregulation. Symp., San Francisco, 1972 (Karger, Basel).

Hammen, C. S. 1969. Metabolism of the oyster, *Crassostrea virginica*. Am. Zool. 9: 309-318.

Hazel, J. R. and C. L. Prosser. 1974. Molecular mechanisms of temperature compensation in poikilotherms. Physiol. Rev. 54: 620-677.

Heath, A. C. and G. M. Hughes. 1973. Cardiovascular and respiratory changes during heat stress in rainbow trout (*Salmo gairdneri*). J. Exp. Biol. 59: 323-338.

Hutchison, V. H. 1976. Factors influencing thermal tolerances of individual organisms, pp. 10-26. In: (G. W. Esch and R. W. McFarlane, eds.). Proc. Symp. Augusta, Ga., April, 1975. ERDA Symp. Ser., CONF-750425, Natl. Tech. Inf. Serv., Springfield, Va.

Kanwisher, J. 1959. Polarographic oxygen electrode. Limnol. Oceanogr. 4: 210-217.

Mangum, C. P. and C. Sassaman. 1969. Temperature sensitivity of active and resting metabolism in a polychaetous annelid. Comp. Biochem. Physiol. 30: 111-116.

Murray, R. W. 1971. Temperature receptors, pp. 121-133. In: (W. S. Hoar and D. J. Randall, eds.) Fish Physiology, Vol. 5. Academic Press, Inc., New York.

Neill, W. H. and J. J. Magnuson. 1974. Distributional ecology and behavioral thermoregulation of fishes in relation to heated effluent from a power plant at Lake Monona, Wisconsin. Trans. Am. Fish. Soc. 103: 663-710.

Neill, W. H. J. J. Magnuson, and G. G. Chipman. 1972. Behavioral thermoregulation by fishes: A new experimental approach. Science 176: 1443-1445.

Newell, R. C. 1966. Effect of temperature on metabolism of poikilotherms. Nature 212: 426-428.

Newell, R. C. 1970. Biology of Intertidal Animals. Logos Press Limited, London. 555p.

Newell, R. C. 1976. Adaptations to Environment: Essays on the Physiology of Marine Animals, R. C. Newell (ed.), Butterworths, London-Boston. 539p.

Newell, R. C. and H. R. Northcroft. 1967. A re-examination of the effects of temperature on the metabolism of certain invertebrates. J. Zool., Lond. 151: 277-298.

Norris, K. S. 1963. The functions of temperature in the ecology of the percoid fish *Girella nigricans* (Ayres). Ecol. Monogr. 33: 23-62.

Olla, B. L. and A. L. Studholme. 1975. The effect of temperature on the behavior of young tautog, *Tautoga onitis* (L.), pp. 75-93. In: (H. Barnes, ed.) Proceedings of the Ninth European Marine Biology Symposium. Aberdeen University Press, Aberdeen.

Olla, B. I. and A. L. Studholme. 1976. Environmental stress and behavior: Response capabilities of marine fish. Second Joint U.S./USSR Symposium on the Comprehensive Analysis of the Environment, Honolulu, Oct., 1975, pp. 25-31, U.S. EPA, Washington, D.C., EPA #600/9-76/024.

Olla, B. L., A. L. Studholme, A. J. Bejda, C. Samet, and A. D. Martin. 1978. Effect of temperature on activity and social behavior of the adult tautog, *Tautoga onitis* under laboratory conditions. Marine Biology 45: 369-378.

Precht, H., J. Cristophersen, H. Hensel, and W. Larcher. 1973. Temperature and Life. Springer-Verlag, Berlin. 779p.

Prosser, C. L. (Ed.). 1973. Comparative Animal Physiology, 3rd ed. W. B. Saunders Company, Philadelphia. 1011p.

Reynolds, W. W. 1977. Temperature as a proximate factor in orientation behavior. J. Fish. Res. Board Can. 34: 734-739.

Reynolds, W. W. and M. E. Casterlin. 1976. Thermal preferenda and behavioral thermoregulation in three centrarchid fishes, pp. 185-190. In: (G. W. Esch and R. W. McFarlane, eds.) Thermal Ecology II. Proc. Symp. Augusta, Ga., April, 1975. ERDA Symp. Ser., CONF-750425, Natl. Tech. Inf. Serv., Springfield, Va.

Reynolds, W. W., M. E. Casterlin, and J. B. Covert. 1976. Behavioral fever in teleost fishes. Nature 259: 41-42.

Reynolds, W. W., M. E. Casterlin, J. K. Matthey, S. T. Millington, and A. C. Ostrowski. 1978. Diel patterns of preferred temperature and locomotor activity in the goldfish *Carassius auratus*. Comp. Biochem. Physiol. 59A: 225-227.

Roberts, J. L. 1957. Thermal acclimation of metabolism in the crab, *Pachygrapsus crassipes* Randall. II. Mechanisms and the influence of season and latitude. Physiol. Zool. 30: 242-255.

Roberts, J. L. 1964. Metabolic responses of fresh-water sunfish to seasonal photoperiods and temperatures. Heogländer wiss. Meeresunters. 9: 459-473.

Roberts, J. L. 1967. Metabolic compensations for temperature in sunfish, pp. 245-262. In: (C. L. Prosser, ed.) Molecular Mechanisms of Temperature Adaptation. American Association for the Advancement of Science, Publ. 84, Washington, D.C.

Roberts, J. L. 1974. Temperature acclimation and behavioral thermoregulation in cold-blooded animals. Fed. Proc. 33: 2155-2161.

Roberts, J. L. 1975. Active branchial and ram gill ventilation in fishes. Biol. Bull. 148: 85-105.

Roy, A. W. and P. H. Johansen. 1970. The temperature selection of small hypophysectomized goldfish (*Carassius auratus* L.). Can. J. Zool. 48: 323-326.

Rozin, P. N. and J. Mayer. 1961. Thermal reinforcement and thermoregulatory behavior in the goldfish, *Carassius auratus*. Science 134: 942-943.

Satinoff, E. 1978. Neural organization and evolution of thermal regulation in mammals. Science 201: 16-22.

Spoor, W. A. 1946. A quantitative study of the relationships between the activity and oxygen consumption of the goldfish, etc. Biol. Bull. 91: 312-325.

Sullivan, C. M. 1954. Temperature reception and responses in fish. J. Fish. Res. Board Can. 11: 153-170.

Sutterlin, A. M. 1969. Effects of exercise on cardiac and ventilation frequency in three species of freshwater teleosts. Physiol. Zool. 43: 36-52.

Tribe, M. A. and K. Bowler. 1968. Temperature dependence of "standard metabolic rate" in a poikilotherm. Comp. Biochem. Physiol. 25: 427-436.

Vernberg, F. J. 1969. Acclimation of intertidal crabs. Am. Zool. 9: 333-341.

Vernberg, W. B. and F. J. Vernberg. 1972. Environmental Physiology of Marine Animals. Springer-Verlag, New York. 346p.

Webb, P. W. 1975. Hydrodynamics and Energetics of Fish Propulsion. Bull. Fish. Res. Board Can. 190, 159p.

Whittow, G. C. (Ed.). 1970. Comparative Physiology of Thermoregulation, Vol. 1, Invertebrates and Nonmammalian Vertebrates. Academic Press, New York. 333p.

Wieser, W. (Ed.). 1973. Effects of Temperature on Ectothermic Organisms. Springer-Verlag, Berlin. 298p.

Withrow, R. B. (Ed.). 1959. Photoperiodism and Related Phenomena in Plants and Animals. American Association for the Advancement of Science, Publ. 55, Washington, D.C. 903p.

MULTIPLE FACTOR INTERACTIONS AND STRESS IN
COASTAL SYSTEMS: A REVIEW OF EXPERIMENTAL
APPROACHES AND FIELD IMPLICATIONS

Robert J. Livingston

Department of Biological Science
Florida State University
Tallahassee, Florida

INTRODUCTION

There has been considerable effort to develop techniques
and approaches for an evaluation of the impact of various
pollutants on aquatic species. The criteria for such experiments have been documented (Cairns and Dickson, 1973; Vernberg
and Vernberg, 1974; Cairns *et al.* 1975; Vernberg *et al.*,
1977). Recently, emphasis has been placed on the nature of
interacting factors as stressors. By definition, studies concerning multiple factor stress reflect a tacit acknowledgment
of the problems involved in the extrapolation of bioassay
results to field situations, where environmental response is
seldom determined by a single driving function. The following
is a critical review of multiple factor interactions, with
particular attention to the nature of stress in estuarine and
coastal systems.

The complexity of the problem of interacting factors is
evident in recent work concerning the fate and effects of
chlorine in coastal systems. The chlorine problem involves
environmental persistence of the pollutant, identification of
degradation products, associated bioassay considerations, the
determination of appropriate water quality parameters, and the
design of experimental approaches that will ultimately lead to
applications at the population and community level (Block *et
al.*, 1977). Although the primary factors that control physiological and behavioral responses of coastal organisms are
generally defined (Vernberg and Vernberg, 1976) and multiple
factor effects of specific organisms have been demonstrated
(Vernberg and Coull, 1975), there has been relatively little

effort to examine the implications of stress with respect to
trophic interactions and population/community response
(Livingston et al., 1978). While such work has been approached in various ways (Cairns et al., 1973; Gaufin, 1973;
Patrick, 1973), demonstration of effect through interactions
at different levels of biological organization with associated
models of environmental response based on experimental manipulation remains relatively undeveloped.

Past work involving multiple factor bioassay has already
been reviewed (Livingston et al.,1974). Various classes of
pollutants demand individual attention. For instance, heavy
metal interactions with various factors have been documented
for estuarine organisms (Bryan, 1976), with temperature and
salinity as the principal modifiers of toxic action (Vernberg
and O'Hara, 1972; Vernberg and Vernberg, 1972; O'Hara, 1973).
Sheline and Schmidt-Nielsen (1977) found that pretreatment of
top minnows (Fundulus heteroclitus) with selenium caused a
redistribution of mercury among organs. Exposure to sub-
optimal conditions of temperature and salinity increased the
level of adverse response of organisms exposed to mercury
(Vernberg et al., 1973). Dillon (1977) found that low
salinity reduced the toxic effects of inorganic mercury on
estuarine clams (Rangia cuneata). Polychaete worms
(Nereis diversicolor) taken from estuarine sediments contaminated with heavy metals were more tolerant of such pollutants than unexposed worms (Bryan, 1976), evidence of
adaptive response of such organisms to heavy metal distribution which remains largely undefined with regard to population
dynamics and distribution. These problems are amplified by
differential response of various developmental stages of
crustaceans such as harpacticoid copepods(Tigriopus japonicus)
to combinations of metals (D'Agostino and Finney, 1974).
Despite a considerable bioassay effort, the environmental
implications of heavy metal contamination of coastal biological systems remain poorly defined.

Experimental demonstration of interactions among
pesticides, polychlorinated biphenyls (PCB's), and various
modifying factors is available; it includes induction and
storage effects, and simultaneous (short-term) interactions
at various levels of complexity (Livingston et al., 1974).
Again, there are significant differences related to the
individual chemical formulation (Vernberg et al., 1977) and
stage of development (Nimmo et al., 1971; Bookhout and
Costlow, 1975) although temperature and salinity appear to be
the primary modifying functions. Nimmo and Bahner (1974)
found that salinity can direct the toxic effects of Arochlor
1254 on Penaeus aztecus well within the range of the natural

environmental distribution of this species. Koenig (1977) found that combinations of mirex and DDT exert synergistic effects on various life stages of the diamond killifish *(Adinia xenica)* as a result of differing toxic mechanisms. The final effect remained dependent on the life-history stage of the fish. Thus, in addition to the persistence, availability, and toxicity of a given compound or compounds, the time of entry of the chemical into the biological system, in addition to corresponding interaction with predominant modifying factors and stage of development of the exposed organisms, must be considered in an evaluation of impact.

Various organisms have been used as stress indicators. Often, it is presumed that such data will allow identification of shifts in species composition or community structure. In view of the high level of variability of a given population in time, the weakness of this approach is manifest. As an alternative, multi-species assemblages can be followed through time and represented by indices which in some way reflect community composition, and the reaction of a specific level of organization to stress. Species diversity, for example, has long been used as an indicator of the effects of pollution (Wilhm and Dorris, 1968; Gaufin, 1973; Patrick, 1973). Cairns *et al.* (1973) have presented the various arguments in favor of the use of diversity as a measurement of impact response in aquatic communities. In estuarine and coastal systems, low species diversity has been equated with reduced structural stability (Boesch, 1972) and species diversity has been used as an indication of the effects of various pollutants on fish assemblages in coastal areas (Bechtel and Copeland, 1970; Holland *et al.*, 1973). However, subsequent studies have established that species diversity cannot be an index of pollution for benthic macrophytes (Zimmerman and Livingston, 1976), epibenthic invertebrates (Hooks, 1973), or fishes (Livingston, 1975). In view of the fact that estuarine assemblages are characterized by high relative dominance of certain species, and that such dominance strongly influences the determination of community indices, it would seem that the indicator concept would be applicable to coastal assemblages. Unfortunately, this is not always the case, because of the complexity of the response of such systems to physical or chemical perturbation.

Temperature

The estuarine environment is often physically stressed (Kinne, 1967) while remaining highly productive (Ketchum, 1967). Short- and long-term variations of salinity, temperature, suspended solids, and other physioco-chemical

functions determine the biological associations in a given
area (Carriker, 1967). These effects are often system-
specific, with individual biological response directed by patterns of energy flow, combinations of dynamic complexes of
forcing functions, temporal stratification, and spatial
limitations determined by various physiographic functions.
Such combination of factors have been used to classify coastal
systems into broadly recognizable units (Odum and Copeland,
1969; Copeland, 1970). This development has led to specula-
tion concerning the biological stability of estuaries, often
characterized by low diversity, reduced trophic specialization,
and high levels of biological variability at the population
and community levels. Of course, the adaptive nature of many
estuarine species is now recognized (Vernberg and Vernberg,
1976) and physico-chemical instability should no longer be
confused with biological instability. Thus stress, whether of
natural or anthropogenic origin, can no longer be viewed in
simplistic terms. Unfortunately, spatial and temporal
variability of coastal systems remain little understood
(Livingston, 1977), especially with respect to long-term
biological changes. Also the mechanisms of trophodynamic
response remain obscured by the magnitude and rapidity of
energy flux in coastal systems. In short, the concept of
stress remains poorly defined with respect to functional and
deterministic interactions.

Temperature is widely recognized as an important
parameter in estuaries. Although various experiments have
been performed with temperature as a principal modifying
factor, little is known concerning the impact of variable
temperature on population response to episodic influxes of
various pollutants. For instance, the time of year of an oil
spill can be critical to actual impact within the context of
site-specific characteristics and the size and nature of the
spill (Michael et al., 1977). This has been demonstrated in
pesticide studies (Koenig et al., 1976) wherein blue crab
(Callinectes sapidus) mortality in a north Florida estuary
contaminated with DDT was seasonally influenced by temperature
variation, and major mortality in the field was associated
with rapid decreases in water temperature as cold fronts moved
through the area (Fig. 1). Relatively high levels of DDT-R
in the crabs together with specific physiological and behavioral stress due to a sudden drop in temperature were hypothesized as the principal agents of mortality. The potential
molecular complexing of DDT with protein components combined
with increased release and mobilization of DDT during temperature stress were thought to produce increased acute toxicity
via DDT-nerve membrane complexing and impaired behavioral

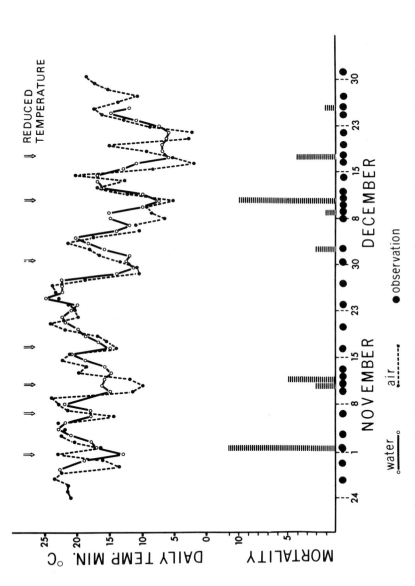

FIGURE 1. Observed (field) mortality of blue crabs (Callinectes sapidus) in a DDT-contaminated marsh system (Alligator Harbor, north Florida, USA) and relationship to reductions of water temperature resulting from movement of cold fronts through the system. (Taken from Koenig et al., 1976.)

reactions. These field observations are now susceptible to laboratory investigation for a determination of mechanism of impact.

There has been a considerable effort to determine the impact of thermal effluents from power plants on estuarine and coastal biota. Undoubtedly, this work will contribute substantially to our knowledge of the influence of temperature on stress reactions in such systems.

Salinity

The interaction of key factors such as salinity with the naturally high levels of productivity of many coastal systems tends to obviate easy generalization. The same parameters which make coastal systems extremely productive (i.e. land runoff, input of nutrients, and detritus) serve also to increase the level of osmoregulatory stress through changes in salinity. Thus, such coastal populations have been termed "monotonous" (Hedgepeth, 1957) and are often dominated by unspecialized, euryhaline populations adapted to a highly unstable physical environment. Gunter (1977) refers to salinity as a "master factor" which determines survival (predator-prey relationships), trophic (nursery) function, optimal spawning and growth conditions, and the movements of various organisms at different levels of development. These influences are particularly important with regard to larval and juvenile stages of various species. It has been shown (Gunter, 1961) that along increasing gradients of salinity, there are often general increases in species richness and decreases in numbers of individuals per species. Variable and extreme conditions of reduced salinity can sometimes be associated with lower numbers of species; however, if such relationships are stable in time, the combination of (sessile) estuarine endemics and various developmental stages of motile euryhaline species can lead to high productivity in oligohaline situations. For example, reduced salinity can enhance the growth of various species such as oysters (*Crassostrea virginica*) as a result of reduced predation pressure and lowered parasitic infestation (Copeland, 1966). Influxes of penaeid shrimp (*Penaeus* spp.)and blue crabs (*Callinectes sapidus*) have been positively correlated with increased river flow (Copeland, 1966) or rainfall (Gunter and Hildebrand, 1954). Increased salinity in estuaries as a result of prolonged drought has been associated with complex changes in species composition due to influxes of stenohaline marine species, reductions of sessile and motile estuarine species, and generally reduced productivity of commercial species (Parker, 1955). Reid (1955a, b; 1956)

found that excavation of a pass in a barrier island and the
resulting introduction of high salinity water into a brackish
Texas bay system reduced the nursery function of the estuary
with associated qualitative and quantitative changes in species
representation. This would indicate that salinity strongly
influences productivity and the composition of species assemblages in coastal systems.

The Apalachicola Estuary: Indicators of Stress

     Studies in the Apalachicola estuary in north Florida
(Livingston *et al.*, 1977) illustrate certain aspects of pro-
ductivity/diversity relationships. Details concerning laboratory and field monitoring techniques have been published
(Livingston, 1976; Livingston *et al.*, 1976) and will not be
presented here. All samples were taken at permanent stations
in the northern Gulf of Mexico (Fig. 2) at monthly intervals
from March, 1972, to the present. Physioco-chemical data were
determined for surface and bottom samples (mean depth: 1-2 m)
taken with a 1-ℓ Kemmerer bottle. Salinity (^{o}/oo) was determined with a temperature-compensated refractometer calibrated
periodically with standard seawater. Epibenthic organisms were
taken with 5-m (16-foot) otter trawls (1.9 cm mesh wing and body;
0.6 cm mesh liner). Repetitive samples (2-minute) were taken
monthly as described by Livingston *et al.* (1975). All organisms were preserved in 10% formalin, sorted and identified to
species, measured and/or counted; biomass transformations and
statistical analyses were carried out using an interactive
computer program designed for analyses of extensive data collections. The Brillouin Index (Brillouin, 1962) was used to
compute species diversity. Cluster analyses were run using the
ρ measure of similarity and a flexible grouping strategy with
β = -0.25 (Livingston *et al.*, 1978). Summed (annual) data
(fishes, invertebrates) were run by station to determine spatial
faunal affinities over the 5-year sampling interval.
     A comparison was made of salinity distribution in the
Apalachicola estuary before (Dawson, 1955) and after the opening of Sike's Cut in the barrier island system. Areas of the
bay contiguous with this artificial pass are now characterized
by increased salinity (Table 1) even though most other stations
showed recent decreases in salinity. Such trends are compati-
ble with the distribution of other water quality parameters
such as turbidity (Fig. 3). Salinity at Sike's Cut is now
relatively high and stable in time in comparison with other
portions of the Apalachicola estuary (Fig. 4).
     Annual species richness (total number of species) and
diversity relationships of epibenthic fishes and invertebrates
are shown in Figures 5 and 6. Areas characterized by stable,
high levels of salinity are associated with relatively **high**

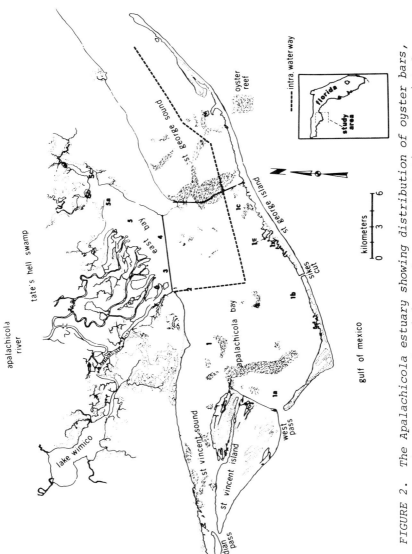

FIGURE 2. The Apalachicola estuary showing distribution of oyster bars,
river entry, and positions of permanent stations associated
with long-term environmental studies of the bay system.

TABLE 1. *Mean (surface) salinities (ppt) in the Apalachicola Bay system before (August, 1953-August, 1954)* [*] *and after (June, 1973-May, 1974) the opening of Sike's Cut. Comparison is made during periods of comparable rainfall and river flow*

Station	(1953-1954)	(1973-1974)
1	14.1	8.6
1A	19.8	20.3
1B	5.0	15.2
1E	16.8	15.0
1C	19.3	16.8
2	6.6	3.1
3	5.7	4.3
4	7.4	4.3
5	7.7	2.8
5A	7.3	4.8
7	2.7	---

[*] *Data published by C. E. Dawson, 1955. Printed with permission of Johnson Reprint Corporation.*

levels of richness and diversity. Although fish species richness is not appreciably greater at 1B than at various upland stations, species diversity is higher in saline areas. The invertebrate data show a different pattern, with high salinity associated with high species richness and diversity. Such trends are also reflected in species distributions in time (Figs. 7, 8). Station clusters for fish and invertebrate associations reflect habitat (grassbeds) and physico-chemical gradients. At stations associated with high salinity (the outer bay), species assemblages were qualitatively distinct from those of the rest of the bay. This difference was particularly pronounced at station 1B.

In upland or river-associated areas of the bay (having variable physico-chemical regimes with low annual mean levels of salinity), there are relatively low richness and diversity values. This distribution is explained by trends of dominance, qualitative species composition, and relative abundance of the various populations. Sike's Cut was characterized by relatively low dominance, high species richness, and low numbers of individuals per sampling effort when compared to areas of low salinity. Relatively high biomass per unit sample was noted in upland (low-salinity) areas. Species such as the blue crab and penaeid shrimp were not numerous at station 1B when compared with other portions of the system. High salinity,

APALACHICOLA BAY SYSTEM
(Bottom salinity, March, 1977)

FIGURE 3. Distribution of salinity (bottom, ppt) and turbidity (surface, J.T.U.) in the Apalachicola Bay system during March, 1977.

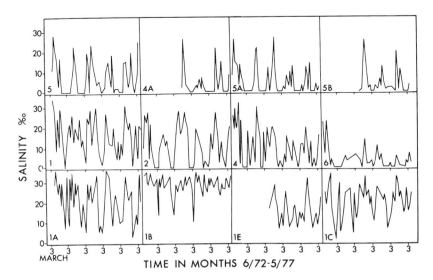

FIGURE 4. Long-term variation of salinity (bottom, ppt) at various stations in the Apalachicola estuary at monthly intervals from June, 1972 to May, 1977. Periodic salinity reductions at most stations reflect winter influxes of freshwater input from the Apalachicola River.

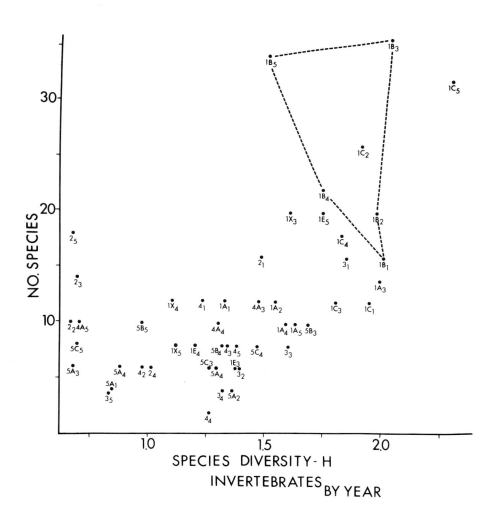

FIGURE 5.   Diversity-richness relationships of epibenthic invertebrates taken monthly at permanent stations in the Apalachicola estuary.   Indices (number of species, Brillouin Diversity Index) were computed from annual total numbers of individuals at each station from year 1 to year 5.   Results at station 1B (Sike's Cut) have been highlighted by dashed lines.

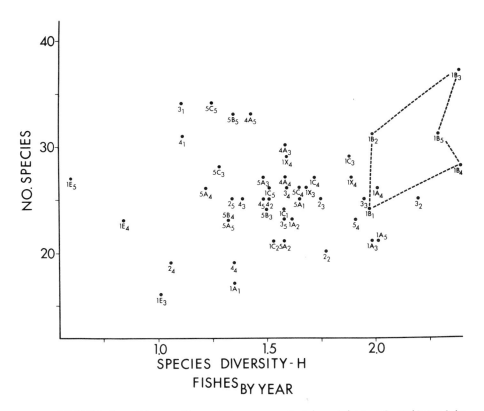

FIGURE 6.   Diversity-richness relationships of epibenthic
fishes taken monthly at permanent stations in the Apalachicola
estuary.   Indices (number of species, Brillouin Diversity
Index) were computed from annual total numbers of individuals
at each station from year 1 to year 5.   Results at station 1B
(Sike's Cut) have been highlighted by dashed lines.

FIGURE 7.   Cluster analysis of summed invertebrate data
from all stations from June, 1972, to May, 1977.   The ρ
similarity coefficient was used in conjunction with a flexible
grouping strategy (β = -0.25) with the station as the cluster
variable and species as the independent variable.

Similarity (Invertebrates)

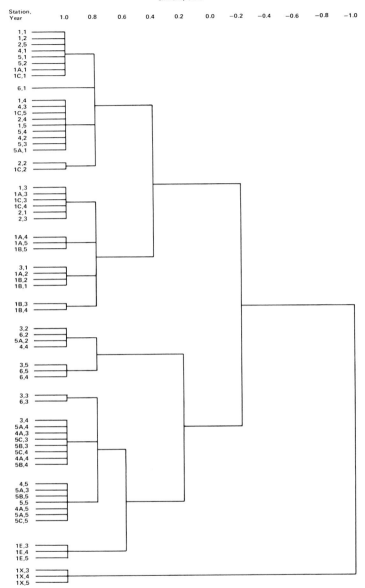

FIGURE 8. *Cluster analysis of summed fish data from all stations from June, 1972, to May, 1977. The ρ similarity coefficient was used in conjunction with a flexible grouping strategy (β = -0.25) with the station as the cluster variable and species as the independent variable.*

stable in time, was thus inversely related to nursery potential
and fishery productivity. From this standpoint, high species
richness and diversity could be viewed as signs of low productivity in the system. Because of specific relationships of
environmental stability, high dominance and productivity, and
the reaction of individual populations to natural stressors
such as salinity, indices such as species richness and diversity may not always serve as reliable indicators of environmental well-being in estuaries. In this case, high species
richness and diversity reflect the altered salinity distribution. The opening of Sike's Cut, while increasing the numbers
of species in this portion of the bay, actually reduced the
productivity of the system. Thus, when viewed in this context,
high species diversity is not necessarily desirable or indicative of favorable water quality conditions.
      The over-simplified use of individual populations or
single indices of community structure may be misleading in an
evaluation of water quality in coastal areas since too much
information is omitted relative to the interaction of the
principal physico-chemical forcing functions, productivity, and
primary biological associations. Such indices are useful only
within the limits of the individual system in question.

APPLICATIONS TO EXPERIMENTAL DESIGN

Synergistic Bioassays and Field Analysis

      Experimental design of so-called "synergistic" bioassays
is directly related to the above observations. Because of a
combination of factors such as high natural productivity and
stress due to a changeable physico-chemical environment, many
estuaries have evolved into relatively stable systems characterized by spatial and temporal progressions of a relatively
small number of dominant species (Livingston *et al.*, 1978).
The description of a given coastal system should be limited to
individual sets of stressor functions within the context of
high productivity, nursery functions, and dynamic trophic
interrelationships and competition. Experimental design should
thus be concerned not only with the acute and chronic effects
of a given pollutant on a given species, but with the impact
of such a stressor on the individual trophic unit (larval,
juvenile, or adult stage) and ultimately with the energy cycling within the system. Such an approach would place more
emphasis on the selection of experimental subjects from a
trophodynamic point of view. It is altogether possible that,
in estuarine and coastal systems, the species concept is
obsolete with respect to impact of a given factor on the

individual organism or the system as a whole. Experimental
approaches to this question should thus be directed by trophic
phenomena within the context of temporal developmental progressions. Analysis of changes in competitive relationships
of populations through time (Wiens, 1977) could help prevent
oversimplification by undue dependence on a given indicator
species or index of community status. Experimental design
could thus be derived from proven interrelationships of the
working (trophic) units of a given system rather than by
construction of field models from laboratory results. As a
result of this line of reasoning, the synergistic bioassay
would then focus on dynamic mechanisms of energy transfer in
a given system with an emphasis on how such changes are
directed by key forcing functions and on how the biological
response is translated into observed distributions of populations in space and time. If field analyses of community
structure in affected areas were also considered, some order
might be realized in the experimental design of the synergistic bioassays.

Microcosms and Simulated Ecosystems

There are various experimental approaches which tend
toward a compromise between straight laboratory experimentation
and field analysis. Artificial communities, as described by
Sigmon et al. (1977), have been used in a variety of ways,
chiefly to determine the movement of pollutants through simplified systems. This method can lead to interpretations of
acute or chronic effects at various trophic levels. These
systems have been used for some time, and have fairly predictable advantages and drawbacks. While there is certain control
over various combinations of physical forcing functions,
laboratory simulations are simplifications of natural ecosystems and consequently reflect inconsistencies or gaps in our
knowledge of real systems. The use of artificial communities
can lead to hypothesis testing (Cripe and Livingston, 1976)
with a particular emphasis on the trophic response of various
components in the experimental systems (Hurlbert et al., 1972;
Hamelink and Waybrant, 1976). As the microcosm becomes more
complex, adequate sampling becomes a problem (as it is in the
field) until a point is reached where it becomes questionable
whether one should simply conduct such tests in the natural
environment. This can be carried out using caged organisms
which are often exposed to a given pollutant under natural
conditions (Lockhart et al., 1977). Such experiments suffer
from problems associated with an isolation of the test organisms from their natural functions, that is, altering their
patterns of movement or feeding while protecting them from
predation. Caging experiments may actually create an entirely

different physical environment as a result of altered sedimentation rates and associated changes of water quality
(P. K. Dayton, personal communication).

Another approach is represented by the Controlled Ecosystem Pollution Experiment (CEPEX) Program (Parsons, 1974;
Takahashi et al., 1975), which is an attempt to understand the
chronic effects of pollutants on combinations of organisms at
various trophic levels (representing a specific food web).
Such experiments are conducted in enclosed water columns
(polyethylene "bags") placed in various natural systems.
Reeve et al. 1976) have noted problems concerning replication
of test populations and difficulties in extrapolation of experimental results to the natural system. Lee et al. (1977) showed
the advantages of such systems in the determination of the fate
of pollutants and their movement from one trophic level to
another. However, in isolating the organisms from the natural
environment, there is a tendency to accept assumptions and weaknesses associated with the usual bioassay approach. Functional
interactions remain unknown and the mechanism of pollutant
interaction at the population (community) level is left unre-
solved. Thus, direct extrapolation to natural systems is weak,
and predictions of impact due to specific pollutants in complex
coastal systems remain unclear.

There is no single solution to the problem of impact evaluation, and a combination of approaches will be necessary if some
sense is to be made from the welter of data concerning the
response of individual populations to a given combination of
forcing functions. The question remains whether it would not be
advantageous, indeed economical in the long run, to design such
experiments from comprehensive, long-term field data so that a
modicum of reality is brought back into the laboratory. If this
is to happen, a basic revision may be necessary of our conceptu-
alization of the species as an ecological unit and how it relates to the almost infinite variability of estuarine and
coastal systems.

SUMMARY AND CONCLUSIONS

Over the past several years, there has been a considerable
effort to develop advanced methodological approaches to bioassay. This effort has been technically successful. However,
efforts to apply bioassay results to environmental situations
have led to serious problems with regard to the actual meaning
of laboratory results. Such application often leads to circular reasoning and major unexplained assumptions are sometimes
accepted without question or even comment. The perfunctory
bow to field implications usually tacked onto the end of
detailed laboratory papers simply does not add to the value of
the experimental data. In view of the innumerable potential
combinations of environmental parameters, the specific choice

of experimental conditions for a synergistic bioassay is often
based on arbitrary sets of variables without particular attention to actual environmental conditions or potential short-
or long-term variation in a given habitat. If a given species
is accepted as an indicator, too little attention is given to
ontological considerations and interspecific competition in
resource utilization.

The current emphasis on individual species (usually adults)
in bioassays of coastal organisms may well be a distortion of
the actual environmental situation, since the functional unit
is often the trophic entity, one of a kaleidoscopic series of
transitory ontological forms utilizing niches which may involve
greater intraspecific differences than interspecific ones
within a given period of time. Thus any one of a series of
larval or juvenile stages of a given estuarine species may,
in its niche requirements or susceptibility to a given pollutant, be more closely related to corresponding representatives
of other species than to its own adult form. Elimination of
a given stage of development through habitat destruction may
alter competitive interactions and the ultimate response of
the system as a whole to a point where natural fluctuations of
key forcing functions cause decreases in useful productivity.
Furthermore, the acceptance of species diversity indices as
indicators of stress due to anthropogenic activity in coastal
systems may be an essential oversimplification due to basic
misconceptions regarding the relationship of environmental and
biological stability, competition, and what may be termed
"useful" productivity. It is entirely possible that once the
basic units of information and the trophic relationships of
such entities within the environmental restraints of a given
system are identified, new approaches can be made in the
experimental design of interacting factors. In other words,
the design of laboratory experimentation from established
field interactions will certainly add depth and significance
to tests of hypotheses.

The concept of stress needs definition. The key functions
of a given habitat ultimately direct specific trends of productivity and energy transfer. Thus, excess nutrients and other
changes in water quality involved in cultural eutrophication
can cause alterations in species composition, disjunct energy
transfer, and alteration (usually a decrease) of the useful
productivity of the affected system. There is some evidence
that toxic agents such as pesticides have a substantial effect
on species composition and energy transfer mechanisms at the
system level. In addition, changes in the physical configuration of an estuary, with consequent alteration of the habitat
with respect to a parameter such as salinity, may completely
alter patterns of competition, productivity, and response to
pollutants. Thus, the potential impact of a specific factor

should be viewed from the standpoint of habitat alteration in time. Analysis of the biological relationships of the system, with particular attention to long-term trends in population and community functions, may eventually make it possible to develop valid hypotheses concerning patterns of resilience. Hopefully, this would then serve as a basis for experimental design and the determination of significant interacting factors in an evaluation of a given pollutant or pollutants.

Establishment of a combination of life-history studies, long-term field analysis, and laboratory and field experimentation may make it possible to bridge the gap between the bioassay approach and actual environmental effects. Unless the experimental approach takes into consideration the short- and long-term variability of coastal systems, there is a definite risk that serious mistakes will be made at the regulatory level. While the recent development of the bioassay technique is an understandable compromise in favor of a rapid solution to pressing environmental problems, further postponement of an integration of the field and laboratory approaches to the determination of the impact of man on aquatic systems can only delay our understanding of the short- and long-term aspects of environmental alteration and pollution.

ACKNOWLEDGMENTS

The author is extremely grateful for the assistance of various graduate students in the field. This includes Mr. G. J. Kobylinski, Mr. F. G. Lewis III, and Mr. P. Sheridan. Mr. G. C. Woodsum aided in retrieval of the information on the Apalachicola estuary. Portions of the field projects were funded by NOAA Office of Sea Grant, Department of Commerce, under Grant No. 04-3-158-43. Data analysis was supported by EPA Program Element No. 1 BA025 under Grant No. R-803339.

LITERATURE CITED

Bechtel, T. J. and B. J. Copeland. 1970. Fish species diversity indices as indicators of pollution in Galveston Bay, Texas. Contr. Mar. Sci. 15: 103-132.

Block, R. M., G. R. Helz, and W. P. Davis. 1977. The fate and effects of chlorine in coastal waters. Ches. Sci. 18: 97-101.

Boesch, D. F. 1972. Species diversity of marine macrobenthos in the Virginia area. Ches. Sci. 13: 206-212.

Bookhout, C. G. and J. D. Costlow, Jr. 1975. Effects of
    mirex on the larval development of blue crab. Water, Air,
    Soil Pollut. 4: 113-126.
Brillouin, L. 1962. Science and Information Theory. New
    York: Academic Press.
Bryan, G. W. 1976. Some aspects of heavy metal tolerance in
    aquatic organisms, pp. 7-34. In: (A. P. M. Lockwood, ed.)
    Effects of Pollutants on Aquatic Organisms. New York:
    Cambridge University Press.
Cairns, J., Jr., and K. L. Dickson (Eds.). 1973. Biological
    Methods for the Assessment of Water Quality. Philadelphia:
    American Society for Testing and Materials.
Cairns, J., Jr., K. L. Dickson, and G. Lanza. 1973. Rapid
    biological monitoring systems for determining aquatic
    community structure in receiving systems, pp. 148-163.
    In: (J. Cairns, Jr. and K. L. Dickson, eds.) Biological
    Methods for the Assessment of Water Quality. Philadelphia:
    American Society for Testing and Materials.
Cairns, J., Jr., K. L. Dickson, and G. F. Westlake (Eds.).
    1975. Biological Monitoring of Water and Effluent Quality.
    Philadelphia: American Society for Testing and Materials.
Carriker, M. R. 1967. Ecology of estuarine benthic inverte-
    brates: A perspective, pp. 442-487. In: (G. H. Lauff,
    ed.) Estuaries. Washington, D. C.: AAAS.
Copeland, B. J. 1966. Effects of industrial waste on the
    marine environment. J. Water Poll. Contr. Fed. 38: 1000-
    1010.
Copeland, B. J. 1970. Estuarine classification and responses
    to disturbances. Trans. Amer. Fish. Soc. 99: 826-835.
Cripe, C. R. and R. J. Livingston. 1976. Dynamics of mirex
    and its principal photoproducts in a simulated marsh
    system. Arch. Env. Contam. Toxicol. 5: 295-303.
D'Agostino, A. and C. Finney. 1974. The effect of copper and
    cadmium on the development of *Tigriopus japonicus*, pp.
    445-464. In: (F. J. Vernberg and W. B. Vernberg, eds.)
    Pollution and Physiology of Marine Organisms. New York:
    Academic Press.
Dawson, C. E. 1955. A contribution of the hydrography of
    Apalachicola Bay, Florida. Publications of the Texas
    Institute of Marine Science 4: 15-35.
Dillon, T. M. 1977. Mercury and the estuarine marsh clam,
    Rangia cuneata. Arch. Env. Contam. Toxicol. 6: 249-256.
Gaufin, A. R. 1973. Use of aquatic invertebrates in the
    assessment of water quality, pp. 96-116. In: (J. Cairns,
    Jr. and K. L. Dickson, eds.) Biological Methods for the
    Assessment of Water Quality. Philadelphia: American
    Society for Testing and Materials.
Gunter, G. 1961. Some relations of estuarine organisms to
    salinity. Limnology and Oceanography 6: 182-190.

Gunter, G. 1977. Salinity, pp. 694-697. In: (J. R. Clark, ed.) Coastal Ecosystem Management. New York: John Wiley & Sons.

Gunter, G. and H. H. Hildebrand. 1954. The relation of total rainfall of the state and catch of the marine shrimp (*Penaeus setiferus*) in Texas waters. Bull. Mar. Sci. Gulf. Caribbean 4: 95.

Hamelink, J. L. and R. C. Waybrant. 1976. DDE and lindane in a large-scale model lentic ecosystem. Trans. Amer. Fish. Soc. 105: 124.

Hedgepeth, J. W. 1957. Ecological aspects of the Laguna Madre. A hypersaline estuary, pp. 408-419. In: (G. H. Lauff, ed.) Estuaries. Washington, D. C.: AAAS.

Holland, J. S., N. J. Maciolek, and C. H. Oppenheimer. 1973. Galveston Bay benthic community structure as an indicator of water quality. Contr. Mar. Sci. Univ. Tex. 17: 169-188.

Hooks, T. A. 1973. An analysis and comparison of the benthic invertebrate communities in the Fenholloway and Econfina estuaries of Apalachee Bay, Florida. M.S. Thesis, Florida State University, Tallahassee.

Hurlbert, S. H., J. Zedler, and D. Fairbanks. 1972. Ecosystem alteration by mosquitofish (*Gambusia affinis*) predation. Science 175: 639-641.

Ketchum, B. H. 1967. Phytoplankton nutrients in estuaries, pp. 329-335. In: (G. H. Lauff, ed.) Estuaries. Washington, D. C.: AAAS.

Kinne, O. 1967. Physiology of estuarine organisms with special reference to salinity and temperature: General aspects, pp. 525-540. In: (G. H. Lauff, ed.) Estuaries. Washington, D. C.: AAAS.

Koenig, C. C. 1977. The effects of DDT and mirex alone and in combination on the reproduction of a salt marsh cyprinodont fish, *Adinia xenica*, pp. 357-376. In: (F. J. Vernberg, A. Calabrese, F. P. Thurberg, and W. B. Vernberg, eds.) Physiological Responses of Marine Biota to Pollutants. New York: Academic Press.

Koenig, C. C., R. J. Livingston, and C. R. Cripe. 1976. Blue crab mortality: Interaction of temperature and DDT residues. Arch. Env. Contam. Toxicol. 4: 119-128.

Lee, R. F., M. Takahashi, J. R. Beers, W. H. Thomas, D. L. R. Seibert, P. Koeller, and D. R. Green. 1977. Controlled ecosystems: Their use in the study of the effects of petroleum hydrocarbons on plankton, pp. 323-344. In: (F. J. Vernberg, A. Calabrese, F. P. Thurberg, and W. B. Vernberg, eds.) Physiological Responses of Marine Biota to Pollutants. New York: Academic Press.

Livingston, R. J. 1975. Impact of Kraft pulp-mill effluents on estuarine and coastal fishes in Apalachee Bay, Florida, USA. Mar. Biol. 32: 19-48.

Livingston, R. J. 1976. Dynamics of organochlorine pesti- cides in estuarine systems: Effects on estuarine biota, pp. 507-522. In: Estuarine Processes, Vol. 1. New York: Academic Press.

Livingston, R. J. 1977. Time as a factor in biomonitoring estuarine systems with reference to benthic macrophytes and epibenthic fishes and invertebrates, pp. 212-234. In: (J. Cairns, Jr., K. L. Dickson, and G. F. Westlake, eds.) Biological Monitoring of Water and Effluent Quality. Philadelphia: American Society for Testing and Materials.

Livingston, R. J., R. L. Iverson, R. H. Estabrook, V. E. Keys, and J. Taylor, Jr. 1975. Major features of the Apalachi- cola Bay system: Physiography, biota, and resource manage- ment. Florida Scientist 37: 245-272.

Livingston, R. J., G. J. Kobylinski, F. G. Lewis III, and P. F. Sheridan. 1976. Long-term fluctuations of epi- benthic fish and invertebrate populations in Apalachicola Bay, Florida. Fish. Bull. U. S. 74: 311-321.

Livingston, R. J., C. C. Koenig, J. L. Lincer, C. D. McAuliffe, A. Michael, R. J. Nadeau, R. E. Sparks, and B. E. Vaughan. 1974. Synergism and modifying effects: Interacting fac- tors in bioassay and field research, pp. 226-304. In: (G. V. Cox, ed.) Marine Bioassays Workshop Proceedings. Washington, D. C.: Marine Technology Society.

Livingston, R. J., P. F. Sheridan, B. G. McLane, F. G. Lewis III, and G. J. Kobylinski. 1977. The biota of the Apalachicola Bay system: Functional relationships, pp. 75- 100. In: (R. J. Livingston and E. A. Joyce, eds.) Proc. of the Conf. on the Apalachicola Drainage Syst.

Livingston, R. J., N. P. Thompson, and D. A. Meeter. 1978. Long-term variation of organochlorine residues and assemblages of epibenthic organisms in a shallow north Florida (USA) estuary. Mar. Biol. 46:355-372.

Lockhart, W. L., D. A. Metner, and J. Solomon. 1977. Methoxy- chlor residue studies in caged and wild fish from the Athabasca River, Alberta, following a single application of blackfly larvicide. J. Fish. Res. Bd. Can. 34: 626-632.

Michael, A. D., D. Boesch, C. Hershner, R. J. Livingston, K. Roos, and D. Straughan. 1977. Benthos, pp. 57-75. In: (G. V. Cox, ed.) Oil Spill Studies: Strategies and Techniques. Montauk: American Petroleum Institute.

Nimmo, D. R. and L. H. Bahner. 1974. Some physiological con- sequences of polychlorinated biphenyl--and salinity--stress in penaeid shrimp, pp. 427-444. In: (F. J. Vernberg and W. B. Vernberg, eds.) Pollution and Physiology of Marine Organisms. New York: Academic Press.

Nimmo, D. R., R. R. Blackman, A. J. Wilson, Jr., and
J. Forester. 1971. Toxicity and distribution of Arochlor
1254 in the pink shrimp *Penaeus duorarum*. Mar. Biol. 11:
191-197.

Odum, H. T. and B. J. Copeland. 1969. A functional classification of the coastal ecological systems, pp. 9-86. In:
(H. T. Odum, B. J. Copeland, and E. A. McMahan, eds.)
Coastal Ecological Systems of the United States. Washington, D. C.: Water Pollution Control Administration.

O'Hara, J. 1973. The influence of temperature and salinity
on the toxicity of cadmium to the fiddler crab, *Uca
pugilator*. Fish. Bull. 71: 149-153.

Parker, R. H. 1955. Changes in the invertebrate fauna,
apparently attributable to salinity changes, in the bays
of central Texas. J. Paleont. 29: 193-211.

Parsons, T. R. 1974. Controlled Ecosystem Pollution Experiment (CEPEX). Env. Conserv. 1: 224.

Patrick, R. 1973. Use of algae, especially diatoms, in the
assessment of water quality, pp. 76-95. In: (J. Cairns,
Jr. and K. L. Dickson, eds.) Biological Methods for the
Assessment of Water Quality. Philadelphia: American
Society for Testing and Materials.

Reeve, M. R., G. D. Grice, V. R. Gibson, M. A. Walter,
K. Darcy, and T. Ikeda. 1976. A Controlled Environment
Pollution Experiment (CEPEX) and its usefulness in the
study of stressed marine communities, pp. 145-162. In:
(A. P. M. Lockwood, ed.) Effects of Pollutants on Aquatic
Organisms. New York: Cambridge University Press.

Reid, G. K., Jr. 1955a. A summer study of the biology and
ecology of East Bay, Texas. Part I. Texas J. Sci. 7:
316-343.

Reid, G. K., Jr. 1955b. A summer study of the biology and
ecology of East Bay, Texas. Part II. Texas J. Sci. 7:
430-453.

Reid, G. K., Jr. 1956. Ecological investigations in a disturbed Texas coastal estuary. Texas J. Sci. 8: 296-327.

Sheline, J. and B. Schmidt-Nielsen. 1977. Methylmercury-
selenium: Interaction in the killifish, *Fundulus hetero-
clitus*, pp. 119-130. In: (F. J. Vernberg, A. Calabrese,
F. P. Thurberg, and W. B. Vernberg, eds.) Physiological
Responses of Marine Biota to Pollutants. New York:
Academic Press.

Sigmon, C. F., H. J. Kania, and R. J. Beyers. 1977. Reductions in biomass and diversity resulting from exposure to
mercury in artificial streams. J. Fish. Res. Bd. Can. 34:
493-500.

Takahashi, M., W. H. Thomas, D. L. Seibert, J. Beers, P. Koeller,
and T. R. Parsons. 1975. The replication of biological events
in enclosed water columns. Archiv. Hydrobiol. 76: 5-23.

Vernberg, F. J., M. S. Guram and A. Savory. 1977. Survival
    of larval and adult fiddler crabs exposed to Arochlor Ⓡ
    1016 and 1054 and different temperature-salinity combina-
    tions, pp. 37-50. In: (F. J. Vernberg, A. Calabrese,
    F. P. Thurberg, and W. B. Vernberg, eds.) Physiological
    Responses of Marine Biota to Pollutants. New York:
    Academic Press.
Vernberg, F. J. and W. B. Vernberg. 1974. Pollution and
    Physiology of Marine Organisms. New York: Academic Press.
Vernberg, W. B. and B. C. Coull. 1975. Multiple factor
    effects of environmental parameters on the physiology,
    ecology, and distribution of some marine meiofauna.
    Cahiers Biol. Mar. 16: 721-732.
Vernberg, W. B., P. J. DeCoursey, and W. J. Padgett. 1973.
    Synergistic effects of environmental variables on larvae
    of *Uca pugilator* (Bosc). Mar. Biol. 22: 307-312.
Vernberg, W. B. and J. O'Hara. 1972. Temperature-salinity
    stress and mercury uptake in the fiddler crab, *Uca
    pugilator*. J. Fish. Res. Bd. Can. 29: 1491-1494.
Vernberg, W. B. and J. Vernberg. 1972. The synergistic
    effects of temperature, salinity, and mercury on survival
    and metabolism of the adult fiddler crab, *Uca pugilator*.
    Fish. Bull. 70: 415-420.
Vernberg, W. B. and F. J. Vernberg. 1976. Physiological
    adaptations of estuarine animals. Oceanus 19: 48-54.
Wiens, J. A. 1977. On competition and variable environments.
    Amer. Sci. 65: 590-597.
Wilhm, J. L. and T. C. Dorris. 1968. Biological parameters
    for water quality criteria. BioScience 18: 477-481.
Zimmerman, M. S. and R. J. Livingston. 1976. Effects of
    Kraft-mill effluents on benthic macrophyte assemblages in
    a shallow-bay system (Apalachee Bay, north Florida, USA).
    Mar. Biol. 34: 297-312.

SEASONAL EFFECTS OF CHLORINE PRODUCED OXIDANTS ON
THE GROWTH, SURVIVAL AND PHYSIOLOGY OF THE AMERICAN
OYSTER, *CRASSOSTREA VIRGINICA* (GMENLIN).

Geoffrey I. Scott

Environmental Protection Agency
National Marine Water Quality Lab
Bear's Bluff Field Station
PO Box 368
John's Island, S. C.

Winona B. Vernberg

School of Public Health &
Belle W. Baruch Institute
University of South Carolina
Columbia, S. C.

INTRODUCTION

Chlorine gas (Cl_2) has been used in industrial, biocidal,
disinfection applications more than 175 years. It was recognized much more recently, however, that discharge of Cl_2 into
estuarine and coastal waters may result in undesirable toxic
effects to many of the organisms residing in these habitats
(Waugh, 1964; Morgan and Stross, 1969; McLean, 1973). While
there is a substantial volume of information on the short term
toxicity of chlorine to marine invertebrates and fishes
(Markowski, 1960; Gameson *et al.*, 1961; Cory and Newman, 1970;
Muchmore and Epel, 1973; and Stober and Hanson, 1974) very
little is known about the long term effects of chlorination on
one of the most common species found along the coast of eastern
North America, the American oyster, *Crassostrea virginica*
(Gmelin).
As early as 1946, Galtsoff reported that adult *C. virginica*,
showed reduced pumping rates at a chlorine concentration of less

than 0.05 mg/l, and at concentrations of 1 mg/l or greater,
oysters closed their valves and ceased to pump. However, the
longest period of exposure was only 48 hours. Recently
Bongers *et al.* (1977) reported little if any mortality in
adult *C. virginica* exposed for 15 days to chlorination (0.35-
0.85 mg/l) and bromine chloride (BrCl) (0.17-0.86 mg/l); BrCl
is an oxidant which has been mentioned as a possible alternative to chlorine in the disinfection process. Sublethal responses revealed that new shell growth was greater in controls
than in either chlorine or BrCl exposed oysters; the amount
of shell deposition decreased with increased toxicant concentrations. No significant difference in the reduction of new
shell deposition occurred in tests comparing chlorinated and
BrCl treated effluents.

   Few data are available on the effects of chlorination upon
larvae of molluscs. Waugh (1964) reported that exposures of
2-20 minutes to various concentrations (less than 10 mg/l) of
chlorine did not induce significant mortality in larvae of
Ostrea edulis. Recent studies of the acute toxicity of chlorination to molluscan larvae by Roberts *et al.* (1975) demonstrated that exposure to chlorination for 48 hours is very
toxic to both oyster (*C. virginica)* and clam (*Mercenaria
mercenaria*) larvae. The 48 hour EC_{50} was estimated to be less
than 0.005 mg/l for larval oysters and 0.006 mg/l for larval
clams. The estimated 96 hour EC_{50} (by shell deposition) for
juvenile oysters was 0.023 mg/l, It also was observed that
oyster larvae survived intermittant chlorination longer than
they did continuous chlorination.

   One of the chief probelms in evaluating chlorination studies
is the number of compounds formed following chlorination of
natural waters. The number and types of compounds are a function of the physical and chemical parameters of the water,
including but not limited to temperature, pH, organic matter,
ammonia, sunlight (UV) and salinity (or the amount of bromine
available as a reaction compound). To complicate matters
even further, bormination rather than chlorination may pre-
dominate as salinities increase, since sea water typically
contains 60 mg/kg bromide. In addition, photolysis may influence the level of bromate produced during the chlorination of
saline waters (Macalady *et al.*, 1977). These compounds that
are formed following chlorination are termed chlorine produced
oxidants (CPO) (Burton, 1976) and may include varying propor-
tions of hypochlorous acid and hypochlorite ion as well as
hypobromous acid and hypobromite ion, depending upon the
ambient salinity. In turn, these CPO compounds may combine
with the naturally occurring organic fractions of sea water to
form halogenated organic compounds (Helz *et al.*, 1978). Thus,
studies in which attempts are made to determine both the
effects of chlorination and of bromination products formed

during the initial treatment as well as the co-occurring halo-genated organic products formed secondarily in sea water, are needed.

The objective of this study was to examine the effects of chronic chlorination on the survival and physiological responses of the adult oyster, *Crassostrea virginica*. These studies were conducted throughout the year to measure any possible seasonal effects, and synergistic interactions of temperature and CPO concentrations were delineated.

MATERIALS AND METHODS

Chronic oyster bioassays were conducted during the fall (45-day exposure), winter (75-day exposure), and spring (60-day exposure) of 1976-1977. Adult oysters (5.0-12.0 cm height) were collected from Leadenwah Creek (coordinates N 32^0 36' latitude, W 80^0 15' longitude), a large tidal tributary of the North Edisto River Estuary in South Carolina. Following collection the oysters were transported immediately to the laboratory. where they were scrubbed and all fouling organisms were removed. After numbering and measuring the height of individual oysters, 50 animals were placed in each of 4 exposure tanks (110 cm x 63 cm x 28 cm) and acclimated in running seawater (>5.0 l/hour/oyster) for 15 days. An additional tank was maintained as a control. Each tank was cleaned at least every other day (daily when possible) during both acclimation and the seasonal exposure periods.

During each seasonal bioassay, the following measurements were used to assess the effects of chronic exposure to CPO:

1. Percentage survival and percentage mortality were recorded daily.

2. Prior to and at the end of each seasonal exposure, shell height was measured using Vernier calipers.

3. Growth was monitored by measuring changes in shell height and weight in 10 oysters/exposure concentration plus controls at day zero and 45 days (fall), 75 days (winter), and 60 days (spring), using the techniques described by Havinga (1928) and Andrews (1961).

4. Condition index measurements were made on 20 oysters per sampling period [day 0, 30, and 60 (spring); day 0, 30, and 75 (winter); and day 45 (fall)] using the method described by Galtsoff (1964):

$$\text{Condition index} = \frac{\text{Dry Body Weight (g)}}{\text{Cavity Volume (ml)}} \times 100. \tag{1}$$

Calculations of cavity volume are usually obtained by subtracting the total volume of the oyster (including its shell)
minus its empty shell volume as determined by the water displacement method. However this method may be subject to
significant manipulative error and is very time consuming.
Andrews (1961) reports that the specific gravity of oyster
tissues is approximately the same as the specific gravity
seawater. Therefore, a direct weight to volume relationship
exists whereby cavity volume may be determined by subtracting
the total weight of the oyster (including its shell) minus
the empty shell weight, since there is approximately a 1:1
ratio of weight to volume in the tissues removed. A comparison of cavity volumes derived by both the weight to weight
method and the volume to volume method, showed no statistically
significant (P < 0.01) differences. Regressional analyses
of cavity volumes derived by both methods on oysters used in
this study and in other studies show that cavity volumes
derived by each method are very closely related (Dr. David R.
Lawrence, personal communication, 1977). The correlation
coefficients (R) ranged from 0.95-0.98. The weight to weight
method is a useful and time-saving modification of the currently
used volume method, and provides a more rapid and precise
manner to determine cavity volume and condition index measurements (Scott, unpublished data).

    5. Gonadal index measurements were made using the techni-
que in Galtsoff (1964):

$$\text{Gonadal Index} = \frac{\text{Dry weight of Gonads (g)}}{\text{Total Dry Body Weight (g)}} \times 100. \qquad (2)$$

    6. Chlorine produced oxidant (CPO) concentrations (free
+ combined chlorine and other residual oxidants) were measured
with a Wallace and Tiernan Titrator. Phenylarsine oxide and
pH buffers were standard Wallace and Tiernan reagents. The
5% KI solution and chlorine (Fisher NaOCl Analytical Reagent,
minimum 5.0% NaOCl) stock solutions were prepared in deionized
water. Stock solutions were kept in lightproof containers to
prevent photo-decomposition. Measurements of CPO concentrations
employed procedures outlined in Standard Methods (American Public Health Association, 1971). All water samples were taken
directly from treatment aquaria at least every 2-3 days during
each seasonal experiment. Agitation of samples was avoided
to reduce the potential for error by photo-decomposition or
flashing-off of CPO (Environmental Protection Agency, 1974).
Nominal amounts of NaOCl (mg/l) and measured ranges and means
of CPO (mg/l) detected in each seasonal bioassay are reported
in Table 1.

    7. Ambient water temperature and salinity were measured
daily in the incoming seawater from the North Edisto River

TABLE 1. Physical and chemical parameters measured during each seasonal bioassay. Parameters measured include temperature (°C), salinity (°/oo), the nominal amounts of chlorine added (nominal NaOCl, mg/l), and measured chlorine produced oxidant (measured CPO, mg/l)

Table 1. Physical and Chemical Parameters

Nominal NaOCl concentration (mg/l)	Fall					Winter					Spring				
	Measured CPO concentration (mg/l)		Temperature (°C)		Mean salinity (°/oo)	Measured CPO concentration (mg/l)		Temperature (°C)		Mean salinity (°/oo)	Measured CPO concentration (mg/l)		Temperature (°C)		Mean salinity (°/oo)
	Range	Mean	Range	Mean		Range	Mean	Range	Mean		Range	Mean	Range	Mean	
5.60	0.98–3.26	1.58	10.5–27.0	17.5	25.4	2.02–4.95	2.98	4.0–20.0	11.2	23.3	0.75–3.82	1.81	20.5–31.5	27.0	26.6
3.20	0.47–1.94	0.91	10.5–27.0	17.5	25.4	0.84–2.09	1.23	4.0–20.0	11.2	23.3	0.39–1.20	0.65	20.5–31.5	27.0	26.6
1.80	0.18–0.49	0.28	10.5–27.0	17.5	25.4	0.22–0.68	0.39	4.0–20.0	11.2	23.3	0.11–0.34	0.24	20.5–31.5	27.0	26.6
1.00	0.08–0.22	0.14	10.5–27.0	17.5	25.4	0.09–0.24	0.16	4.0–20.0	11.2	23.3	0.08–0.17	0.12	20.5–31.5	27.0	26.6
Control	—	—	10.5–27.0	17.5	25.4	—	—	4.0–20.0	11.2	23.3	—	—	20.5–31.5	27.0	26.6

Estuary in each seasonal experiment. Mean ambient seawater
temperatures ranged from 17.5^0C (fall) to 11.2^0C (winter) to
27.0^0C (spring). Mean salinities during the study ranged
from 25.4o/oo (fall) to 23.3o/oo (winter), to 26.6o/oo (spring).
Periodic dissolved oxygen measurements showed that levels were
always near saturation. The pH ranged from 7.4 to 8.3 during
the study.
    8. Fecal production measurements were made every 2-3 days
in both control and experiments by observing the number of
oysters producing feces.
    9. Respiration was measured of excised mantle tissues
removed from adult oysters (height 7-10 cm) which had been
exposed to 3 nominal concentrations of NaOCl (3.20, 1.80, and
1.00 mg/l) for 30 days during the winter (ambient water temperature 5^0C) and spring (ambient water temperature 28^0C)
experiments. Oxygen consumption rates of tissue from control
animals (NaOCl-free water) were also determined. Mantle tissues
were extracted in a manner similar to that described by Wilbur
& Jodrey (1957), but modified as follows. The section of
the mantle dissected, was that tissue adjacent to the abductor
muscle, beginning at the anus, continuing posteriorly past
the fusion of the left and right mantle and terminating at
the end of the abductor muscle on the side opposite the anus.
This technique allows a more uniform section of mantle tissue
to be removed. Respiration determination were made using a
Gilson Differential Respirometer. In all experiments 10 ml of
0.45 µ filtered seawater (22.5% salinity) and 0.4 ml of 20%
KOH were added to each flask and side-arm of each flask,
respectively. Two empty flasks (containing only filtered
seawater and KOH) were maintained as thermobarometers in each
experimental run.
    10. Statistical treatment included determination of mean,
standard deviation, and standard error of each experimental
measurement in control and exposed oysters. T-tests were used
to determine if measured differences between controls and
experimentals were statistically significant. The 95% confidence
level ($p < 0.05$) was the minimum statistical significance
accepted.

RESULTS

Physical and Chemical Measurements

    Measurements of physical [temperature (^0C) and salinity
(o/oo)] and chemical [nominal NaOCl and measured CPO (mg/l)]
parameters are recorded in Table 1, for each seasonal bioassay.
Measured CPO concentrations varied considerably from season

to season at each nominal CPO concentration. This variation is expected since differing seasonal water temperatures and dissolved and suspended organic loads in seawater result in variable reaction rates with chlorine. Measured CPO concentrations are highest during the winter exposure period, decreasing during both the fall and spring. Overlaps in the range between various nominal CPO concentrations possibly occur because of the continually changing organic load in the ambient seawater. This changing organic load causes occasional shifts in measured CPO concentrations in all nominal concentrations in response to seasonal changes in water temperature (and hence organic load) and in response to sudden changes in water temperature and salinity generally associated with passing weather fronts (usually accompanied by increased water turbulence and rainfall).

Seasonal survival. The survival of oysters to different measured CPO concentrations during each seasonal bioassay is shown graphically in Figures 1 (fall), 2 (winter), and 3 (spring).

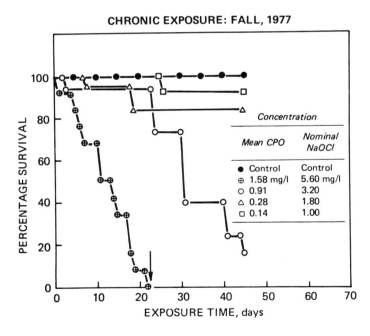

FIGURE 1. *Survival of adult oysters exposed to different concentrations of CPO during the fall exposure. Time intervals are days of exposure (maximum 45 days). The arrow (↓) denotes time of 0% survival.*

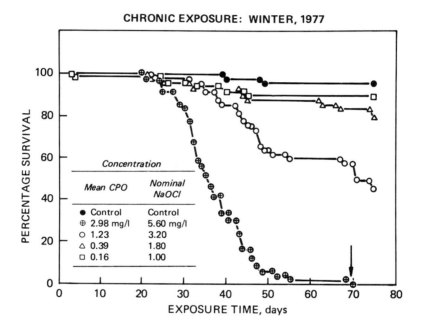

FIGURE 2. *Survival of oysters exposed to different concentrations of CPO during the winter exposure. Time intervals are days of exposure (maximum 75 days). The arrow (↓) denotes time of 0% survival.*

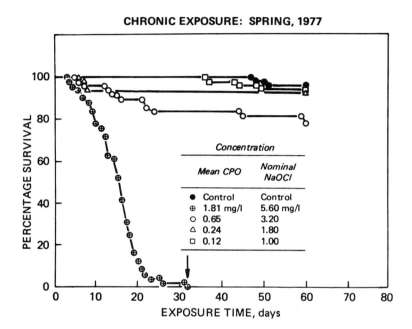

FIGURE 3. *Survival of oysters exposed to different concentrations of CPO during the spring exposure. Time intervals are days of exposure (maximum 60 days). The arrow (↓) denotes the time of 0% survival.*

In the nominal 5.60 mg/l NaOCl concentration, there was a delay in mortality, with 0% survival by day 70 (winter) versus day 22 (fall) and day 32 (spring). This delay in mortality during the winter bioassay occurs when mean measured CPO concentrations are at their highest [2.98 mg/l (winter) versus 1.58 mg/l (fall) and 1.81 mg/l (spring)].

Seasonal survivability varied in the nominal 3.20 mg/l NaOCl concentration. Lowest survival occurred during the fall bioassay (12% survival after 45 days). Survivability was higher in the winter (44% survival after 75 days) and much higher in the spring (78% survival after 60 days). Very little mortality occurred in the winter and spring bioassays when compared to the fall exposure. Survival in the nominal NaOCl concentrations of 1.80 mg/l and 1.00 mg/l was much higher in the spring than in the fall and winter bioassays (Figures 1 and 2). Control survival was always above 94% in all seasonal exposures.

Fecal Production

Measurements of seasonal mean fecal production are listed in Table 2. Concentrations of 0.12-0.16 mg/l of CPO resulted in a 50% or more reduction in fecal production regardless of season. Fecal production in controls was always greater than 83% during each seasonal exposure.

Condition index and gonadal index measurement. Sublethal effects of CPO concentrations on adult oysters as measured by condition index and gonadal index measurements are listed in Table 3 and depicted graphically in Figures 4, 5, and 6.

TABLE 2. Summary of mean fecal production in adult oysters (seasonally) exposed to CPO. Measured CPO values (mg/l) are seasonal means. Fecal production in control oysters was significantly (p < 0.05) greater than in exposed oysters during each bioassay

Table 2. Seasonal Oyster Fecal Production

Nominal NaOCl concentration (mg/l)	Fall		Winter		Spring	
	Mean measured CPO concentration (mg/l)	Mean percentage of oysters producing feces	Mean measured CPO concentration (mg/l)	Mean percentage of oysters producing feces	Mean measured CPO concentration (mg/l)	Mean percentage of oysters producing feces
5.60	1.58	0	2.98	0	1.81	0
3.20	0.91	0	1.23	1	0.65	10
1.80	0.28	6	0.39	10	0.24	26
1.00	0.14	19	0.16	39	0.12	49
Control	—	85	—	83	—	93

TABLE 3. *Mean condition index and gonadal index measured in oysters exposed to CPO. All measurements of condition index and gonadal index in oysters exposed to CPO are significantly (P < 0.05) smaller than those measurements in control (unexposed) oysters, during each seasonal bioassay. N = 20 (number of oysters/measurement) in each mean value reported. Reported CPO values (mg/1) are seasonal means*

Table 3. Seasonal Condition Index and Gonadal Index Measurements

	Fall			Winter							Spring								
Nominal NaOCl concentration (mg/l)	Measured CPO concentration (mg/l)	45-Day condition index X̄	S.d.	Measured CPO concentration (mg/l)	30-Day condition index X̄	S.d.	75-Day condition index X̄	S.d.	75-Day gonadal index X̄	S.d.	Measured CPO concentration (mg/l)	30-Day condition index X̄	S.d.	30-Day gonadal index X̄	S.d.	60-Day condition index X̄	S.d.	60-Day gonadal index X̄	S.d.
5.60	1.58	—	—	2.98	—	—	—	—	—	—	1.81	—	—	—	—	—	—	—	—
3.20	0.91	1.74	0.799	1.23	5.60	1.346	2.66	1.015	2.38	1.252	0.65	2.75	0.603	9.14	9.976	3.29	1.049	12.87	4.317
1.80	0.28	2.52	0.500	0.39	5.69	1.669	3.99	1.194	3.61	2.028	0.24	3.91	1.203	19.32	10.639	3.59	0.755	12.04	6.161
1.00	0.14	2.23	0.576	0.16	5.71	2.761	4.01	1.036	5.10	2.821	0.12	3.94	0.962	20.44	8.554	3.00	0.966	9.33	5.443
Control	—	3.21	0.738	—	7.38	2.468	4.92	1.201	10.38	5.580	—	5.08	1.465	30.83	9.644	4.60	1.112	23.65	10.180

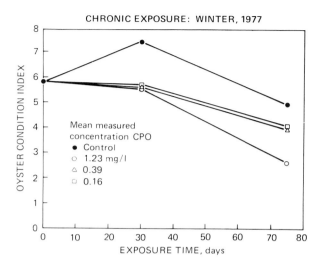

FIGURE 4. *Chronic exposure: Winter, 1977. Mean condition index measurements in adult oysters exposed to different CPO concentrations (mg/l) during the winter bioassay (75 days exposure). All values of CPO exposed oysters at day 30 and day 75 were significantly (p < 0.05) different from control values.*

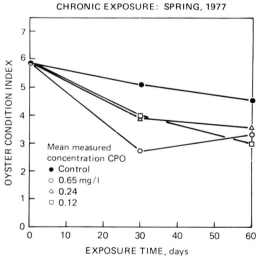

FIGURE 5. *Chronic exposure: Spring, 1977. Mean condition index measurements in adult oysters exposed to different CPO concentrations (mg/l) during the spring bioassay (60 days exposure). All values of CPO exposed oysters at day 30 and day 60 are significantly (p < 0.05) different from control values.*

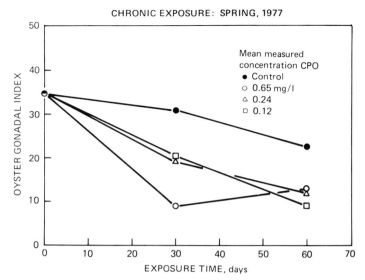

FIGURE 6. *Chronic exposure: Spring, 1977. Mean gonadal index measurements in adult oysters exposed to different CPO concentrations (mg/l) during the spring bioassay (60 days exposure). All values of CPO exposed oysters at day 30 and day 60 are significantly (p < 0.05) different from control values.*

Conditions indices measurements in controls were always significantly (p < 0.05) higher than those in measured CPO concentrations during each seasonal exposure. There was a general decrease in condition index with increased exposure time at all measured CPO concentrations. Seasonal fluctuations in condition indices in controls are obviously related to the storage and release of gonadal material.

Trends in gonadal index values closely paralleled those observed in condition index measurements. Control gonadal indices were significantly greater (p < 0.05) than those measured at each CPO concentration during each seasonal exposure. There was a general decrease in mean gonadal index with increasing exposure time in all seasonal exposure concentrations of CPO. Peak control gonadal indices measurements occurred during the spring exposure (May) followed by a general decrease during late spring (early June).

Growth Measurements

Growth measurements are not reported in this study. The discovery of a growth layer inside the shell of CPO exposed

oysters made interpretation of results difficult if not impossible. Electron microscopy studies to characterize this
growth layer are presently under way at the University of
South Carolina. As oysters are exposed to CPO, the mantle
apparently becomes irritated and retracts farther back into
the foliated layer of the shell away from the prismatic
edge of the shell. At its new position farther inside the
shell, the mantle then lays down new shell growth upon the
older foliated layer of the shell. This new shell growth
appears to be different from the normal growth noted in either
the foliated or prismatic portions of the shells in controls.

Mantle Respiration

Measurements of mantle respiration (µl/g/hr) are depicted
graphically in Figure 7. Oxygen uptake is much lower during
the winter (5°C; 30 days exposure) than during the spring
(28°C; 30 days exposure). During the winter experiment, there

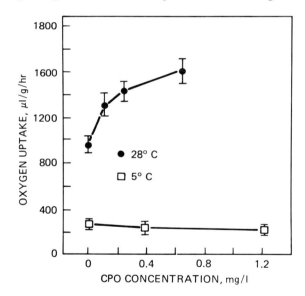

FIGURE 7. *Mean seasonal oyster mantle respiration rates
(µl/g/hr) in control oysters and oysters exposed to 3 different
NaOCl concentrations (3.20, 1.80, and 1.00 mg/l) for 30 days
at 5°C (winter) and 28°C (spring). At 28°C, O_2 consumption
rates of CPO exposed oysters were significantly (p < 0.05)
higher than control values. At 5°C there was no significant
difference between CPO exposed oysters and controls. N = 15
(number of oysters/measurement) in each value reported. CPO
values are seasonal means. Bars (I) indicate the standard
error of each mean.*

was no significant difference between the oxygen uptake in
control and exposed oysters. In the spring test, there were
statistically significant differences ($p < 0.05$) in the oxygen
consumption in dosed oysters. The observed higher rates of
oxygen consumption in dosed oysters, were noted when CPO concentrations were lower than those measured during the winter
test, but when water temperatures were much higher.

DISCUSSION

There are obvious seasonal differences in the survival
of oysters exposed to the various CPO concentrations, especially the nominal 5.60 and 3.20 mg/l NaOCl levels. The
delay in mortality in the nominal NaOCl 5.60 mg/l level during
the winter bioassay when measured CPO concentrations were the
highest (2.98 mg/l) suggests that other factors, such as
temperature and the physiological conditions of the oyster,
are important in determining survival and mortality patterns.
The high survival during the initial 30 days of the winter
bioassay when measured CPO concentrations were greatest
(2.98 mg/l) (see Fig. 2), corresponds with a time period in
which water temperatures were 5°C or below. According to
Galtstoff (1964) oxygen uptake decreases markedly at temperatures below 5°C, and the valves are often closed at this low
temperature and ventilation of the gills is stopped. This
closure of the valves and lowered metabolic rate may prevent
the oyster from being exposed to the CPO until water temperatures increase, causing elevated metabolic rates, increased
gill ventilation, and more direct exposure to the toxicant.
A previous study by Middaugh et al. (1976) has shown
exposures to CPO caused increased toxicity with increasing
temperature in juvenile spot, Leiostomus xanthurus. Thatcher
et al. (1976) have also shown the possible synergistic effects
of increasing temperature and chlorine on the survival of
juvenile brook trout, Salvelinus fontinalis.
Although synergism has been defined differently by various
authors (i.e. Sprague, 1970; Warren, 1971) by such terms as
joint toxicity or additive toxicity, it is usually used in
reference to the co-action of 2 or more factors. In the
present study, the multiple co-occurring CPO (i.e. hypochlorous acid, hypochlorite ion, hypobromous acid, hypobromite
ion, monochloramine and monobromamine) and possible halogenated organics are generated from the chlorination of seawater. The combined interactions of these various chlorination by-products on toxicity should be a function of the CPO
concentrations measured during each seasonal experiment as
well as the physical factors of seawater, such as water

temperature. Although synergistic effects must be quantified
statistically by response curves and associated statistical
analysis, inference of the possible synergistic interactions
between CPO concentration and water temperature can be made
on a qualitative basis by comparing seasonal survival and
seasonal sublethal physiological measurements (i.e. respiration, condition index).

Such a qualitative synergistic effect between seasonal
CPO concentrations and seasonal water temperatures is
obvious if the time to 0% survival is compared in the winter
bioassay, 70 days, to that found in fall and spring, 22 days
and 32 days respectively in the nominal 5.60 mg/l NaOCL
concentration. A comparison of the mean measured CPO concentrations reveals highest concentrations were measured during
the winter 2.98 mg/l (versus 1.58 mg/l (fall) and 1.81 mg/l
(spring)) when time to 0% survival was delayed. These data
strongly support our earlier conclusion that seasonal factors,
such as water temperature and the physiological condition
of the oysters, are contributing factors which interact with
the CPO concentration to affect mortality levels.

Differences in the percentage survival observed in the
nominal 3.20 mg/l concentration were variable and again appear to relect the seasonal fluctuations in mean measured
CPO concentrations (0.65 mg/l (spring), 0.91 mg/l (fall), and
1.23 mg/l (winter)), water temperature, and physiological
condition of the oysters tested. Comparisons of time versus
percent survival revealed striking seasonal differences
(Figures 1, 2, and 3). In the fall bioassay only 12% of the
oysters survived to day 45; during the winter and spring
survival rates were 76% and 80% respectively. The decreased
survival in the fall was probably related to the physiological
state of the oysters during this season of the year. Mean
condition index of control oysters during the fall was 3.21
versus 7.38 during the winter and 5.08 during the spring.
The lower fall condition index measurement is related to
post-spawning effects. Our data suggest that this lowered
physiological state makes the oyster more susceptible to the
toxic effects of chlorination.

The percent survival of oysters at nominal NaOCl concentrations of 1.80 and 1.00 mg/l was much greater than that
observed at the higher exposure concentrations. Although
seasonal differences in percent survival did occur, these
differences were not significant. Bongers et al. (1977)
reported 92-100% survival in C. virginica exposed to Cl_2
concentrations of 0.35-0.85 mg/l for 15 days. Our data show
a similar survival rate after only 15 days exposure but, as
exposure time increased, the percent survival substantially
decreased during each seasonal bioassay. For example, during
the winter experiment, the percent survival was 94% after 15

days exposure to a mean measured CPO concentration of 0.39 mg/l, compared to 80% survival after 75 days. Control survival during this same time period was 96%.

Results of seasonal measurements of fecal production indicate significant reductions in fecal production with increasing CPO concentrations. However, there are no observable seasonal differences in oysters exposed to different CPO concentrations. Our data suggest that the presence of CPO products in the seawater alters the production of fecal material, which in turn infers effects upon feeding. Recent studies by Gentile et al. (1976) have shown that a CPO concentration of less than 1.00 mg/l was responsible for a complete mortality of all phytoplankton species in Narragansett Bay waters. The reduction in the ability of the oyster to produce feces (hence feed) may be related to its preference for algae not exposed to CPO concentrations. Reductions in tissue production appear to be directly related to increasing lengths of exposure and increasing concentrations of CPO. The reduction in tissue production is probably directly related to the increased inability to produce feces (feed) with increasing CPO concentrations. Our data indicate that the exposed oysters close their valves and rely upon body food reserves.

Seasonal gonadal index measurements firmly support seasonal condition index measurements. There are significant (p < 0.05) reductions in seasonal gonadal indices with increasing lengths of exposure and increasing CPO concentrations when compared to control values. The most severe effect was found during the spring when gonadal development was at its peak. Gonadal index data from the spring bioassay study (Fig. 6) confirm observations of reduced tissue production in our spring condition index measurements (Fig. 5) and supports the idea that tissue reduction must occur at the expense of glycogen reserves in the gonads.

From the respiration studies there are indications of a synergistic effect between water temperature and CPO concentration. During periods of low temperature (winter; 5°C) and high CPO concentrations (0.17-1.23 mg/l) there are only slight (nonstatistical) differences between the respiration rates of control and exposed mantle tissue (see Figure 7). During periods of higher temperature (spring; 28°C) and lower concentrations (0.12-0.65 mg/l) however, there are significantly (p < 0.05) higher respiration rates in mantle tissue from exposed versus control oysters. It would appear, then, that the combination of high temperature and seasonal CPO concentrations produces a synergistic physiological stress. These results tend to confirm the work by Hoss et al (1977) which showed that the combination of thermal shock and chlorine had synergistic effects on the survival of flounder (Paralichthys lethostigma) and mullet (Mugil cephalus).

There has been considerable evidence suggesting that
oysters are able to avoid exposure to unfavorable conditions
by closing their valves and shifting to anaerobic pathways
(Lund, 1957; Galtsoff, 1964). Oysters exposed to CPO's showed
a concomitant reduction in tissue production and lessened
glycogen reserves, which points to a shift from aerobic to
anaerobic metabolism. Further evidence of such a shift in
CPO-exposed oysters can be found in results of scanning electron microscopy examinations. Glycolysis, as measured in
the mantle tissue, produces more succinic acid than lactic
acid (Simpson and Awapara, 1966). Hammen (1969) concluded
that during anaerobiosis, pyruvate which results from the
degradation of glucose by glycolysis, is converted through a
reversal of the citric acid cycle to fumarate and succinate.
The production of more succinate than lactate may be advantageous in that succinate is the weaker acid of the two and
would cause less shell dissolution. Recent scanning electron
microscopy examinations of CPO-exposed oyster shells have
revealed rearrangement of crystals in the prismatic layer,
showing that shell dissolution has occurred (Norimitsu Watabe,
personal communication), and provides further indirect evidence for the shift to anaerobic metabolism during exposure
to CPO.

In summary, then, chronic exposure of Crassostrea
virginica to CPO products is highly toxic at high concentrations; severe sublethal effects occur at low concentration
levels. The degree of toxicity, however, varies from season
to season and is related to seasonal changes in measured CPO
concentrations, water temperature, and the physiological condition of the oyster. The combination of various CPO concentrations and high water temperatures significantly decreased
survival and increased respiration rates in exposed oysters,
probably due to a sublethal synergistic physiological stress
and a lethal synergistic effect on survival due to the interaction of these factors. Other sublethal effects appear to be
related to reductions in feeding and increased avoidance of
CPO which results in reduced levels of tissue production. Decreased tissue production is reflected in severe reductions in
the amount of gonadal tissue accumulated; implying an increased
dependence upon gonadal glycogen reserves due to a possible
shift to anaerobic pathways during exposures to CPO.

LITERATURE CITED

Andrews, J. D. 1961. Measurement of shell growth in oysters
    by weighing in water. Proc. Nat'l. Shellfish. Assoc. 52:
    1-11.

BEYOND THE LC$_{50}$: AN OPINION ABOUT RESEARCH ACTIVITIES AND NEEDS CONCERNING PHYSIOLOGICAL EFFECTS OF POLLUTANTS IN THE ENVIRONMENT

Carl J. Sindermann

U. S. Department of Commerce
National Oceanic and Atmospheric Administration
National Marine Fisheries Service
Northeast Fisheries Center
Sandy Hook Laboratory
Highlands, New Jersey

INTRODUCTION

With the completion of the third highly successful symposium on physiological effects of pollutants in the marine environment, it would seem to be a reasonable endeavor to reflect on the past and to consider the status and near-term future of this dynamic and highly relevant research area.

The scientific reports in this symposium series (Vernberg and Vernberg, 1974; Vernberg *et al.*, 1977) have added significantly to the core of available experimental information about contaminant effects on marine/estuarine species and ecosystems. Pollutants of major present concern -- heavy metals, halogenated hydrocarbons, and petroleum components -- have properly been foci of attention. Symposium participants have provided an admirable mix of university, government, and industry research people -- including many outstanding contributors to the expanding literature on pollutant effects. The symposium has also provided an appropriate mix of those very real but essentially undefinable aspects of research called "fundamental" and "applied". The needs of government agencies -- Environmental Protection Agency, Food and Drug Administration, Corps of Engineers, and state environmental conservation agencies -- have been considered, but the scientific "need to understand" aspects have also prevailed. Rallying points cluster around

437

physiology, biochemistry, bioassay, and population impacts. A satisfying repetition of such terms as "sublethal effects" and "metabolic pathways" characterized the discussions.

The stimulation and excitement of the interactions generated by the symposia are difficult to convey in printed volumes, and impossible to distill into a single overview paper. Accepting this impossibility as fact, I can then address myself freely to issues, pertinent and otherwise, that have emerged or crystallized during the sessions and in the discussions that followed.

THE CONTRIBUTIONS OF PHYSIOLOGICAL STUDIES

One fundamental but probably not startling insight is that continuous attention to sublethal physiological/biochemical effects of pollutants is an essential ingredient in any broad examination of environmental pollution. As Jan Prager of EPA has put it "Death is too extreme a criterion for determining whether a substance is harmful to marine biota or not". Sublethal effects -- many of them physiological -- can provide understanding of causes of toxicity -- an understanding that is rarely gained from acute tests. Of course, physiological studies are but part of a matrix of interlocking research efforts, having especially critical synapses and anastamoses with animal behavior, population genetics, pathology, and ecosystem analyses. Frequently during the discussions at this symposium there seemed to be a tangible need for expertise just beyond the extended range of the discussants. No doubt this will always be so, since research with marine organisms and populations is never simple; since variables are many, and often difficult to deal with; and since research groups, however large they may seem, can rarely include every needed specialization. Similarly, every scientific meeting cannot and probably should not become a babel of voices from many disciplines.

Another insight is the repeated observation (now approaching the status of an axiom) that early life history stages of marine animals tend to be most susceptible to environmental contaminants. Excellent data from studies of invertebrates as well as vertebrates document LC_{50}'s and sublethal effects on larvae and postlarvae. These studies should be extended back in the life cycle to include mutagenic effects of contaminants on the earliest stages -- meiotic divisions of the egg, and early post-fertilization mitotic division stages. Preliminary results of such studies (Longwell, 1975, 1976a, 1976b) suggest significant genetic events here, before embryos are formed or hatched.

Associated cytotoxic effects of contaminants may also be important in very early life history stages; here the biochemist
and the geneticist may find common interests.

Still another frequently-expressed but not startling
insight relates to the continuing need for field experimentation. Too often the results of well-conceived laboratory
experiments must be relegated to the category of circumstantial
evidence when confronted with the inevitable question "How do
these findings relate to events in polluted natural waters?"
This does not imply, of course, that every laboratory study
must have a field trial before receiving an "approved" stamp,
or must necessarily satisfy legal requirements of regulatory
agencies, but it does seem that where facilities and personnel
permit, field experimentation should be part of the total
experimental design. If this premise is accepted, the emergence
of genuine understanding about particular pollutant effects
could assume a flat spiral configuration, often beginning in
the center with a puzzling or unusual field observation, moving
then to the laboratory for controlled experimentation, then to
the field for extension of experimental findings, then back
to the laboratory tanks and chromatographs for second generation
experiments, possibly with related or unrelated species and
with different pollutant stresses. There seems to be a
reasonable migration toward field experimental programs in
recent years, but the extent and scale of this movement is
still not overwhelming.

This avenue of thought leads obviously to more complex
experimental studies of contaminant effects on populations,
communities, and ecosystems. Some painstaking attempts have
been made to bring pieces of the marine environment into the
laboratory, or at least onto the laboratory grounds, in
various tanks and ponds. Other attempts have been made to
enclose and manipulate pieces of the natural environment *in
situ* -- in plastic bags or floating nets. While such approaches
often confront the investigators with horrendous logistic problems, and often fail to control enough variables for the satis-
faction of purists, there is much to be learned with these
methods, if they are supported by adequate analytical and
laboratory experimental components.

The technological base for use of marine/estuarine animals
in bioassays has progressed remarkably in the past decade --
so that *Artemia* no longer must represent "marine" animals.
However, we still use a heterogeneous assemblage of physiologically (and in some cases genetically) diverse animals that
we blithely refer to as a test species. Oysters, mussels,
crabs, fish -- all have sufficient intraspecies geographic
variability to make us yearn for genetically described, disease
resistant, easily transportable and easily maintained standard
reference stocks of bioassay species, available at various
life history stages. This piece of Utopia is still remote

from us, even for species like the oyster, which may have
simultaneous utility as aquaculture **crops**. In the interim,
physiologists must continue to characterize such species in
terms of ranges of physiological adaptations to local environmental factors; pathologists must be involved in examining
physiological as well as morphological effects of disease;
and geneticists must be involved in examining the nature and
plasticity of resistance to contaminants.

While on the subject of bioassays, some good word should
be said for interactive or multiple factor bioassays, since
these more closely approximate real world situations. The
possibilities are limited only by the extent of available
facilities, size of staff, and of course funding. Approaches
inclue predator-prey relationships, microcosm perturbations
(especially of benthic assemblages), interaction of modulated
and fixed environmental variables, and use of contaminant-
adapted populations.

An exciting new (or at least under-exploited) aspect of
environmental assessment, of interest to physiologists and
biochemists, can be described as "effects monitoring", in
which standardized physiological-biochemical criteria can be
applied to assessment of the relative state of health of
coastal/estuarine waters. This approach would augment but
not replace the more traditional but often inadequate criteria
such as species distribution, abundance and diversity, and
contaminant analyses of sediments, water column, and biota.
Test organisms may be obtained at sampling stations, or standard species may be transported by ship and tested with water
from the sampling stations.

A recent detailed evaluation of the possible utility of
this approach was completed and published by a working group
of the International Council for the Exploration of the Sea
(ICES, 1976). The general conclusion was that physiological/
biochemical criteria, as well as behavioral, morphological,
and genetic criteria, were eminently feasible, but that the
methodology would require substantial development by appropriate
specialists. The United States has taken the lead in a program of effects monitoring in the NOAA initiative called
"Ocean Pulse" in the western North Atlantic (Pearce, 1977).
A new era of sea-going physiologists may soon be upon us.

THE MAJOR CHEMICAL CONTAMINANTS

If a list were made of the 25 most heavily contaminated
coastal/estuarine areas of the world, some, such as Escambia
Bay in Florida, would be included because of prolonged or persistent effects of single contaminants such as PCB's; others,

such as Raritan Bay in the New York Bight, would be included
because of a mind-boggling mixture of pollutants, but particularly halogenated hydrocarbons, petroleum components, and
heavy metals -- all toxic in some way to the biota. Effects
of each of these major classes of contaminants have been
accorded reasonable respect and attention throughout this
symposium series.

Halogenated hydrocarbons -- notably DDT and PCB's --
have dominated a principal part of the contaminent scene for
almost a decade, and will be persistent problems in the years
ahead -- partly because of their continued use and their survival time in the marine environment. Second generation
pesticides -- such as organophosphates, and more recent formulations such as insect growth inhibitors and chitin synthesis
inhibitors--have appeared and are appearing. Their effects
on marine organisms are still poorly understood, but there are
some disquieting suggestions that their persistence in the
marine environment may be underestimated, and that effects on
early life stages of crustaceans may be severe, especially
where improper applications are made near coastal/estuarine
waters, or when high runoff occurs soon after application.
Often early warning of unexpected harmful effects of these
newer compounds may come from laboratory experiments. Confirmed results of such experiments should be made known to
the scientific community, to regulatory agencies and to the
public as quickly as possible.

Consideration of chlorinated hydrocarbon contaminants
leads logically to the activities of chlorine in the marine
environment. Use of chlorine has long been a highly effective
control measure to prevent fouling in electric generating
stations and for water purification. Such use can impinge on
coastal/estuarine animals in ways that are slowly being
elucidated. Chlorine-produced oxidants (CPO) can cause mortalities of oysters and can produce a variety of sublethal
effects on growth (as reported by Scott and Vernberg at this
symposium). Chlorinated and brominated organic compounds
may be formed *in situ*; their possible effects on marine biota
have been discussed, but little experimental data exists.

Effects of petroleum hydrocarbons in marine waters have
been subjects of an unrelenting flood of literature which
began in earnest in the 1950's, and accelerated geometrically
in the 1970's -- urged along by a number of major and minor
oil spills coincident with increasing ocean transport of oil.
Because of the often extreme visibility of oil spills, and
the obvious impact of resulting shoreline fouling, public as
well as scientific concern about less obvious impacts has
been expressed. Long-term effects of spills depend in part
on the nature of the oil, on environmental temperatures, and

CARL J. SINDERMANN

on sediment characteristics in the spill area -- since release
of petroleum components into interstitial water constitutes
an important route of exposure to benthic animals.

A coherent research program on physiological effects of
petroleum hydrocarbons should contain the following:

(1) certainly a consideration of long-term chronic as well
as short-term effects;

(2) probably some experimental work with animals which
have adapted to chronically-contaminated areas as well as those
from reasonably clean sites;

(3) undoubtedly some attention to effects of particular
purified components as well as of complex mixtures;

(4) definitely some attempt to describe the metabolic pathways involved in observed responses; and

(5) possibly some examination of the morphological as well
as the physiological responses.

Effects of oil spills on living resources and the marine
environment at different places around the world have been
reported to include an entire spectrum from catastrophic, with
mass mortalities, to minimal, with few immediate effects and
no permanent evironmental damage. Of course, as Blumer (1972)
pointed out, there have been remarkably few long-term studies
of the impact of spills, and most of these have been qualitative rather than quantitative. Interest in and funding for
such studies seems inversely proportional to time elapsed
after the event.

Major impacts of oil pollution are probably not in the
acute oil spill events that create great public attention and
lead to generation of hastily-formed scientific task force
operations -- but rather in the long-term chronic oil contamination of estuarine/coastal waters from sewage plant
effluents and sludges, other land runoff, deliberate dis-
charges from vessels, and atmospheric fallout. Petroleum
hydrocarbon residues in contaminated sediments may reach levels
which could be considered hazardous to the biota; Raritan Bay
(N.J.) sediments, for example, have been reported to contain
up to 3672 ppm total saturated and aromatic hydrocarbons
(Thomas, J. P., personal communication, 1977). Long-term
changes in some parts of the fauna of that bay have been observed, particularly the disappearance of some Crustacea
important in food chains of fishes (McGrath, 1974). However,
except in the immediate vicinity of urban areas and major industrial sites, even this chronic contamination probably produces little if any measurable general effects on fish and
shellfish species. Local effects have certainly been demonstrated -- and localized changes in species distribution and
abundance reported -- but no major reductions in species abun-
dances have been directly ascribed to oil pollution.

Although crude oil contains many potentially harmful components, it seems that low boiling point water soluble fractions are most toxic to marine animals, whereas some of the
components with higher boiling points can be carcinogenic to
man and other mammals -- possibly even to fish and shellfish.
These are the primary areas of concern to physiologists and
pathologists.

A literature explosion comparable to that seen for petroleum hydrocarbons has also occurred for effects of heavy metals.
This has been (as Ron Eisler pointed out in this symposium)
partly because of the increasing availability to marine scientists of complex instrumentation, but also because of public
health concerns and because of possible effects on coastal/
estuarine biota. Results of surveys of heavy metal concentrations in fish and shellfish have been published, and acute toxic
effects have been demonstrated. We are now deeply involved in
more fundamental aspects of heavy metal toxicity, as illustrated by papers in this symposium on hematologic effects of mercury
in fish, retardation of fin and limb regeneration in fish and
crustaceans after sublethal exposure to certain metals, and the
induction of cadmium-binding protein in molluscs. Important
additional work still ahead concerns such topics as the methylation potential of all metals, multiple factor modification
of heavy metal toxicity, the detoxification pathways for metals
in fish and shellfish, the nature and extent of adaptation to
high environmental heavy metal content, and the factors which
determine and modify availability of metals to marine biota.

The much-deserved emphasis of this symposium series has
been physiological effects of the chemical contaminants already
considered, but in the process an entire complex of chemical
interactions and physiological stress has receded into the
background. This complex involves effects on marine animals
of anoxia and related hydrogen sulfide poisoning. In my
opinion one of man's greatest impacts on coastal/estuarine
waters is organic enrichment from sewage treatment plant
effluents and agricultural runoff. There are indications here
and there in coastal and estuarine waters -- principally in
the form of recurring phytoplankton blooms -- of gradually
increasing eutrophication. The consequences include anoxic
episodes, particularly in estuarine and sub-thermocline coastal
waters.

There are of course numerous studies of responses of marine
species, particularly benthic ones, to oxygen depleted environments (Hochachka and Mustafa, 1972; Mangum, 1973; Rossi and
Reish, 1976; Taylor, 1976), but the possible synergistic effects
of imposition of this stress in chemically contaminated habitats have received little attention. Anoxia clearly can

eliminate large segments of benthic populations; ultimate
effects on abundance of species in coastal/estuarine habitats
may outweigh those of any other class of marine pollutant.

SURVIVAL IN DEGRADED ENVIRONMENTS

It is easy to wonder if the combined impact of natural and
man-made factors on physiology, reproduction, and survival of
coastal/estuarine species can ever be unraveled. This oppres-
sive thought occurs to me in quiet moments especially after
listening to excellent experimental papers reporting severe
effects of particular contaminants on survival of larvae, or
on success of spawning -- and then looking out my office win-
dow toward Raritan Bay in the New York Bight Apex, where
humans have for more than a century done their damnedest to
contaminate, modify, and degrade, but where large stocks of
shellfish persist, and where recreational fishing is still
carried on. We have seen data in this symposium series that
demonstrate the dramatic modification that a variable such as
salinity or temperature can have on physiological effects of
specific contaminants; we have heard of the protective action
of selenium and now cadmium against mercury toxicity; and we
have seen data on removal of heavy metals from the water column
by adsorption on particulates. Is it too extreme or too
simplistic to ask if the sum of all the pluses and minuses
for all contaminants and natural factors -- some canceling,
some augmenting, some complexing, some adsorbing, some being
metabolized, some being stored -- could not result in some
peaceful near-neutrality (or at least an environmental impact
that is much reduced from that which might be inferred from
laboratory experiments)? And what of the often rapid dilution
of contaminants in coastal waters and the often brief residency
of many motile species in the more toxic environments? The
whole impact seems to be clearly less than the sum of its
dissected parts, judged by the ultimate criteria of survival
and abundance of species. This is where large-scale, long-term
multidisciplinary field experiments could be done -- by selecting
some of our most degraded coastal estuarine waters and attempting to demonstrate what is present, what is available to the
organisms, and what the physiological effects are over a number of generations.
     To do this properly, we should have better data on the
extent of the natural environment which contains toxic or
even above-background levels of particular contaminants in
forms that would be available to and toxic to marine animals.
Often the figures given for environmental contamination levels
in impacted waters represent the highest levels found; these

may occur in extremely restricted space (such as in oil spills)
and the averages outside this zone may quickly become very low.
This would suggest that populations of marine organisms would
be at risk only in extremely localized parts of their ranges --
measured in areas of a few square kilometers rather than hundreds or even thousands of square kilometers. Within these
localized areas sedentary animals could of course be affected,
as could migratory species if they paused long enough.

Survival of animals in contaminated habitats is obviously
related also to the extent of resistance of the individual or
species to toxic effects of the form of contaminant it en-
counters directly. The nature of resistance and the degree of
plasticity of resistance are both worthwhile research objectives. Some observational and experimental evidence supports
the conclusion that resistance of populations of animals to
contaminants can be induced or increased by chronic exposure to
the contaminants. Several mechanisms for detoxification of
contaminants were discussed during this symposium. Richard
Lee referred to studies with the polychaete *Capitella capitata*
in which detoxifying mixed function oxidases were induced by
10-week exposures to crude oil or benzanthracene. Jack
Anderson referred to as yet unpublished results of Rossi (in
press) in which exposure of three generations of polychaetes
in a chronic oil spill area increased resistance of survivors
to acute exposures. David Engel reported induction of cadmium-
binding protein by the oyster after two-week exposures, and
considered this response to be an important defense against
heavy metal toxicity. Bryan (1974, 1976) and Bryan and Hummerstone (1971, 1973) published findings of increased resistance
to acute contaminant toxicity in chronically exposed *Nereis*
populations.

There is at present an obvious need for contributions from
population geneticists to an understanding of this very important adaptive phenomenon -- which may be only selection of
appropriate phenotypes from highly polymorphic populations, or
it may be genetic selection.

Possibly when we begin to sort out the *net* toxic effects
of contaminants plus natural factors, and then superimpose
increased population resistance in areas of chronic exposure,
we may begin to understand how populations persist where intui-
tion and data from experimental exposure studies tell us they
should not survive, or at least should be dwindling.

THE FEDERAL AGENCIES

A logical thread woven through discussions of pollutants
in coastal/estuarine waters and organisms is that of the pervasive and necessary activities of the federal agencies. Two

of them -- the National Oceanic and Atmospheric Administration
(NOAA) and the Environmental Protection Agency (EPA) -- have
had significant involvement with this symposium series, and
are both concerned with effects of pollution on marine organisms.
NOAA has a primary responsibility for living marine resources
and the ecosystems of which the resource species form a part;
EPA is concerned with water quality, development of bioassay
methods, and development of regulations to protect the environment. Both agencies have multiple "legislative mandates" to
support activities related to marine pollution.

NOAA, through three of its components (The National Marine
Fisheries Service, the Office of Sea Grant, and the Environmental Research Laboratories), has supported and is supporting
research on ocean ecosystems and the effects of pollutants on
food chains and resource species. With the gradual awareness
that humans can disturb ocean ecosystems, and that it is
possible to contaminate resource species even before they are
caught and marketed, the impetus for such studies is growing --
with the result that more (though still only a small fraction)
of NOAA's budget is invested in pollution-related work. There
does seem to be a growing concern within the agency for the
possible impacts of ocean pollution, but increased funding to
reflect this concern must compete with the numerous other high-
priority missions of NOAA.

Through its in-house research and extensive university
research contracts, the EPA performs its role as a sorcerer's
apprentice, trying to maintain adequate examination of physiological and other effects of known toxic substances already
in the aquatic environment, and at the same time trying to
assemble data on effects of the many new potentially hazardous
chemical formulations. Despite some feeling that certain
funding actions of the agency seem capricious, and that its
research support is confined to short-term crisis-response
proposals, its regulatory arm is probably one of the principal
user groups for data on pollution effects, and its research
arm an important contributor to and supporter of relevant
research.

Speaking from personal experience, one distinct advantage of
the presence in our midst of representatives of an agency with
combined research and regulatory functions is to encourage many
of us to reexamine the rigor of our experimental design, the
validity of our sampling methods, and the nature of our conclusions. These things EPA includes under the concept of "legal
viability". The concept can exert a salutary influence on
scientific investigations, but it is too narrow to encompass all
of the complex studies with multiple variables that constitute
important aspects of research on pollution effects.

Furthermore, the active and continuing participation of
EPA scientists in this symposium series insures that the best
and most current research data on physiological/biochemical
effects of contaminants will be immediately available for the
complex machinations involved in such processes as registration
of chemicals, and issuance of restricted use labels. We hope
that some of the data presented can provide early warning to
the perceptive, so that we may in the future avoid some of the
gross contamination horror stories (such as DDT, PCB, Kepone,
and others) that have characterized the past decade. John
Costlow's description during this symposium of lethal effects
of Dimilin on 5 species of crab larvae over an extended period
and in extreme dilution may be an example of this. Every
attempt should be made to encourage and coddle this essential
but fragile free interaction, unrestricted communication, and
reasonable cooperation among government, university and
industry scientists.

INTERNATIONAL OCEAN POLLUTION UPDATE

Coastal/estuarine pollution is not a uniquely American
phenomenon, and some infusion of thinking and research information from Europe and the Far East to this symposium could
be useful. Europeans in particular have been cognizant of
deterioration of coastal waters in the Baltic, the North Sea,
the Mediterranean, and elsewhere. Research on effects tends
to be pragmatic, but equal in quality and often in quantity
to that done in the Western Hemisphere. Annual reports to
the International Council for the Exploration of the Sea and
to the European Symposium on Marine Biology constitute principal outlets for scientific findings in Europe. Reports of
Prefectural and Regional Laboratories of Japan contain much
relevant information about pollution effects. The 1970 FAO
Conference on Marine Pollution and Sea Life (Ruivo, 1972) did
much to bring global pollution studies into some common focus;
in my opinion it is now time to prepare for a 1980 world
conference with the same title, in view of the great advances
in our understanding of the effects of marine pollution achieved
during the decade of the 1970's.
Demonstration that contaminants such as PCB's and DDT may
spread beyond national waters has made ocean pollution a
proper concern of the United Nations and its recently created
Environmental Program. The 1972 Stockholm Conference on the
Human Environment produced some worthwhile residues in the
form of international actions to control and even reduce ocean
pollution (Sindermann, 1973). Additionally, some international

cooperative pollution monitoring efforts, such as "mussel watch", have been developing slowly and painfully. Monitoring the levels of pollutants in the ocean is of course a worthwhile national and international initiative, provided (1) that a focus on coastal/estuarine waters is maintained, and (2) that monitoring efforts remain balanced with studies of *effects* of the pollutants being monitored.

SUMMARY

Physiological studies of pollutants in the coastal/estuarine environment -- heavy metals, petroleum components, and halogenated hydrocarbons--have provided much information about effects on the biota. Data from laboratory experiments, combined with field observations and results of field experiments, reduce intuitive conclusions and expand areas of genuine understanding. There is still, however, a large area of uncertainty about multiple pollutant effects on marine organisms, and the critical problem of resistance to contaminant toxicity is only now beginning to be elucidated. A satisfying shift of emphasis toward understanding sublethal effects of long-term exposure to contaminants and away from artificial short-term acute exposures has taken place. Additionally, greater attention has been focused on metabolic pathways, induction of resistance mechanisms, and possibilities for monitoring using physiological/biochemical criteria.

LITERATURE CITED

Blumer, M. 1972. Oil contamination and the living resources of the sea. pp. 476-481. In: Ruivo, M. (ed.). Marine pollution and sea life. Fishing News (Books) Ltd., London, 624 pp.

Bryan, G. W. 1974. Adaptation of an estuarine polychaete to sediments containing high concentrations of heavy metals. pp. 123-135. In: Vernberg, F. J. and W. B. Vernberg (eds.). Pollution and physiology of marine organisms. Acad. Press, N. Y., 492 pp.

Bryan, G. W. 1976. Some aspects of heavy metal tolerance in aquatic organisms. pp. 7 -34. In: Lockwood, A. P. M. (ed.) Effects of pollutants on aquatic organisms, Vol. 2, Camb. Univ. Press, 193 pp.

Bryan, G. W. and L. G. Hummerstone. 1971. Adaptation of the polychaete *Nereis diversicolor* to estuarine sediments containing high concentrations of heavy metals. I. General observations and adaptation to copper. J. Mar. Biol. Assoc. U. K. 51: 845-863.

Bryan, G. W. and L. G. Hummerstone. 1973. Adaptation of the polychaete *Nereis diversicolor* to estuarine sediments containing high concentrations of zinc and cadmium. J. Mar. Biol. Assoc. U. K. 53: 839-857.

Hochachka, P. W. and T. Mustafa. 1972. Invertebrate facultative anaerobiasis. Science 178: 1056-1060.

ICES. 1976. ICES Working Group on Pollution Baseline and Monitoring Studies in the Oslo Commission and ICNAF Areas. Report of the Subgroup on the Feasibility of Effects Monitoring (A. D. McIntyre, Convenor). Internat. Council Expl. Sea, Doc. CM1976/E:44, 36 pp.

Longwell, A. C. 1975. Mutagenicity of marine pollutants as it could be affecting inshore and offshore marine fisheries. Middle Atlantic Coastal Fisheries Center, Natl. Mar. Fish. Ser., Rept. No. 79, 72 pp.

Longwell, A. C. 1976a. Chromosome mutagenesis in developing mackerel eggs sampled from the New York Bight. Amer. Soc. Limnol. Oceanog. Spec. Sympos. 2: 337-339.

Longwell, A. C. 1976b. Chromosome mutagenesis in developing mackerel eggs sampled from the New York Bight. NOAA Tech. Memo. ERL-MESA-7, 61 pp.

Mangum, C. 1973. Responses of aquatic invertebrates to declining oxygen conditions. Amer. Zool. 13: 529-541.

McGrath, R. A. 1974. Benthic macrofaunal census of Raritan Bay -- preliminary results. Proc. Third. Sympos. Hudson River Ecol. (1973), Paper No. 24, 23 pp.

Pearce, J. B. 1977. A report on a new environmental assessment and monitoring program, "Ocean Pulse". Internat. Counc. Expl. Sea, Fish. Imp. Comm. Doc. CM1977/E:65, 11 pp.

Rossi, S. S. and D. J. Reish. 1976. Studies on the *Mytilus edulis* community in Alamitos Bay, California. VI. Regulation of anaerobiasis by dissolved oxygen concentration. Veliger 18(4):357-360.

Ruivo, M. (ed.). 1972. Marine pollution and sea life. Fishing News (Books) Ltd., London, 624 pp.

Sindermann, C. J. 1973. A biologist's view of the Stockholm Conference on the Human Environment. Proc. Third Food-Drugs from the Sea Conf., Marine Tech. Soc. (1973), pp. 11 - 16.

Taylor, A. C. 1976. Burrowing behaviour and anaerobiasis in the bivalve *Arctica islandica* (L.). J. Mar. Biol. Assoc. U. K. 56: 95-109.

Vernberg, F. J. and W. B. Vernberg (eds.). 1974. Pollution
    and physiology of marine organisms. Acad. Press, N. Y.
    492 pp.
Vernberg, F. J., A. Calabrese, F. P. Thurberg and W. B.
    Vernberg (eds.). 1977. Physiological responses of marine
    biota to pollutants. Acad. Press, N. Y., 462 pp.

SUBJECT INDEX

A

Abiotic modifiers, 115
Acclimation temperature, 373
Accumulation, 189, 190, 197, 200, 201, 208, 209, 211, 212
Active metabolism, 370, 372, 374
Altosid, 264
American oyster, 373
Annelids, 114
Anthropleura, 293
Antimony, 113
Aroclor 1254, 271, 273, 276–281, 283–286
Aromatic hydrocarbons, 23, 26, 40, 41, 49, 50
Arsenic, 113, 116, 188, 197
Astacus fluviatilis, 315
ATPases, 330, 331

B

Behavior, 8–10, 117
Behavioral conditioning, 379
Behavioral fever, 379
Behavioral thermoregulation, 365, 366, 372, 382
Benthic bioassay, 341, 349
Beryllium, 118
Bioassay, 439, 440
Bioindicator, 207, 209–213
Biological half-life, 194, 195
Biotic modifiers, 115
Bivalves, 70, 80, 183–185, 190, 195–200, 203, 206, 209, 211–214, 242
Bluefish, 380
Boron, 113
Breathing, 39–43, 46, 47, 49, 50
Breathing rates, 39, 40, 42, 43, 46, 47, 49, 50
Bryozoa, 114

C

Cadmium, 113, 116, 120, 151, 152, 157, 158, 161, 162, 165, 185, 188, 190, 192, 193, 195–199, 205, 207–209, 212, 213, 239–241, 245, 249–253
Calcium, 112
Callinectes sapidus, 223–225, 309, 320–327, 329, 331, 332, 356–361
Cambarus (Orconectes) virilis, 315
Cancer irroratus, 322
Carcinus maenas, 322
Cerium, 113
Cesium, 113
Chaetognaths, 114
Chemorecptor, 223–226, 231
Chlamys operularis, 199
Chlorine, 415, 416
Chlorine produced oxidants, 418
Choromytilus meridionalis, 192
Chromium, 113, 116
Cichlasoma bimaculatum, 328
Cobalt, 113, 199
Coelenterata, 114, 293
Condition index, 69, 70, 73, 75, 79, 417, 424, 427, 430, 431
Copper, 113, 116, 120, 187, 189, 190, 192, 195–199, 204, 208, 209, 212, 213, 223, 234, 235, 239–244, 246–253
Copper sulfate, 223, 224, 226, 231, 234
Coughing rates, 39
Crabs, 151, 157, 161, 164, 165
Crangon crangon, 315
Crassostrea angulata, 206
Crassostrea commercialis, 186, 187, 190, 195, 196
Crassostrea gigas, 185, 188, 204
Crassostrea virginica, 185, 187–197, 202, 205, 207
CSF, 272, 273, 286, 287
Cyanide sensitive factor, 272

D

Daphnia magna, 315
Daphnia pulex, 315

451